# Transplant Production in the 21st Century

## Scientific and Editorial Advisory Board



Each paper was refereed by Scientific and Editorial Advisory Board members and only papers accepted after the authors' revision according to corrections and suggestions by the referees are included in this book.

# Transplant Production in the 21st Century

## Proceedings of the International Symposium on Transplant Production in Closed System for Solving the Global Issues on Environmental Conservation, Food, Resources and Energy

Edited by

**C. Kubota**

and

**C. Chun**

*Chiba University,*
*Matsudo, Japan*

Springer-Science+Business Media, B.V.

A C.I.P. Catalogue record for this book is available from the Library of Congress.

ISBN 978-90-481-5570-5 ISBN 978-94-015-9371-7 (eBook)
DOI 10.1007/978-94-015-9371-7

*Printed on acid-free paper*

# Contents

Preface ix

**1. Closed transplant production systems**

Necessity and concept of the closed transplant production system
*Toyoki Kozai, Chieri Kubota, Changhoo Chun, Fawzia Afreen and Katsumi Ohyama* ---- 3

Closed transplant production system at Chiba University
*Changhoo Chun and Toyoki Kozai* ---- 20

Electric energy, water and carbon dioxide utilization efficiencies of a closed-type transplant production system
*Katsumi Ohyama, Toyoki Kozai and Keita Yoshinaga* ---- 28

Microprecision irrigation system for transplant production
*Haruhiko Murase* ---- 33

Design concepts of computerized support systems for large-scale transplant production
*Takehiko Hoshi, Yasumasa Hayashi and Toyoki Kozai* ---- 38

**2. Technology in transplant production**

***Modeling, measurement and environmental control for transplant production***

Modeling and simulation in transplant production under controlled environment
*Chieri Kubota* ---- 47

Object-oriented analysis and modeling of closed plant production systems
*David H. Fleisher and K.C. Ting* ---- 53

Estimating cuticle resistance of seedling shoot tips based on the Penman-Monteith model
*Hiroshi Shimizu and Royal D. Heins* ---- 59

Measurement of pH in guard cells using a confocal laser scanning microscope
*Masahiro Yabusaki, Yasuomi Ibaraki, Kenji Kurata and Keiko Iwabuchi* ---- 63

Does electrolyzed-reduced water protect plants from photoinhibition?
*Keiko Iwabuchi, Kazuyuki Seyama, Kenji Kurata and Yukiko Hiruta* ---- 67

Environmental control for improved plant quality within controlled environment plant production systems
*Stephen T. Kania and Gene A. Giacomelli* ---- 72

Environmental engineering for transplant production
*Chalermpol Kirdmanee and Kriengkrai Mosaleeyanon* ---- 78

Effects of air current on transpiration and net photosynthetic rates of plants in a closed plant production system
*Yoshiaki Kitaya, Johshin Tsuruyama, Masayuki Kawai, Toshio Shibuya and Makoto Kiyota* ---- 83

Effects of air temperature, relative humidity and photosynthetic photon flux on the evapotranspiration rate of grafted seedlings under artificial lighting
*Yong Hyeon Kim* ---- 91

Growth of tomato (*Lycopersicon esculentum* Mill.) plug transplants in a closed system at relatively high air current speeds - A preliminary study -
*Watcharra Chintakovid and Toyoki Kozai* ---------------------------------- 98
Advances and current limitations of plug transplant technology in Korea
*Byoung Ryong Jeong* ---------------------------------------------------- 102

***Lighting strategies for transplant production***
A review on artificial lighting of tissue cultures and transplants
*Wei Fang and R.C. Jao* ------------------------------------------------- 108
Light emitting diodes (LEDs) as a radiation source for micropropagation of strawberry
*Duong Tan Nhut, Takejiro Takamura, Hiroyuki Watanabe and Michio Tanaka* ---------------------------- 114
Application of red laser diode as a light source for plant production
*Aya Yamazaki, Hiroshi Tsuchiya, Hirofumi Miyajima, Takayoshi Honma and Hirofumi Kan* ------------------------------------------------------------ 119
Effective vegetable transplant production programs for closed-type systems under different lighting regimes
*Toru Maruo, Masahiro Tsuji, Hitomi Kida, Yutaka Shinohara and Tadashi Ito* ---------------------------- 125
Photoautotrophic micropropagation in a natural light environment
*Jeff Adelberg, McNair Bostick, David Bishop and Robert Pollock* ---------------------------- 131

***High-quality transplant production***
Production of value-added transplants in closed systems with artificial lighting
*Hyeon-Hye Kim and Toyoki Kozai* ------------------------------------------ 137
High quality plug-transplants produced in a closed system enables pot-transplant production of pansy in the summer
*Yoshitaka Omura, Changhoo Chun, Toyoki Kozai, Kei Arai and Katuyoshi Okabe* ------------------------------------------------------------------------ 145
Yield and growth of sweetpotato using plug transplants as affected by their ages and planting depths
*A.F.M. Saiful Islam, Changhoo Chun, Michiko Takagaki, Kosuke Sakami and Toyoki Kozai* ------------------------------------------------------------ 149
Yield and growth of sweetpotato using plug transplants as affected by cell volume of plug tray and type of cutting
*Dongxian He, Yee Hin Lok, Changhoo Chun and Toyoki Kozai* -------- 154
Production of medicinal plant species in sterile, controlled environments
*Susan J. Murch, Sankaran KrishnaRaj and Praveen K. Saxena* -------- 160
Effect of air temperature on tipburn incidence of butterhead and leaf lettuce in a plant factory
*Ki Young Choi, Kee Yoeup Paek and Yong Beom Lee* ------------------- 166
Evaluation of lettuce cultivars suitable for closed plant production system
*Masahisa Ishii, Toru Maruo, Yutaka Shinohara and Tadashi Ito* ------- 172
Root growth subsequent to transplanting in plug-grown cabbage seedlings
*Satoshi Yoshida* ------------------------------------------------------------ 178

Effective storage conditions for subsequent growth enhancement of *Ficus carica* L. cuttings
*Michiko Takagaki, Yoshihiro Murata, Naoto Sakurai, Hideo Enomoto and Yuji Udagawa* ---------- 183

## 3. Biotechnology for transplant production

***Biotechnology for woody plants***

Characterization of transformed poplar formed by the inhibition of peroxidase
*Noriyuki Morohoshi* ---------- 191
Micropropagation of Canadian spruces (*Picea spp*)
*Trevor A. Thorpe and Indra S. Harry* ---------- 197
In vitro culture of Japanese black pine (*Pinus thunbergii*)
*Katsuaki Ishii and Emilio Maruyama* ---------- 205
Control of the development of somatic embryo of Japanese conifers by the density of embryogenic cells in liquid culture
*Shinjiro Ogita, Hamako Sasamoto and Takafumi Kubo* ---------- 209
A preliminary experiment on photoautotrophic micropropagation of *Rhododendron*
*Carmen Valero-Aracama, Sayed M.A. Zobayed and Toyoki Kozai* ----- 215
Mass clonal propagation of *Artocarpus heterophyllus* through in vitro culture
*Shyamal K. Roy, P.K. Roy, P. Sinha and M.S. Haque* ---------- 219
Photoautotrophic growth of *Pleioblastus pygmaea* plantlets in vitro and ex vitro as affected by types of supporting material in vitro
*Yumiko Watanabe, Yoshiaki Sawa, Nobuharu Nagaoka and Toyoki Kozai* ---------- 226

***Transplant production using micropropagation techniques***

Evolution of culture vessel for micropropagation: from test tube to culture room
*Sayed M.A. Zobayed, Fawzia Afreen, Chieri Kubota and Toyoki Kozai* ---------- 231
Physiology of in vitro plantlets grown photoautotrophically
*Fawzia Afreen, Sayed M.A. Zobayed, Chieri Kubota and Toyoki Kozai* ---------- 238
Enhanced growth of in vitro plants in photoautotrophic micropropagation with natural and forced ventilation systems
*Quynh Thi Nguyen, Toyoki Kozai and Jeongwook Heo* ---------- 246
Micropropagation of ornamental plants using bioreactor system
*Kee Yoeup Paek, Eun-Joo Hahn, Jeongwook Heo and Seong Ho Son* ---------- 252
Effects of medium sugar on growth and carbohydrate status of sweetpotato and tomato plantlets in vitro
*Sandra B. Wilson, Chieri Kubota and Toyoki Kozai* ---------- 258
Practical sugar-free micropropagation system using large vessels with forced ventilation
*Yulan Xiao, Jiacong Zhao and Toyoki Kozai* ---------- 266
Growth and acclimatization of chrysanthemum plantlets using bioreactor and hydroponic culture techniques
*Eun-Joo Hahn, Sun Ja Kim, Kee Yoeup Paek and Yong Beom Lee* ------ 274

Mass propagation of pineapple through in vitro culture
*Shyamal K. Roy, M. Rhaman and S. Hauqe* ---------------------------------- 279
Microbial contamination under photoautotrophic culture system
*Nazrul Islam and Sayed M.A. Zobayed* --------------------------------------- 284

Author Index 289

# Preface

We are facing global issues concerning environmental pollution and shortages of food, feed, phytomass (plant biomass) and natural resources, which will become more serious in the forthcoming decades. To solve these issues, immeasurable numbers of various plants and huge amounts of phytomass are required every year for food, feed and for the improvement of amenities, the environment and our quality of life. Increased phytomass is also required as alternative raw material for producing bio-energy, biodegradable plastics and many other plant-originated industrial products. Only by using phytomass as a reproducible energy source and raw material, instead of fossil fuels and atomic power, we can save natural resources and minimize environmental pollution. To increase phytomass globally, we need billions of quality transplants (small plants) to be grown yearly, in the field or in the greenhouse, under various environmental conditions. However, these high quality transplants can be produced only under carefully controlled, rather than variable environmental conditions.

Recent research has shown that the closed transplant production system requires considerably small amounts of electricity, water, fertilizer, $CO_2$ and pesticide to produce value-added transplants as scheduled with minimum release of environmental pollutants and minimum loss of transplants. The closed or closed-type transplant production system is defined as a transplant production system covered with opaque walls with minimized or controlled ventilation rates, using artificial lighting. With this system, photoperiod, light intensity and quality, air temperature, humidity, $CO_2$ concentration and air current speed can be controlled as desired. The closed transplant production system can be cost-competitive against the open-type transplant production systems such as used in greenhouses and nurseries.

Considering the facts mentioned above, International Symposium on Transplant Production in Closed System for Solving the Global Issues on Environmental Conservation, Food, Resources and Energy was held at Chiba University, Japan during February 28 to March 2, 2000. The Symposium was sponsored by Japan Society for Promotion of Science, Chiba University, Japanese Society of Environment Control in Biology, Japanese Society of High Technology in Agriculture, International Association of Biotechnology Applications, International Plant Propagators Society - Japan Region, Japanese Society for Horticultural Science and Chiba Convention Bureau. We thank Professor Atsushi Komamine who played a vital role as Chairpersons of Organizing and Advisory Committees.

This book contains selected papers presented at the Symposium. The papers included in this book were reviewed, accepted and edited for publication. We would like to thank all the participants of the Symposium, the Committee members, the secretaries and sponsoring organizations, who contributed to making the Symposium fruitful and enjoyable. We hope that this book serves as a source of information on transplant production in closed systems for solving the global issues in the 21st Century.

*Chairperson of the National Committee*
Toyoki Kozai
*Editors*
Chieri Kubota and Changhoo Chun

# 1. CLOSED TRANSPLANT PRODUCTION SYSTEMS

# NECESSITY AND CONCEPT OF THE CLOSED TRANSPLANT PRODUCTION SYSTEM

Toyoki Kozai, Chieri Kubota, Changhoo Chun, Fawzia Afreen and Katsumi Ohyama
Faculty of Horticulture, Chiba University, Matsudo, Chiba 271-8510 Japan. E-mail: kozai@midori.h.chiba-u.ac.jp

**Abstract.** We are requested to develop a concept, a methodology and an industry to solve the global issues on environmental pollution and shortages of food, feed, phytomass (plant biomass) and natural resources including fossil fuels and usable water. These issues are considered to become more and more serious on a larger scale in the forthcoming decades. In order to solve those issues in the 21st Century, billions of plants are required every year not only for food, feed and environment conservation, but also for alternative raw materials to produce energy, bio-degradable plastics and many other industrial products. By using plant-derived products, we can minimize the environmental pollution and the use of fossil fuels and atomic power. Then, we need billions of quality transplants (small plants) every year to be grown in the fields with maximum use of solar energy and minimum use of resources under harsh environmental conditions. These quality transplants can be produced only under carefully controlled environments. Bioengineering is expected to provide a useful concept and a methodology to develop the bioindustry for solving the above global issues substantially. In bioengineering, the global and local flows of energy, mass (or materials) and information are analyzed with special attention to the organic and inorganic metabolisms of plants, animals including humans and microorganisms. Concept of 'closed-type or closed production systems' is essential to develop a production system with minimum use of resources and with minimum environmental pollution. This concept can be applied to develop a closed-type transplant production system with artificial lighting for producing billions of quality transplants with minimum use of resources and with minimum environmental pollution. Research and development of the closed transplant production systems will create a new field of bioengineering and bioindustry.

## 1. Introduction

The world population in the year 2000 is over 6 billions and has been predicted to reach about 10 billions by the middle of the 21st Century. Recent annual rate of population increase is nearly 3 % in Asian, African and South American countries. In those countries, the environmental pollution and the shortages of food, feed, phytomass (plant biomass) and natural resources including fossil fuels and usable fresh water will become more and more serious on a larger scale in the forthcoming decades. Increase in phytomass in those countries is essential also to stabilize their climates and to conserve their ecosystems or environments.

The difficulty with solving these global issues on food, energy and environments is that we need to solve these issues concurrently based upon one common and innovative concept and methodology created from broad and long-term viewpoints, and to develop an industry, called 'bioindustry' hereafter. The bioindustry is strongly related to agriculture, horticulture, forestry and aquaculture and also to other manufacturing industries, but is not the same as those industries (Kozai et al., 1997).



*C. Kubota and C. Chun (eds.), Transplant Production in the 21st Century,* 3–19.

The reason why we need to solve those issues concurrently is that solving only one issue separately from the other two often makes the situations of the other two even worse. This type of conflicting situations is called by Yoda (1993) a global 'trilemma', a difficult choice between three possibilities. For example, the worldwide spread of 'advanced' agricultural technology for increasing 'crop yield' based on more energy and resources will make the environmental pollution worse, increase the atmospheric $CO_2$ concentration, and cause shortages of fossil fuels and other natural resources. This is because the modern agricultural technology is heavily dependent upon the oil-derived products such as chemical fertilizers, chemical pesticides, plastics, and fuels for machines.

In order to solve the global trilemma, we are requested to develop the bioindustry by working internationally and interdisciplinary, based upon a 'bioengineering' concept, paradigm, doctrine and/or methodology (Fig. 1). In this article, the perspectives of 'bioengineering' and 'bioindustry' are discussed in relation to the global trilemma and other major problems to be solved in the 21st Century.

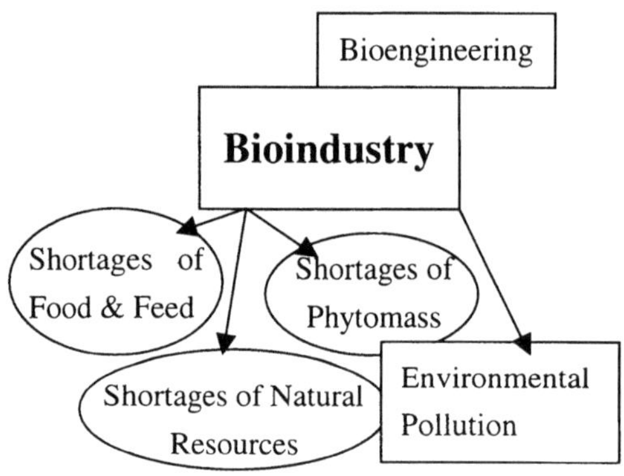


*Fig. 1* Bioindustry and bioengineering contribute to solving the global issues on environmental pollution, shortages of food, feed, natural resources and phytomass (plant biomass) concurrently.

Apart from the global trilemma, bioengineering and bioindustry need to be developed to innovate the paradigms of other fields of production engineering and production industries. The bioindustry and bioengineering will be a key production industry and production engineering, respectively, in any field of production industries, engineering and sciences in the 21st Century (Fig. 2). A crucial concept there is 'closed production systems', as is discussed later in this article.

## 2. Flows of Energy and Carbon on the Earth

Figure 3 shows the outline of energy and carbon flows on the Earth. The solar energy is the sole energy source received by the Earth from its outside. A small portion (ca. 0.1%) of the solar energy is converted into chemical energy in the form of carbohydrates and their derivatives by photosynthesizing or growing plants on the Earth.

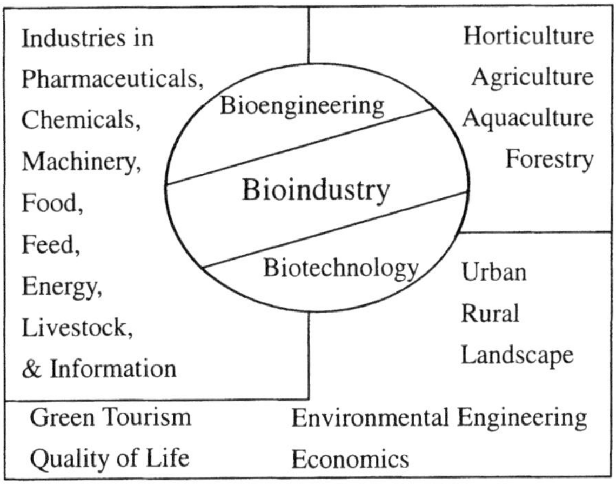


*Fig. 2* Bioindustry and bioengineering in relation to the other fields of industries and engineering.

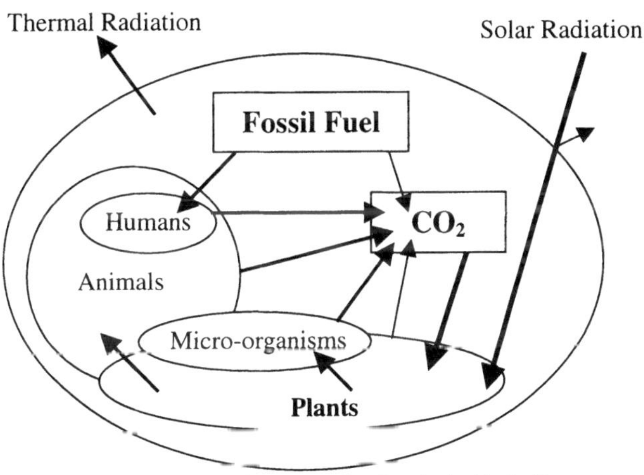


*Fig. 3* Schematic diagram showing the flows of energy and carbon in the Earth. Plants are only organisms that can absorb $CO_2$ from the air and fix the solar radiation via photosynthetic activities. By using plants as industrial raw material, we can reduce the consumption of fossil fuel.

The rest of solar energy is mostly (ca. 85%) converted into heat energy and emitted back as longwave or thermal radiation energy into the space, and is partly (ca. 15%) reflected back to the space.

The growing plants absorb $CO_2$ in the atmospheric air as carbon source for photosynthesis. Namely, they grow photoautotrophically using only $CO_2$, water, light and some inorganic nutrients. On the other hand, animals including humans and microorganisms cannot convert solar energy into chemical energy by themselves. They produce $CO_2$ into the air by respiration, and totally depend upon plants directly or indirectly with respect to their food. Namely, they grow heterotrophically. In addition, humans have been increasingly producing $CO_2$ into the air by use of fossil fuels.

In the 20th Century, the world human and domestic animal populations increased substantially, whereas amounts of global phytomass and fossil fuels decreased substantially. As a result, atmospheric $CO_2$ concentration became higher than 360 $\mu$mol $mol^{-1}$ (or ppm) in 1999, which was lower than 300 $\mu$mol $mol^{-1}$ in 1900. World oil consumption was 500 million tons in 1900 and was 3 billion tons in 1999. World natural gas consumption was almost zero in 1900.

In the 21st Century, we are requested to increase the phytomass, with the aids of microorganisms and animals, and by doing so, we can decrease $CO_2$ concentration in the atmosphere and consumption of fossil fuels.

## 3. Transplant Production as a Field of Bioindustry and Bioengineering

As described above, billions of plants are required every year not only for food, feed and environment conservation, but also for alternative raw materials to produce energy, bio-degradable plastics and many other industrial products. By using plant-derived products, we can minimize the environmental pollution and the use of fossil fuels and atomic power.

Then, we need billions of quality transplants (small plants) every year to be grown in the fields with maximum use of solar energy and minimum use of resources under harsh environmental conditions (Fig.4). These quality transplants can be produced only under carefully controlled environments.

Those huge demands of transplants will create new fields of bioindustry and bioengineering in the 21st Century. The followings are two examples of transplants which demands will be increased dramatically in the 21st Century.

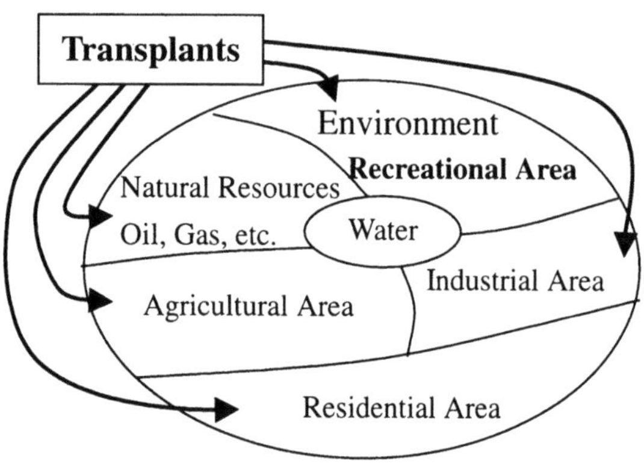


*Fig. 4* A large number of quality transplants are needed for environmental conservation and for producing food, feed, bio-energy and industrial raw materials.

### 3.1 Production of Forest Tree Transplants

The total area under forest in tropical countries has been decreasing at an annual rate of 0.8% according to the survey by FAO. It was 1,910 million hectares in 1981, but it decreased to 1,756 million hectares in 1990, i.e., decreased by 15.4 million hectares per year. On the other hand, annual afforestation and re-afforestation areas in those countries are estimated to be 1.8 million hectares during the period 1981-1990, resulting in the yearly net decrease in forest area of 13.6 (=15.4-1.8) million hectares.

The forest area in temperate and cold climate regions has also been decreasing. In addition, the decrease in biomass of woody plants due to desertization in arid regions is significant. Such local and global decreases in forest area and thus plant biomass are a factor causing recent climate changes at different geographical scales.

As is indicated above, in the 21st Century, a huge quantity of transplants of woody plant species such as eucalyptus, acacia, rattan, teak, bamboo and pine trees will be required in the pulp, paper, timber, plantation, horticulture and furniture industries, in the re-forestation, forestation and dessert rehabilitation for environment conservation. Use of plants and plant-derived products reduces the consumption of fossil fuels for manufacturing plastics, and lowers the atmospheric $CO_2$ concentration by fixing atmospheric $CO_2$ through photosynthesis of plants, and stabilizes the local and global climates (Kozai et al., 1999b).

It should be noted that use of wooden materials for furniture, houses and other constructions decrease the atmospheric $CO_2$ concentration and at the same time decrease the use of metals and plastics derived from oils. Furniture and houses made of woods consists of mostly carbon and hydrogen and oxygen. In this sense, forestation, re-forestation and sustainable use of forest trees contribute significantly to solve the global issues.

### 3.2 Production of Sweetpotato Transplants

One crop which will be of importance in tropical and subtropical countries in the 21st Century is sweetpotato (*Ipomoea batatas* (L.) Lam.), which has been used as starch and health food rich in vitamins and antioxidant elements such as beta-carotene and ascorbic acid as well as fibers.

In the 21st Century, it will be potentially used in large quantity as feed and raw materials for producing bio-degradable plastics and hydrogen gas (Kozai et al., 1998a; Kozai et al., 1998b). The hydrogen gas will be used as clean energy source for charging the batteries of automobiles and other electricity-driven machines repeatedly.

Potential yield of sweetpotato in tropical and subtropical countries is 1.5 times higher than that of rice and maize, and 2 times higher than that of potato. Sweetpotato needs less water, nitrogen fertilizer and labor to be grown than rice, which is advantageous for environmental conservation and efficient plant production in the 21st Century. It should be noted that we will be lack of usable fresh water for agriculture, especially in many Asian countries in the 21st Century.

Thus, the transplant production for forestation, re-forestation, dessert rehabilitation, sweetpotato cultivation and many other purposes can be a field of bioindustry and of bioengineering in the 21st Century. Furthermore, developing production systems of quality transplants of different plant species with minimum use of energy, material and human resources will make a significant contribution in solving the global issues concurrently (Kurata and Kozai, 1992; Aitken-Christe et al., 1995).

Quality transplants means transplants having superior genetic, physiological, morphological and physical properties and being free from pathogens. They are also requested to be transported and transplanted easily by hands or by a mechanical means and to grow vigorously and give high yield and quality after transplanting in the open fields or in the greenhouses with minimum use of the resources even under unfavorable environmental conditions.

Again, we need to grow plants in the large open fields at different geographical locations. Then, we also need to increase the biomass on the Earth to conserve our environment, to obtain more food, to use more biomass as raw materials for industrial products in order to minimize the use of oil and other fossil fuels.

Use of quality transplants can reduce agrochemicals, labor, and other resources to be used in the fields or in the greenhouse during cultivation, and gives high yield and quality of the final crops at harvest (Fig. 5).

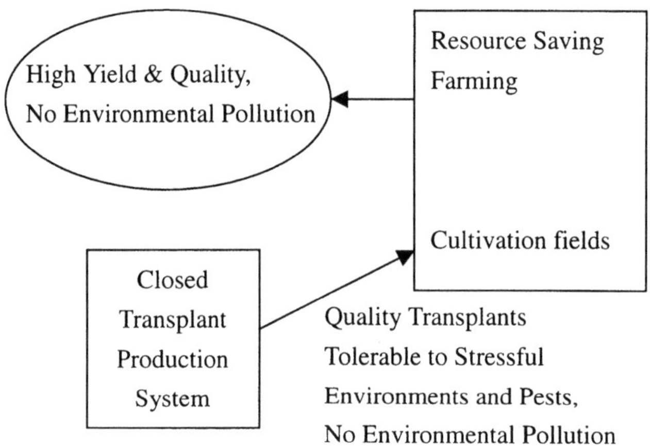


*Fig. 5* Use of quality transplants can save agrochemicals, labor, and other resources to be used in the fields or in the greenhouse during cultivation, and gives high yield and quality of crops.

## 4. Necessity of Closed Production Systems in Agriculture and Other Industries

In any production system, some resources are brought in from the environment of the production system and some products are brought out to the environment of the production system. The environment of the production system is defined here as the surroundings of the production system affecting or being affected by the production system.

In addition to the products, some by-products are produced in the production system. A large portion of the by-products is often released to the environment. Those by-products are called environmental pollutants when they are unfavorable or harmful for the environment. The rest of the by-products or their derivatives are accumulated in the production system and some of them are recycled in the closed system. Their accumulations often decrease the productivity of the production system.

In the 20th Century, the production systems were often designed to maximize its productivity or cost performance. As a result, large amounts of resources were consumed and large amounts of environmental pollutants were released to the environment. This type of production system can be called 'one-way production system' (Yoshikawa, 1998) or 'open production system' (Fig. 6).

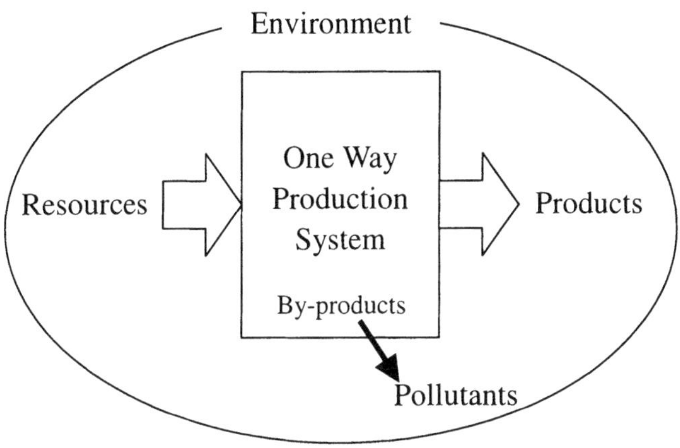


*Fig. 6* Schematic diagram of an one-way (or open) production system which tend to produce pollutants.

This is the case even in agriculture. Generally speaking, the crop yield increases with increasing the amount of nitrogen fertilizer supplied to the field. For example, approximately 200 kg per hectare of nitrogen fertilizer is usually supplied to the crop field annually. However, only about 25 % of nitrogen fertilizer is absorbed by the crops and the rest (75 %) becomes wastes and eventually, environmental pollutants. Micropropagation may be a good example mode in agriculture where various kinds of wastes are released to the environments. Table 1 shows rough estimates of percent wasted for each supply in micropropagation.

In the 21st Century, basic requirements of a production system would be to minimize the consumption of resources and the release of environmental pollutants while keeping the productivity of a production system at high levels. In order to achieve this goal, it is required to develop 'closed production systems' (Yoshikawa, 1998). In closed production systems (Fig. 7), the production process is not 'one-way', but is 'closed, looped or recycled'.

*Table 1* Kinds of wastes and percent of wasted for each supply in micropropagation.

| | Percent |
|---|---|
| Heat | 100 |
| Damaged vessels | 5 |
| Damaged vessel caps | 5-50 |
| Sugar | 60-80 |
| Other nutrients and plant growth regulators | 50-60 |
| Agar | 100 |
| Plants (Dead and unused) | 30-40 |
| Polluted water | 100 |
| Broken lamps | 0.1 |
| $CO_2$ | 100 |

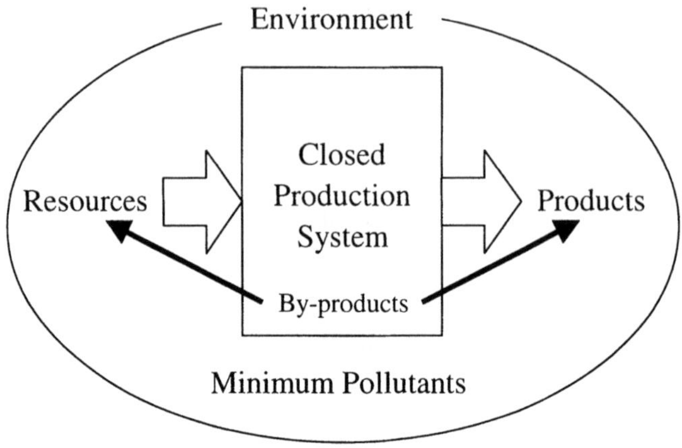


*Fig. 7* Schematic diagram of a closed (or recycling) production system. In this system, by-products are recycled and used as resources and/or products, resulting in reduction of resources and pollutants and increase of products.

With the closed production system, the consumption of resources and the release of environmental pollutants can be minimized (Fig. 8). 'Zero emission production system' meaning 'production system with no emission of waste or pollutants' is another way of expressing this concept. More than several zero emission factories have been established in Japan recently. In those closed production systems, recycling uses of by-products and production of value-added products from by-products are essential to minimize the release of environmental pollutants and to improve the productivity of the production system.

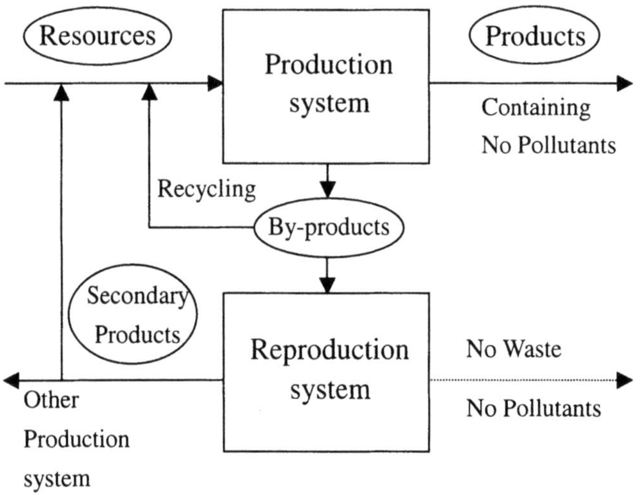


*Fig. 8* Schematic diagram of a closed (or recycling) production system with a reproduction subsystem for no emission of pollutants. This system is often called a zero emission system.

## 5. Closed Transplant Production Systems with Artificial Lighting

Closed transplant production system is one application of closed production systems. In case of agriculture, we 'grow' plants as products rather than we 'make' plants. To grow plants, a large amount of light energy is required. Thus, we need to use solar or natural light as light source, which is free of charge, to grow crops. However, under natural light, light intensity, daylength, wind velocity, rainfall and other environmental factors vary with time of day, season and years, and often unfavorable to crop growth.

On the other hand, only a small amount of light energy is required to produce 'transplants' from seeds or cuttings. Thus, we may produce transplants in a closed transplant production system with artificial lighting and with minimum use of electricity for lighting and air conditioning. The outer structure of the closed transplant production system does not transmit solar light and the exchanges in mass (materials such as air) and energy between inside and outside the closed system are restricted or controlled. With the closed transplant production system, quality transplants are produced at high cost performance with minimized release of environmental pollutants.

Kozai et al. (1999a) hypothesized that in the 21st Century "closed production systems with artificial lighting" would become more favorable to transplant production including seedling and cutting production than "open or one-way production systems with natural lighting" in many aspects and in many cases. Ohyama and Kozai (1998) showed that the consumption of electricity for lighting and cooling in transplant production in the closed system was about 0.35 MJ (= 0.1 kWh) per transplant or less and its cost is about 1.5 yen in Japan and 0.5 cent in USA. This electricity consumption decreases with increasing plant density with respect to each production area.

Most transplants are currently produced in open production systems such as greenhouses under natural light, with a notable exception of production of disease-free transplants in micropropagation via tissue culture, which is mostly conducted under artificial light in closed environments. In either case, almost all the transplants are moved to and transplanted in the open fields or in the greenhouses. (Some major crops such as maize, soybean, etc. which seeds are large in size and are sown directly in the open fields are the notable exceptions, and are beyond the scope of this article.)

Concept, basic design requirements and advantages of the closed transplant production systems with artificial lighting over the open or open transplant production systems with natural lighting have been discussed in detail using numerous data with respect to the initial investment costs, operation costs, transplant quality and environmental pollution due to the use of agro-chemicals, water consumption (Kozai, 1999a). A pilot scale model of the closed transplant production system has been constructed for research purposes at Chiba University, Japan (Chun and Kozai, 2000).

## 6. Fusion of Artificial Production Systems with Natural Systems

Nowadays people tend to prefer the concept "natural" to the concept "artificial" and "open" to "closed". However, of course, "natural" and "open" are not always better than 'artificial" and "closed", respectively. The word "artificial" implies a human art or effort to make something better than "natural".

On the contrary, in reality, "natural" is often better than "artificial" with respect to recycling, symbiosis, and intelligent information processing if considered from broader and longer-term view points. Nonetheless, we should do our best to make our society better than before from broader and longer-term view points with human wisdom. This kind of positive human action can be called “artificial”. In fact, agricultural fields are ‘artificial ecosystems’ which bring about more food with less labor than ‘natural ecosystems’.

Likewise, “closed transplant production system” is better than “open transplant production system”, provided that the closed transplant production system does not produce any pollutants to its environment, while it produces quality transplants using less energy and resources.

In order to solve the global issues mentioned above, we need to change our concept of transplant production systems, considering the possibility of closed transplant production system with artificial lighting from logical points of view. In the 21st Century, development and application of closed artificial systems which do not disturb natural recycling will become important (Yoshikawa, 1998).

In the 21st Century, artificial systems can be acceptable to human societies only when they do not pollute the environments significantly. In other words, they must be designed, constructed and used as ‘zero emission systems’. Only in this way, artificial systems can be fused or integrated favorably with natural systems.

## 7. Reasons for Consideration of the Closed System with Artificial Lighting

Natural light is often considered to be a free and ideal light source for producing crops. At least, natural light seems to be much more economical as a light source than artificial light. However, this statement is not considered valid concerning the transplant production. Use of natural light for transplant production is risky in respect of stable and rapid production of quality transplants. In order to reduce its risk, a considerable investment is needed for environment control facilities, labor and artificial energy (fuels and/or electricity), resulting in the relatively high production costs of transplants. Surprisingly, it is considered that transplant production under natural light is more costly than that under artificial light, if considered all the related costs and economic value of the products (transplants).

Main reasons why the closed system with artificial lighting can be considered more appropriate for transplant production in many cases than the open system with natural lighting are described below. It should be noted that the reasons are applicable for transplant production only, not for plant production in general. In the followings, the open system means the one-way transplant production system using natural light, such as the greenhouse. Of course, the degree of ‘open’ differs in different types of greenhouses.

### 7.1 Environmental Control in General

Transplants are required to grow vigorously after transplanting in the open fields or in the greenhouses under natural and often harsh environmental conditions.

Germinated seeds, cuttings and explants are weak and sensitive to environmental stresses, and the physiological and morphological properties of transplants are influenced considerably by environmental conditions during transplant production.

Thus, their environment during the transplant production period should be carefully controlled to grow them favorably, and to produce quality transplants giving high yield and quality after transplanting even under harsh environmental conditions. Then, it is worth to invest some resources during transplant production period to obtain transplants giving high yield and quality with minimum investments during crop production period in the fields or in the greenhouses. It is assumed that in this article that transplants are produced somewhere and transferred to the fields or greenhouses with natural light and are grown there until harvest.

Air temperature, $CO_2$ concentration, relative humidity and air current speed can be controlled easily in the closed system, regardless of the outside weather. Thus, relationship between transplant growth and the environment can be simplified, and the growth of transplants can be predicted and controlled easily without using plant growth regulators and other sophisticated or risky techniques. Then, quality transplants can be produced easier and faster in the closed system than in the open system. Also, global standardization of transplant production can be realized because the system operation is almost independent of the local weather which differs from place to place.

On the other hand, in the open system, air temperature, relative humidity, $CO_2$ concentration, air current speed are affected considerably by the outside weather, especially, by the fluctuating solar radiation with time. Thus, it is almost impossible to maintain the environmental factors at optimum ranges throughout the transplant production period and to control the growth and development of transplants as desired. Under such conditions, experience and knowledge on the relationship between those environmental factors and the plant growth are difficult to integrate systematically.

Moreover, we can produce many kinds of value added transplants in the closed system with environment control equipment (Chun and Kozai, 2000, Kim and Kozai, 2000). For example, a) inhibition and promotion of flower bud development, bolting and stem elongation by controlling the photoperiod, air temperature and/or light quality, b) addition of microorganisms to the substrate in the plug trays to increase disease resistance, to promote growth and development, and to control plant morphology, and c) production of well-acclimatized transplants which can grow fast and vigorously in the open fields or in the greenhouses after transplanting. All-year round production of transplants can be realized more easily with the closed system than with the open system. The period required for transplant production is shortened by 20-40 % and percentage of marketable transplants is higher in the closed system than in the open system. Thus, annual capacity of transplant production increases and the cost of initial investment per transplant decreases.

The closed system can offer the workers a more comfortable working environment compared to the open system. Then, the workers can perform more efficiently in the closed system than in the open system in most of their working time and thus resulting in the reduction of labor costs.

### 7.2 Light Environment Control

In the closed system, light environment can be controlled easily in terms of PPF (photosynthetic photon flux), photoperiod, spectral composition or distribution (light quality), and lighting direction, regardless of the outside weather. Thus, vegetative, reproductive and morphological (stem elongation, flowering, bolting, etc.) developments of transplants can be manipulated easily. On the other hand, in the open system, daylength, spectral composition, PPF and lighting direction (solar azimuth and altitude) vary with season, time of the day and/or geographical location, which are beyond human capacity to control. Daily and monthly variations in these factors also change annually depending upon the annual changes in weather.

The optimum light intensity or PPF is significantly lower for transplant production (ca. 250-350 $\mu mol\ m^{-2}\ s^{-1}$) than for crop production (ca. 500-1,000 $\mu mol\ m^{-2}\ s^{-1}$), because the PPF saturation point of net photosynthesis is lower in young plant communities than in mature plant communities. The main reason is that the mutual shading of leaves is not so heavy in the young plant community. The PPF of 250-350 $\mu mol\ m^{-2}\ s^{-1}$ at the transplant community surface can be easily obtained by using the fluorescent lamps which are least expensive lamps.

In the open system, PPF in the greenhouse on sunny days is nearly 1,000 $\mu mol\ m^{-2}\ s^{-1}$ around noon in summer, while PPF in the greenhouse on rainy and cloudy days is often lower than 200 $\mu mol\ m^{-2}\ s^{-1}$ even around noon in summer. Thus, in nature, the total time period with a PPF of 250-350 $\mu mol\ m^{-2}\ s^{-1}$ is limited.

Daily integrated PPF on a sunny day in September at the north latitude of 36$^{\circ}$ , for example, is typically about 15 mol $m^{-2}d^{-1}$, which can be obtained by using a PPF of 260 $\mu mol\ m^{-2}\ s^{-1}$ with a photoperiod of 16 h $d^{-1}$; this PPF can be obtained easily by using fluorescent lamps.

The fluorescent lamps are a line light source and can give a relatively uniform horizontal PPF distribution at the transplant community level even though they are placed close to the transplants.

Then, the vertical distance between shelves can be about 60 cm: 5 cm for plug tray, 10-25 cm for transplants, 20-25 cm for air flow above the transplant communities, 2.5 cm for fluorescent lamps and 7.5 cm for the space between the lamps and the bottom surface of the upper shelf.

Wavelength of solar radiation ranges between 300 and 2,500 nm, but the PAR (photosynthetically active radiation) with wavelength ranging between 400 and 700 nm accounts for less than 50 % of the solar radiation energy. The radiation with a wavelength ranging between 800-2,500 nm belongs to thermal or infra-red radiation, the action of which is simply to raise the air and crop temperatures. Thus, during the summer, the temperatures are often too high around noon on fine days for optimum growth of transplants.

Morphological development of plants is influenced not only by daylength (or dark period) and light intensity, but also by the spectral composition of light, especially by the lights with wavelength regions of 300-400 nm (ultraviolet), 600-700 nm (red) and 700-800 nm (far-red). When natural light is used as light source, morphological development cannot be controlled independent of photosynthetic growth of plants which is influenced by PPF with wavelength region of 400-700 nm.

In the closed system, the spectral compositions of lamps can be controlled with time as desired, and thus, morphogenesis of plants can be controlled considerably independent of photosynthetic growth of plants.

### 7.3 Protection and Recycling in the Closed System

Insect pests and pathogenic microorganisms can be easily prevented from entering to the closed system. Thus, practically, there is no need to use pesticides. Protection from the outside disturbance by weather and human violence is also easier in the closed system than in the open system. Besides, the closed system produce minimum environmental pollutants such as unused fertilizer to the outside, and can produce transplants free from pathogens with no use of pesticides.

Evapotranspirated and humidified water is condensed at the cooling coils of air conditioner and the condensed water is recycled for irrigation. Thus, actual (net) water consumption in the closed system, a sum of water contained in the transplants and the substrate and water released with outgoing air, is less than 20 % of that in the open system. $CO_2$ released by plants due to respiration during the dark period can be re-used as $CO_2$ source for photosynthesis during the photoperiod. Percentage of $CO_2$ released from inside to outside is considerably smaller (percent released: <5%) in the closed system than in the open system (percent released: 30-70%) when $CO_2$ concentration is higher inside than outside. Thus, $CO_2$ can be enriched efficiently regardless of the outside weather.

### 7.4 Amounts of Resources and Light Energy Required in Transplant and Crop Production

Transplants are produced for planting them in the fields or in the greenhouses. Thus, transplants need to be produced in such a way that minimum waste is emitted during transplant production and crop production and that the transplants require minimum resources to grow and give high yield and quality. If we can produce quality transplants in such a system with the use of small amounts of resources, and can save large amounts of resources in the crop production process, this way of transplant production should be reasonable and acceptable. Therefore, an optimum transplant production system must be determined considering the resources required both in transplant production and crop production systems and the pollutants emitted from those systems.

The space required for transplant production can be at least more than 2 times greater in the open system than in the closed system, because the multi-shelves can be used in the latter. In other words, space utilization efficiency is at least more than 2 times higher in the latter than in the former. Thus, handling and transportation of transplant trays can be minimized, saving the energy and labor costs.

The space required for transplant production is only 1-2 % of the space required for crop production in most cases due to much higher planting density in transplant production. Also, culture period is generally shorter in transplant production than in crop production. The total light energy required for plant growth is a product of PPF, photoperiod per day, the number of days and the area for planting. Thus, the total light energy per plant required for growing plants, which is the total light energy divided by the number of plants, is much lower in transplant production than in crop production. In other words, the light energy required to produce a transplant which fresh (dry) weight is typically 1 - 2 g (100 - 200 mg) is much lower (1 - 2 %) than that to produce harvestable crops which fresh (dry) weight is typically over 100 - 1000 g (10 - 100 g).

### 7.5 Initial and Operation Costs Considering Recent Technical Advancements

The initial and operation costs for producing quality transplants can be lower with the closed system than with the open system if the closed system is designed and operated carefully, based upon the advanced concept and technology.

Main components of the closed system with artificial lighting are the outer structure covered with optically opaque and thermally insulated materials, artificial lighting system, cooling system, $CO_2$ enrichment system, irrigation/fertilizing system and multi-shelves for placing the transplant trays. In the closed system, heating system is not required even in a cold region or climate, provided that the walls are thermally insulated. Fluorescent lamps can be purchased in bulk (1,000 to 10,000 tubes or more) at a large discounted price.

On the other hand, main components of the open system with natural lighting are the greenhouse covered with glass or plastic sheets, shading and/or supplemental lighting system, heating system, ventilation and/or evaporative cooling system, bench/bed system, irrigation system and fertilizing system (and $CO_2$ enrichment system). The shading system is used mainly during the mid-day in summer and supplemental lighting system is used in winter. The $CO_2$ enrichment system can be operated during daytime mostly when roof and side ventilators are closed.

Facilities required in the closed system but not required in the open system are an artificial lighting system, a cooling system with a refrigerator and multi-shelves. Floor area of the closed system with several-shelves should be 25-40 % that of the open system with the same production capacity. Thus, the cost of outer structure for the closed system should be lower than or comparable to that for the open system, i.e., the greenhouse equipped with a transplant production system. Appearance of the structure of the closed system would be a warehouse with thermally insulated walls.

A major component of the operational cost in the closed system is the electricity cost, which is often negligible in the open system. On the other hand, the costs for heating and ventilation in the open system is significant in many cases, while it is negligible in the closed system. Cost for $CO_2$ enrichment in the closed system is negligible if compared to the electricity cost (Ohyama and Kozai, 1998), while it is not in the open system because most of $CO_2$ supplied is released to the outside from the open system.

The lamps can be turned on mainly during the night and in the morning/evening when the outside temperature is lower than that during the daytime, so that the electricity consumption for cooling can be reduced by achieving higher C.O.P. In Japan and some other countries, the generation of electricity is surplus at night and the price of electricity during the nighttime is about 50 % of that at the daytime. Total electricity consumption in the closed system is a sum of electricity consumption of lamps, air conditioners and other electric equipment such as air circulation fans, humidifiers and electric control units.

In the closed system covered with thermal insulation walls, heat energy generated by lamps and other electric units is called the cooling load which must be removed by air conditioners or cooling units. The cooling load due to the lamps accounts for nearly 90 % of the total cooling load. The cooling load due to the heat transmission through the thermal insulation walls and air infiltration accounts for less than 5 % of the total cooling load when the ventilation fans are not operated.

The electricity consumption of air conditioners is determined by the total cooling load divided by the C.O.P. (Coefficient Of Performance) of the air conditioners for cooling. The C.O.P. for cooling is defined as the ratio of cooling capacity in Watt to electric energy consumption of the air conditioner in Watt. Namely, the electricity consumption of air conditioners can be estimated by that of lamps and other electric equipment divided by the value of C.O.P. Therefore, it is important for saving the total electric consumption to reduce the electric consumption of lamps using efficient ones with a reasonable design of lighting system, to keep the C.O.P of air conditioners as high as possible, and to minimize the electric consumption of the other electrical equipment. The electricity cost can be further reduced by use of a co-generation system and the use of exhaust heat. For the discussion in more detail, refer to Kozai (1999).

Computerization and flexible automation for production management can be realized more easily with the closed system than with the open system.

From the discussion given above, It would be worth to conduct research and development to clarify whether or not the operational cost per transplant is really lower in the closed system than in the open system.

People tend to think that the combinative use of natural and artificial light is a reasonable compromise and economical. However, this statement is not valid with respect to transplant production. Because, its initial investment and operational costs should be higher than those of open system, but improvements of quality and quantity of transplants should not be so significant compared with those produced in the closed system.

## 8. Ratio of Chemical Energy Accumulated by Photosynthesis to Electric Energy Consumed for Lighting

Increase in dry mass of transplants starting from seeds or cuttings by photosynthesis under artificial light can be recognized as a conversion process of electric energy to chemical energy of transplant dry mass. For efficient use of electric energy and for reducing the electricity cost per transplant, we need to examine this energy conversion process carefully (Kozai, 1998).

The ratio (R) of chemical energy accumulated in marketed transplants as carbohydrates by photosynthesis (C) to the electric energy consumed by the artificial light source (E) is expressed as:

$$R = C / E = a \cdot b \cdot c \cdot d$$

where a: conversion factor from electric energy to photosynthetically active radiation (PAR) energy, b: ratio of PAR energy received by the transplants to the PAR energy emitted by the light source, c: ratio of chemical energy accumulated as carbohydrates by net photosynthesis to the PAR energy received by the transplants, and d: ratio of the number of marketed transplants to the number of transplants produced.

Numerical examples of factors a, b, c and d are : a: 0.25 for fluorescent lamps and 0.30 for high-pressure sodium lamps, b: 0.2 - 0.5 depending upon the planting density or spacing, c: 0.15 - 0.25 depending upon the photosynthetic characteristics of transplants and transplant canopy structure, and d: 0.7 - 0.9

Thus, a product of those factors ($a \cdot b \cdot c \cdot d$), R, ranges between 0.006 and 0.0338. It is noted that the value of R is relatively low and varies widely; the latter value being 5.6 times higher than the former one. For reducing the initial and operational costs for artificial lighting, the value of R must be maximized. It is technically difficult to increase the value of 'a' for fluorescent and high-pressure sodium lamps drastically in a few years. The maximum possible values would be 0.3 and 0.4 for fluorescent lamps and high pressure sodium lamps, respectively. It is also difficult to increase the value of 'c' greatly. However, The value of 'b' can be increased considerably by improving the time-dependent spacing of transplants and lighting methods. It should be noted that 10 % of improvement for each factor (a, b, c, and d) will result in 46 % increase in 'R' value ($1.1^4 = 1.46$).

## 9. Concluding Remarks

Developments of 'closed production systems' in bioindustry will give significant impacts in the 21st Century on the bio-production systems in agriculture and its related fields. 'Closed production system' with no emission of environmental pollutants and minimum use of resources will be a key concept for designing and operating the production systems.

Understanding the roles of plants, microorganism, insects and animals in the local and global ecosystems and artificial bio-production systems is crucial to solve the global issues in the 21st Century. Understanding and control of the flows of mass and energy in the ecosystems and artificial bio-production systems are essential to develop intelligent systems for producing biomass for food, feed, energy and for environment conservation.

The consumption of artificial energy (electricity and/or fossil fuels) and resources (labor, facility, supplies, etc.) and the release of environmental pollutants can be decreased in the closed transplant production system with artificial lighting than in the open or one-way transplant production system (i.e., greenhouse) with natural light.

The quality of transplants will be much higher when produced in the closed system with artificial light than in the open system with natural light, resulting in higher yield and quality of harvests and less use of agro-chemicals in the fields or greenhouses. However, many engineering and eco-physiological problems remain unsolved, which can be solved by interdisciplinary and international collaborative work in the near future. Microorganisms, insects and small animals will play important roles in the future closed plant production systems.

By using the closed transplant production system as a model of plant production systems in the fields, we may develop semi-closed plant production systems, in a broad sense, using natural light in the open fields. In this case, the plant production systems are almost closed on a large scale as possible as they can be with respect to recycling of water, fertilizers, biomass and others.

**References**

Aitken-Christie, J., T. Kozai and M.A.L. Smith (eds.) 1995. Automation and Environmental Control in Plant Tissue Culture, Kluwer Academic Publishers, Dordrecht, The Netherlands. pp. 574.

Chun, C. and T. Kozai. 2000. Closed transplant production system at Chiba University. In: C. Kubota and C. Chun (eds.) Proceedings for the International Symposium on Transplant Production in Closed System.

Kim, H-H. and T. Kozai. 2000. Production of value-added transplants in closed systems with artificial lighting. In: C. Kubota and C. Chun (eds.) Proceedings for the International Symposium on Transplant Production in Closed System.

Kozai, T., C. Kubota and Y. Kitaya. 1997. Greenhouse technology for saving the earth in the 21st century. (In: E. Goto, K. Kurata, M. Hayashi and S. Sase (eds.) Plant Production in Closed Ecosystems, Kluwer Academic Publishers, Dordrecht, The Netherlands. pp. 139-152.

Kozai, T., C. Kubota, J. Heo, C. Chun, K. Ohyama, G. Niu and H. Mikami. 1998a. Towards efficient vegetative propagation and transplant production of sweetpotato (*Ipomoea batatas* (L.) Lam.) under artificial light in closed systems. Proc. of International Workshop on Sweetpotato Production System toward the 21st Century, Miyazaki, Japan. 201-214.

Kozai, T. 1998b. Transplant production under artificial light in closed systems. In: H-Y. Lu, J-M. Sung and C. H. Kao (eds.) Asian Crop Science 1998 (Proc. of the 3rd Asian Crop Science Conference), Taichung, Taiwan. 296-308.

Kozai, T, K. Ohyama, F. Afreen, S. Zobayed, C. Kubota, T. Hoshi and C. Chun. 1999a. Transplant production in closed systems with artificial lighting for solving global issues on environmental conservation, food, resource and energy, Proc. of ACESYS III Conference, Rutgers University, CCEA (Center for Controlled Environment Agriculture). 31-45.

Kozai, T., C. Kubota, S. Zobayed, Q.T. Nguyen, F. Afreen-Zobayed, J. Heo. 1999b. Developing a

mass-propagation system of woody plants. In: K. Watanabe and A. Komamine (eds.) Proc. of the 12th Toyota Conference: Challenge of Plant and Agricultural Sciences to the Crisis of Biosphere on the Earth in the 21st Century", Landes Company. 293-307.
Kozai, T., (ed.), 1999. Development and Application of Closed-type Transplant Production System for Solving the Global Issues. Yokendo Co., Tokyo. pp. 191 (in Japanese).
Kurata, K. and T. Kozai (eds.) 1992. Transplant Production Systems. Kluwer Academic Publishers, Dordrecht, The Netherlands. pp. 335.
Ohyama, K and T. Kozai. 1998. Estimating electric energy consumption and its cost in a transplant production factory with artificial lighting: A case study. J. High Technology in Agriculture. 10: 96-107. (in Japanese with English summary and captions).
Yoda, N. (ed.) 1993. Trilemma. Mainichi Shinbunsha, Tokyo, Japan. pp. 270. (in Japanese)
Yoshikawa, H. 1998. Problems facing with current engineering (Chapter 2), Takahashi, H. and S. Kato (eds.) Modern Science & Technology and Global Environmental Science, Iwanami Shoten Co., Tokyo. pp. 45-70. (in Japanese).

# CLOSED TRANSPLANT PRODUCTION SYSTEM AT CHIBA UNIVERSITY

Changhoo Chun and Toyoki Kozai
Department of Bioproduction Science, Faculty of Horticulture, Chiba University, Matsudo, Chiba 271-8510, Japan. E-mail: changhoo@midori.h.chiba-u.ac.jp

**Abstract.** A closed system for transplant production was developed at the Matsudo campus of Chiba University, in order to study biological and engineering aspects of transplant production in closed systems and to test the feasibility of closed systems for production of transplants. In this system, virus-free transplants of sweetpotato (*Ipomoea batatas* (L.) Lam.) are vegetatively propagated and produced under sterilized conditions. Many new concepts and technologies are applied to this system to achieve the goal: to produce high-quality transplants at minimum usage of resources, and as scheduled. In the present paper, these concepts and technologies are introduced. A few examples of studies related to transplant production in closed systems using this facility are also described.

**Key index words.** Artificial lighting, automation, production management, sweetpotato.

## 1. Introduction

A closed system for transplant production (Fig. 1) was developed at the Matsudo campus of Chiba University, Japan in February 2000. This system is a research facility to study biological and engineering aspects of transplant production in closed systems and to test the feasibility of closed systems for production of transplants. The basic concept of closed transplant production systems was proposed by Kozai et al. (1999a,b).

*Fig. 1* The closed transplant production system at Chiba University, Japan.

In these systems, high quality transplants, defined as transplants that are physiologically and morphologically superior and show vigorous growth after being transplanted into fields and greenhouses, can be produced regardless of the weather conditions outside.

Although this system can produce transplants of any species, it is presently being used to produce virus-free transplants of sweetpotato (*Ipomoea batatas* (L.) Lam.). Sweetpotato is used as a fresh vegetable, a feed for animals and a raw material for producing biodegradable plastics and hydrogen gas, and vast amounts of quality transplants are expected to be needed in the near future (Kozai et al., 1999a). In this system, the transplants are vegetatively propagated and produced under sterilized condition. This system can be thought of as scaled-up culture vessels of photoautotrophic micropropagation system (Kozai, 1991), in which plantlets are grown using sugar-free medium with enriched $CO_2$ and high PPF).

For sweetpotato production, vine cuttings are conventionally used as transplants. Because the vine-cutting transplants do not have roots, almost 100% of them wilt after



*C. Kubota and C. Chun (eds.), Transplant Production in the 21st Century,* 20–27.

transplanting. Even though large portions of the transplants recover from the wilting in a few weeks, the delay of the early growth caused by the wilting after transplanting may decrease the final yield and quality of sweetpotato. In the newly developed system, however, sweetpotato transplants with root balls are produced in multi-cell trays for direct transplanting into fields (Islam et al., 2000). In this case, transplants keep growing without wilting and a higher yield can be expected. Machines can be used for transplanting the sweetpotato transplants with root balls, which is difficult with vine-cutting transplants.

## 2. Sterilized area

This system has a floor area of 500 square meters and consists of a sterilized area and a non-sterilized area. The sterilized area consists of a transfer room, three production rooms (micropropagation room and production rooms 1 and 2) and a low-temperature storage room 2.

To enter the sterilized area, workers must pass through air showers and materials must be sterilized and transferred though double-door systems which have UV lamps. To maintain a positive pressure in the sterilized area, air is supplied from outside through a ventilation system that has filters and air heaters. The amount of air supplied to the production rooms is kept at a minimum by maintaining the production rooms as closed as possible, while the amount of air supplied to the transfer room is adjusted to meet worker's needs.

### 2.1 Transfer room

Plantlets grown in the production rooms, each with 4-6 unfolded leaves, are "transferred" into the transfer room. In this room, workers prepare explants (single-node leafy cuttings) from the plantlets and place them on multi-cell trays. The newly prepared trays with explants are placed on conveyers and "transferred" back to the production rooms. Plantlets that have grown in the production rooms can also be "transferred" into low-temperature storage room 2 through the transfer room to slow down their growth for a few days or a few weeks in order to meet production schedules.

Two plant growth measuring systems are located in the transfer room. Each system consists of two digital cameras and a computer. By locating individual explants or a multi-cell tray with transplants in the system, the length, the width and the area of leaves of individual explants, and the maximum and height and volume of the transplant canopy can be estimated. These systems also determine the change in leaf color of explants and transplants. The quality of explants that are used for propagation and the quality of transplants that are consigned to consumers can be evaluated with these systems.

### 2.2 Production rooms

The production rooms are where the sweetpotato plantlets are grown to be propagated and/or to be consigned as transplants. There is no difference in the facilities or operational methods among these rooms. The only differences among the rooms are in the degree of cleanness or sterilization and the numbers of basic modules that they contain. The degree of cleanness is highest in the micropropagation room and lowest in the production room 2. Plantlets can be moved from the rooms with a higher degree of cleanness to one with a lower degree of cleanness, but not in the opposite direction. Two basic modules are installed in the micropropagation room and four are installed in each of production rooms 1 and 2. All of the rooms have a tray transporting system and an irrigation system.

### 2.2.1 Basic Module

The basic module is a bookshelf-like structure located in each production room in which plantlets in multi-cell trays are placed and cultured. Each basic module has a lighting unit, an air conditioning unit, a control unit and seven shelves (2670 mm x 685 mm). The vertical spacing between shelves is 520 mm. Eight multi-cell trays (approx. 300 mm x 600 mm) can be accommodated on each shelf. Each shelf is lighted with sixteen 32-W fluorescent lamps and three 16-W fluorescent lamps. The output of each lamp (excluding the 16-W lamps) can be controlled with pre-decided lighting patterns to achieve the desired photosynthetic photon flux (PPF) uniformly distributed at the surface level of empty trays. The PPF of each shelf is automatically checked once a day to find any malfunctioning lamps using a PPF sensor installed on a tray transporting unit (described later). The PPF on the empty tray surface of each shelf can be controlled between 140 to 300 $\mu$mol $m^{-2}$ $s^{-1}$ when the lamps are turned on.

The air conditioning unit consists of three home-use air conditioners, three mixing fans and a humidifier that are installed on the top of each module. These devices are operated with pre-decided operating patterns designed to minimize energy consumption for eliminating the heat produced by the basic module. The conditioned air is distributed by the mixing fans to each shelf through an air duct and through holes in the backside panel of the shelves. Air temperature of the basic module can be controlled in the range between 22 and 30 °C, relative humidity can be controlled up to 80 %, and air current speed can be controlled in the range between 0 and 1 m $s^{-1}$.

A Local Operating Network (LON) unit, which is a remote control/monitoring system, independently operates the devices of each basic module. The devices of each basic module such as air conditioners, fans, lamps, a humidifier and environmental measuring devices are operated by selecting an operation pattern from ten patterns memorized in the LON unit. Maximum, minimum and averaged values of environmental parameters of each basic module and important incidents such as errors and warnings are reported to the gateway computer. These decentralized functions of each module give a great amount of freedom for future scale-ups of this closed system.

During the dark period, if necessary, the air of each room circulates through a filtering system to remove dust, pathogens, insects and debris. An air sampler is installed in the air circulation line for monitoring microbes. The air pressure of the air circulation line is monitored to know when to change the filters.

### 2.2.2 Tray transporting system

Since workers are not permitted to enter the production area for handling the trays with plantlets, the transportation of trays needs to be automated. A tray transporting system is installed in each room of the production area. This system consists of a tray transporter, conveyers, and a double-door system.

The tray transporter (Fig. 2) was developed by modifying a transporter that is used for inventory management in warehouses. It is mainly for transporting trays with plantlets from one location to another in the same room and/or from a room of the production area to the transfer room. The tray transporter has another function. It carries sensors and a digital camera for measuring environmental factors at any location in the basic module and for capturing continuous images during its operation.

A conveying system is installed between the transfer room and each room of the production area. It is used for transferring trays both ways. A double-door system is installed on the conveyer to separate each room of the production area from the transfer room. A door opens when trays are passed on the conveyer, but two of the doors never open at the same time to maximize the independence of each room.

*Fig.2* A tray transporter installed between the basic modules in each production room.

### 2.2.3 Irrigation system

An automatic irrigation system is installed on each tray transporter. The irrigation system was developed based upon the concepts of microprecision irrigation (Murase, 2000).

Only the proper amount of nutrient solution for a particular plant (or for a particular cell of a tray) is delivered from this irrigation system. Therefore no draining or recycling process is needed and a very high efficiency of water usage can be expected. It is also possible to adjust the amount of water for a certain plant (or cell).

When a command for irrigating a particular tray is given, the transporting/irrigation system moves to the designated tray. It lifts the tray, weighs it, and moves it to the irrigation device. Then, the irrigation device with 72 needle-type nozzles moves up and down for injecting nutrient solution. An irrigation pattern including injection positions and amounts of nutrient solution for each injection can be chosen out of 64 patterns. After injection, the tray is weighed again and then returned to the original location.

A device that automatically prepares nutrient solutions is located in the machine room (a non-sterilized area). This device dilutes stock solutions to pre-set concentrations and supplies them to the nutrient tank (1 l) of each irrigation system. The nutrient solution is delivered to each cell by gas pressure supplied by a tank of compressed nitrogen gas. Solenoid valves are installed in each of the 72 lines. The amount of nutrient solution injected into a cell is controlled by opening a valve for a specified time. A pan for catching nutrient solution and/or media that drip from a tray is installed at each irrigation system.

## 2.3 Low temperature storage room 2

Low temperature storage room 2 (2942 x 1592 x 2230 mm(H)) is located in the sterilized area. It is used for storing plantlets for a few days or a few weeks when it is desired to maintain their growth at a minimum. This room is dimly lighted and kept at a temperature in the range between –3 and 15 °C. An air conditioner (1.5 kW) and a humidifier (1.0 l $h^{-1}$) are installed. Two movable racks, each with four shelves, are placed in this storage room. Four trays with plantlets can be accommodated on each shelf. A specially designed lighting panel (32 W) is installed in each shelf, and the PPF of each shelf can be controlled with a light controller.

**3. Non-sterilized area**

The areas for transplant production in the non-sterilized area include a cleaning room, a preparation room, low temperature storage room 1 and a control room.

**3.1 Cleaning room**

Reused trays, supporting materials (media) and other materials are first cleaned in a washing area in this room. Then, they are packed in plastic bags and sterilized by dipping in hot water (about 70 ºC) for a few hours. The water is heated mainly by a solar heater installed on the roof of this system, and a boiler is used as an auxiliary heater. Materials that have been cleaned and sterilized are transferred to the preparation room. This room has a door of the low temperature storage room 1 and the transplants are transferred to this room to be packed and shipped.

**3.2 Preparation room**

Peoples and materials must pass through this area to enter the transfer room. It has a door to the air shower room and a pass-box. There is an autoclave in this room to sterilize the tools and materials that are used in the sterilized area.

**3.3 Low temperature storage room 1**

All transplants to be shipped must pass this area through a pass box between the transfer room. Transplants can be directly shipped out or stored in this storage room. The transplants can be stored here for anywhere from a few hours to a few weeks to adjust to shipping schedules. It is important to have a buffering capacity to adjust to sudden changes in the amounts of orders and/or scheduled or nonscheduled incidents in the production process. To provide more degrees of freedom to the production facility, it is important to have not only a greater capacity in terms of space but also the ability to hold transplants for a longer duration. Currently, sweetpotato transplants can be stored for about two weeks without any decrease in their economic value.

This room has six movable racks with lamps for dim lighting, similar to those used in low temperature storage 2. Four air conditioners (1.1 kW) and two humidifiers (1.5 l $h^{-1}$) are installed for controlling air temperature and relative humidity of this storage room.

**3.4 Control room**

Computers for the production managing system, a plant operating system and a data base system are located in the control room.

**3.4.1 Production managing system**

The production managing system is the system that manages the whole processes of this facility. It consists of a few subsystems: a production planning subsystem, a production simulating subsystem, a process managing subsystem and a research supporting subsystem.

The production planning subsystem finds the best plan to produce and ship out the transplants as the orders are received. The plan takes into consideration the resources, such as spaces, labor, plant materials, and time that are available. This subsystem develops a production schedule that specifies how many explants should be prepared on a given day, if a particular order is found to be within the production capacity of the system.

The production simulating subsystem is a kind of virtual factory that simulates the production with several models. The growth of trays of plantlets is simulated based upon the environmental conditions where the trays are located. The concepts of decentralized and object-oriented systems (Hoshi, 1992) are used for simulating the actual production activities in this system.

If the production plan is found to be feasible by the production simulating subsystem, the process managing subsystem gives a various instructions related to tray transporting, irrigation, manual operations by workers and others. Each worker carries a handy terminal (a small communicating device) to receive instructions from the subsystem. The workers only need to follow the instructions that are shown on the small monitor of the handy terminal.

The research supporting subsystem is used to assist researchers with their experiments. The researchers can manage their plantlets independently from the regular production processes with the support of this subsystem.

### 3.4.2 Plant operating system

The plant operating system is the system that actually operates the facilities and monitors the actions of the facilities. It consists of a few subsystems: a facility operating subsystem, a plant monitoring subsystem and an alarm subsystem.

The facility operating subsystem gives an order of particular actions to the facilities with the commands from the process managing subsystem of the production managing system. The operation of each facility and the results by the action are monitored by the plant monitoring subsystem. They can be monitored on the computer monitors. If problems are detected by the plant monitoring subsystem, the alarm subsystem takes several steps to protect the facilities. For major alarms, this subsystem sends a message to the cellular phones of the management staff, and thus the staff can check the facilities and reset them. The history of alarms and warnings are stored in a data base system.

### 3.4.3 Data base system

Data related to the growth of transplants and the plant operation (which are produced and/or retrieved by the production managing system, the plant operating system and the plant growth measuring system) are saved and managed in the computer of the data base system. The data saved in this system can be searchable and compatible with other software. Not only numerical data but also texts and images can be saved in this data base.

## 3.5 Research area

The research area has three laboratories (a biological lab, a physical lab, and a basic module lab) for research and other work that is done to support production. The biological lab is used to analyze plant samples and microorganisms in the air on a regular basis. The physical lab is used to examine samples of the nutrient solutions and supporting materials. The basic module lab is used for examining environmental details and conducting other experiments in the basic module.

## 4. Concluding remarks

Using the closed system at Chiba University, the authors and their research group conduct various studies relating to transplant production. Energy and mass analyses in the production area are important for improving system performance. These analyses can help to decrease energy consumption and production costs. From preliminary experiments with similar closed systems, the electricity consumed for producing a transplant is about 0.35 MJ (= 0.1 kWh), which costs 1.5 Japanese yen (Ohyama and Kozai, 1998). We believe that both the initial and operational costs of the closed system can be decreased by selecting facilities that have better performances and proper capacities and by optimizing their operation.

The performance of each unit operation should be optimized. The energy, material, time and labor required in a unit operation should be first carefully analyzed to maximize the performance. Then the mutual relations among multiple operations need to be inspected. Studies of system engineering are essential to optimize whole processes and finally maximize the performance of the whole system.

In this closed system, many new concepts and technologies are introduced. Some of the pre-existing technologies and concepts are applied to this system, and several concepts and technologies have been newly developed for this system. It gives us new research areas to work on. Each technique or concept can be applied to other industries in agriculture, while this new concept of a closed system for transplant production itself can be utilized as a new agricultural business.

Storing transplants is different from storing fruits and flowers. The growth of transplants needs to be retarded but their growth abilities should be maintained during storage, because they have to perform the best growth after transplanting (Kubota and Kozai, 1995). Storage technique for plantlets that are in the middle of the propagation process and that need to be moved back to the production is a new research topic.

By manipulating the environmental conditions during transplant production, production of high quality transplants or highly value-added transplants becomes possible. Chun et al. (2000a,b) and Kim et. al (2000) demonstrated that the quality of transplants and their future performances after being transplanted can be improved by manipulating the environment during transplant production. To broaden the application of this idea to closed systems, many studies related to plant physiology are needed, too.

The authors believe that the research results from the closed system at Chiba University will broaden the new agricultural business and provide new research areas. And these results can be utilized for further improvement of the systems and for development of small-sized systems that farmers can use for production of their own transplants.

### References

Chun, C., T. Kozai, C. Kubota, and K. Okabe. 2000a. Manipulation of bolting and flowering in a spinach (*Spinacia oleracea* L.) transplant production system using artificial light. Acta Hort. 515:201-206.

Chun, C., A. Watanabe, H.-H. Kim, T. Kozai, J. Fuse. 2000b. Bolting and growth of Spinach (*Spinacia oleracea* L.) can be altered by using artificial lighting to modify the photoperiod during transplant production. HortScience 35:(4) (in press).

Hoshi, T. 1992. Objected oriented software development support system of environmental controller in plant growth factories (in Japanese with English summary and captions). J. High Technology in Agriculture 3(2):129-136.

Islam, A.F.M.S., C. Chun, M. Takagaki, K. Sakami, and T. Kozai. 2000. Yield and growth of sweetpotato using plug transplants as affected by their ages and planting depths. In: C. Kubota and C. Chun (eds.) Proceedings for the International Symposium on Transplant Production in Closed System (in

press).
Kim, H-.H., C. Chun, T. Kozai, and J. Fuse. 2000. The potential use of photoperiod during transplant production under artificial lighting condition on floral development and bolting, using *Spinacia oleracea* L. as a model. HortScience 35:43-45.
Kozai, T. 1991. Photoautotrophic micropropagation. In Vitro Cell. Biol. 27P:47-51.
Kozai, T., K. Ohyama, F. Afreen, S. Zobayed, C. Kubota, T. Hoshi, and C. Chun. 1999a. Transplant production in closed systems with artificial lighting for solving global issues on environmental conservation, food, resource and energy, Proc. of ACESYS III Conference, Rutgers University, CCEA (Center for Controlled Environment Agriculture) , 31-45.
Kozai, T., C. Kubota, S. Zobayed, Q.T. Nguyen, F. Afreen-Zobayed, J. Heo. 1999b. Developing a mass-propagation system of woody plants. In: K. Watanabe and A. Komamine (eds.) Proc. of the $12^{th}$ Toyota Conference: Challenge of Plant and Agricultural Sciences to the Crisis of Biosphere on the Earth in the 21st Century", Landes Company, 293-307.
Kubota, C. and T. Kozai. 1995. Low-temperature storage of transplants at the light compensation point: air temperature and light intensity for growth suppression and quality preservation. Sci. Hortic. 61:193-204.
Murase, H. 2000. Microprecision irrigation system for transplant production. In: C. Kubota and C. Chun (eds.) Proceedings for the International Symposium on Transplant Production in Closed System (in press) .
Ohyama, K. and T. Kozai. 1998. Estimating electric energy consumption and its cost in a transplant production factory with artificial lighting: A case study (in Japanese with English summary and captions). J. High Technology in Agriculture, 10:96-107.

# ELECTRIC ENERGY, WATER AND CARBON DIOXIDE UTILIZATION EFFICIENCIES OF A CLOSED-TYPE TRANSPLANT PRODUCTION SYSTEM

Katsumi Ohyama[1], Toyoki Kozai[1] and Keita Yoshinaga[2]
[1]Faculty of Horticulture, Chiba University, Matsudo, Chiba 271-8510, Japan. E-mail: ohyama@green.h.chiba-u.ac.jp
[2]Nepon Inc., Shibuya, Tokyo 150-0002, Japan.

**Abstract.** The electric energy utilization efficiency ($E_E$), water utilization efficiency ($E_W$) and $CO_2$ utilization efficiency ($E_C$) of a closed-type transplant production system (CTPS) were estimated. Sweetpotato (*Ipomoea batatas* (L.) Lam. cv. Beniazuma) plants were grown for 15 days in the CTPS. The $E_E$, $E_W$ and $E_C$ of the CTPS were 0.006, 0.93 and 0.92, respectively. These results indicate that $E_W$ and $E_C$ are more than several times higher in the CTPS than in an open-type transplant production system such as a greenhouse. The $E_E$ of the CTPS was relatively low compared to $E_W$ and $E_C$ and can be further improved by more efficient lighting and air conditioning systems. Therefore, the CTPS is suitable for transplant production with reduced consumption of energy and resources.

**Key index words.** $CO_2$ balance, electric energy consumption, energy balance, Sweetpotato (*Ipomoea batatas* (L.) Lam.), water balance.

## 1. Introduction

Development of a closed-type transplant production system (CTPS) with minimum uses of energy and resources will contribute to solving the global issues on environmental pollution and shortages of food, energy and natural resources (Kozai et al., 1999). The CTPS is defined as a transplant production system covered with opaque and thermally-well insulated walls, where the energy and mass transfer between inside and outside the system is controlled and/or restricted, and thus, using artificial light as a light source for growing transplants is required. Environmental conditions inside the CTPS can be controlled as desired independent of the outside weather. Ohyama and Kozai (1998) reported that the electricity cost for transplant production in the CTPS was 5-10% of the transplant market price in Japan. In the CTPS, water and $CO_2$ are easily reused, and thus the consumption of water and $CO_2$ can be decreased.

Transplants are generally produced in an open-type transplant production system (OTPS) such as a greenhouse. The OTPS is defined as a transplant production system covered with transparent and thermally insulated walls, although the insulation is poor, and the energy and mass transfer between inside and outside the system is not well controlled. Natural light is mainly used as a light source for transplant production in the OTPS. Environmental conditions inside the OTPS are affected by the outside weather. Water can escape from the OTPS relatively easily because the number of air exchanges is higher in the OTPS than in the CTPS. Therefore, much water is consumed for transplant production. This is similar for $CO_2$, when atmospheric $CO_2$ is enriched in the OTPS.

High quality transplants can be produced in the CTPS with an efficient use of electric energy. $E_W$ and $E_C$ are more than several times higher in the CTPS than in the OTPS. It is necessary to further increase the $E_E$, $E_W$ and $E_C$ of the CTPS to minimize energy and resource consumption. However, little information is available on these



C. Kubota and C. Chun (eds.), *Transplant Production in the 21st Century*, 28–32.

utilization efficiencies of the CPTS. The objective of this study was to estimate the $E_E$, $E_W$ and $E_C$ of the CTPS.

## 2. Materials and Methods

The CTPS used in the present experiment is shown in Fig. 1. This system was covered with thermally insulated walls, which were composed of 5-cm Styrofoam and wooden panels. White fluorescent lamps were used as a light source. A home-use air conditioner was used for control of air temperature inside the system. A centrifugal humidifier and a humidistat were used for relative humidity control.

Sweetpotato (*Ipomoea batatas* (L.) Lam., cv. Beniazuma) plants were grown for 15 days in the CTPS. Single node cuttings were planted in 38-cell plug trays. PPF at the tray surface level were set at 100, 200 and 300 μmol $m^{-2}$ $s^{-1}$ on days 0-2, 3-11 and 12-14, respectively. The photoperiod was 16 h. Air temperature inside the CTPS was set at 30°C. Atmospheric $CO_2$ concentration and relative humidity inside the CTPS during the photoperiod were set at 1000 μmol $mol^{-1}$ and 80%, respectively, and those in the dark period were uncontrolled.

The electric energy, water and $CO_2$ balances are schematically shown in Fig 2. The

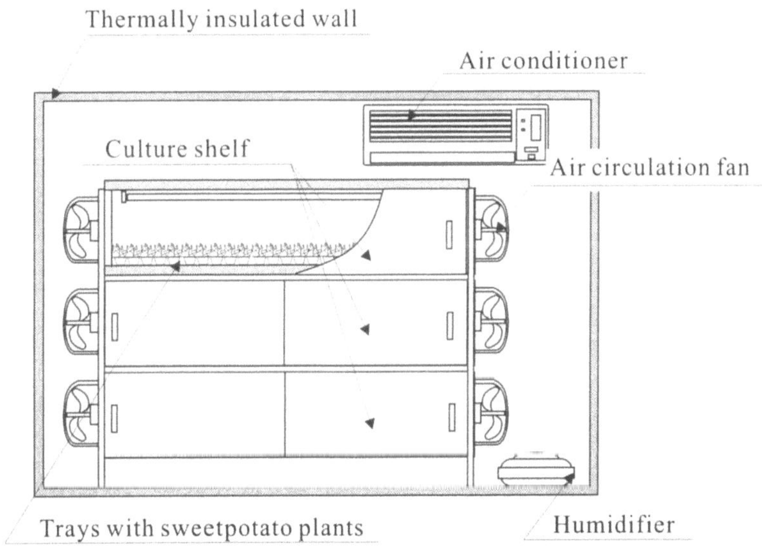


*Fig. 1* Schematic diagram of closed-type transplant production system used in the present experiment.

$E_E$ of the CTPS was estimated by using the following equation:

$$E_E = \frac{Q}{W} \quad \cdots (1)$$

where $Q$ is the chemical energy contained in the plants, which was fixed by photosynthesis, and $W$ is the electric energy consumption of the CTPS during the experiment (MJ $m^{-2}$). $Q$ was estimated by the product of the dry weight increase and a conversion factor (0.02 MJ $g^{-1}$, Larcher, 1995). $W$ was measured with a watt meter.

The $E_W$ of the CTPS was estimated by using the following equation:

$$E_W = \frac{D + C_p + C_s}{I + H} \quad \cdots (2)$$

where $D$, $I$ and $H$ are the amounts of dehumidified, irrigated and humidified water during the experiment (kg $m^{-2}$), respectively; $C_p$ and $C_s$ are the increases in water contents of plants and substrate during the experiment (kg $m^{-2}$), respectively. $D$ is the cumulative amount of drained water from the air conditioner. $C_P$ was estimated by the difference between increases in fresh and dry weights during the experiment. $C_s$ was estimated by subtracting the fresh weight of plants and the dry weight of the tray with substrate from the tray weight at the end of the experiment. $I$ is the cumulative value of the tray weight difference before and after irrigation. $H$ is the cumulative decrease in the

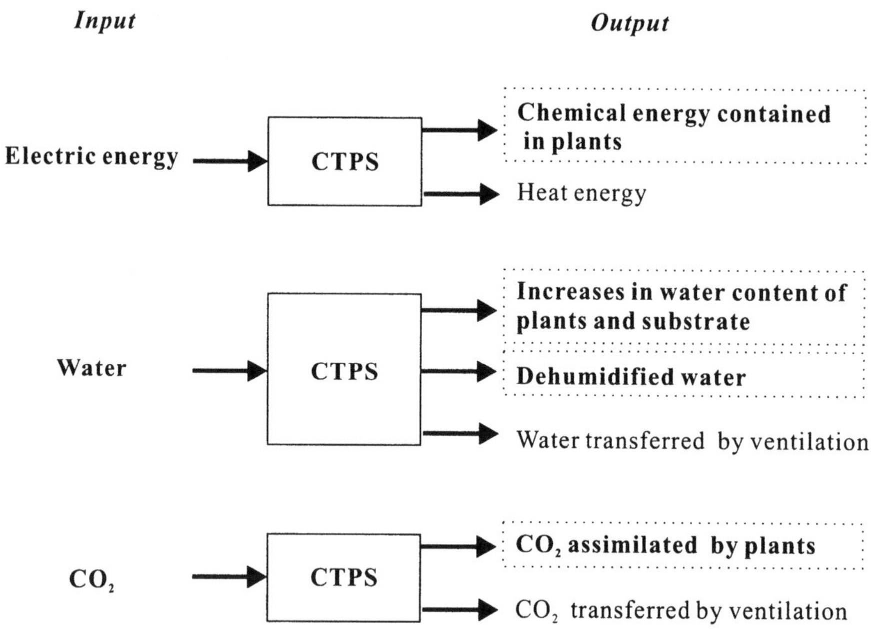


*Fig. 2* Schematic diagram of electric energy, water and $CO_2$ balances of the closed-type transplant production system. The boxes drawn with dotted line indicate usable products.

humidifier tank. We assumed that the dehumidified water could be reused for irrigation and humidification in this experiment.

The $E_C$ of the CTPS was estimated by using the following equation:

$$E_C = \frac{N}{S} \quad \cdots (3)$$

where $N$ and $S$ are the amounts of $CO_2$ assimilated by the plants through net photosynthesis, and supplied during the experiment (mol $m^{-2}$), respectively. $N$ was estimated by the product of dry weight increase and a conversion factor (1.57 mol $gDW^{-1}$, Šetlík and Šesták, 1971). $S$ was estimated by the product of the $CO_2$ supply rate ($2.5 \times 10^{-4}$ mol $s^{-1}$) and the integrated time of the solenoid valve at the open position during the experiment.

**3. Results and Discussion**

The cumulative electric energy consumption of lamps, air conditioner and equipment other than the lamps and air conditioner during the experiment (15 days) were 376, 81 and 41 MJ $m^{-2}$ (497 MJ $m^{-2}$ in total), respectively. The increase in dry weight during the experiment was 158 g $m^{-2}$. The chemical energy contained in the plants was estimated to be 3 MJ $m^{-2}$. The $E_E$ of the CTPS was estimated to be 0.006. It may be possible to further increase $E_E$ of the CTPS, e.g., by 1) improving the lighting system through changes in the geometry of the plants-lamps and the use of reflective materials (e.g., Ikeda et al., 1992), and 2) using an air conditioner with a high coefficient of performance (C.O.P.). It should be mentioned that many technical improvements in fluorescent lamps and home-use air conditioners have been reported recently. The ratio of integrated luminous flux to the electric energy consumption rate of fluorescent lamp (lm $W^{-1}$) has been doubled during the last two decades (Yamamoto, 1998). The energy consumption rate of home-use air conditioner has been reduced to 1/2 in resent years (Ohyama, 1998). Hence, the electric energy consumption of the CTPS would be reduced, and thus, the $E_E$ of the CTPS would be increased when newly developed fluorescent lamps and air conditioners were introduced.

The $E_E$ of the CTPS was low compared to the $E_W$ and $E_C$ of the CTPS. The current market price of sweetpotato plant is 1 to 1.5 US $. The major operational cost in the CTPS is the cost of electricity. In this experiment, electricity cost per plant was 0.05 US $, which represent 3-5% of the market price. Although the $E_E$ of the CTPS is relatively low, the cost of electricity is a relatively small percentage of the total plant production costs, therefore the feasibility of commercialization of the CTPS is not limited by this resource.

The cumulative amounts of irrigated, humidified and dehumidified water during the experiment were 49.5, 24.3 and 55.9 kg $m^{-2}$, respectively. The increases in the water content of the plants and substrate during the experiment were 1.6 and 10.8 kg $m^{-2}$, respectively. The $E_W$ of the CTPS was estimated to be 0.93. Hence, transplants can be produced in the CTPS with minimum consumption of water. It should be noted that the $E_W$ of the CTPS is increased with decreasing number of air exchanges. The $E_W$ would be increased when dehumidified water was reused for irrigation and humidification. Usually, water recycling system does not exist in the OTPS and the number of air exchanges of the OPTS is higher than that of the CTPS. The $E_W$ of the OTPS was estimated to be less than 0.17 when dehumidified water did not reused. The $E_W$ of the CTPS was 5 times higher than that of the OTPS. Therefore, the water consumption for transplant production was lower in the CTPS than in the OTPS.

The cumulative amounts of $CO_2$ supplied to the system and assimilated by plants during the experiment were 6.1 and 5.6 mol $m^{-2}$, respectively. The $E_C$ of the CTPS was estimated to be 0.92. These results show that transplants can be produced in the CTPS with a high efficient use of $CO_2$. Decreasing the number of air exchanges can also increase the $E_C$ of the CTPS. Yoshinaga et al. (2000) reported that the maximum value of $E_C$ of the OTPS was 0.90 when the number of air exchanges was 0.1 $h^{-1}$. Usually, the number of air exchanges of the OTPS is higher than 0.1 $h^{-1}$ in order to maintain the air temperature inside the OTPS during daytime. The $E_C$ of the CTPS was higher than that of the OTPS, and thus, $CO_2$ consumption for transplant production was lower in the CTPS than in the OTPS.

Introducing a ventilation system to the CTPS in the winter season would decrease electric energy consumption for the air conditioner (Ohyama and Kozai, 1998). On the other hand, the $E_W$ and $E_C$ of the CTPS would be decreased. Our results show that 7% of the input water and 8 % of the input $CO_2$ into the CTPS were not used for plant production. This water and $CO_2$ were released from inside to outside the CTPS by ventilation. It is necessary to decrease the amounts of water and $CO_2$ released in order to decrease the environmental impact of the system. Therefore, ventilation of the CTPS should be considerably restricted.

## 4. Conclusions

The above results indicate that the closed-type transplant production system (CTPS) is suitable for transplant production with reduced consumption of energy and resources. The electric energy utilization efficiency ($E_E$), water utilization efficiency ($E_W$) and $CO_2$ utilization efficiency ($E_C$) of the CTPS were 0.006, 0.93 and 0.92, respectively. The $E_W$ and $E_C$ were more than several times higher in the CTPS than in an open-type transplant production system (i.e., a greenhouse). In this experiment, the electricity cost per plant was 3-5% (approximately 0.05 US $) of the market price per plant (1-1.5 US $). Although the $E_E$ of the CTPS is relatively low, the cost of electricity is a relatively small percentage of the total plant production costs, therefore the feasibility of commercialization of the CTPS is not limited by consumption of electric energy.

**References**

Ikeda, A., Y. Tanimura, S. Nakayama and K. Iwao. 1992. Study on lighting design in plant factory using fluorescent lamps. Environ. Control in Biol. 30 (1):9-15 (in Japanese with English abstract).

Larcher, W. 1995. Carbon utilization and dry matter production. In Physiological plant ecology. Springer, Berlin. pp. 57-166.

Kozai, T., K. Ohyama, F. Afreen, S. Zobayed, C. Kubota, T. Hoshi and C. Chun. 1999. Transplant production in closed systems with artificial lighting for solving global issues on environment conservation, food, resource and energy. Proc. of ACESYS III conference, Rutgers University. pp. 31-45.

Ohyama, K. and T. Kozai. 1998. Estimating electric energy consumption and its cost in a transplant production factory with artificial lighting: A case study. J. of High Technology in Agriculture.10 (2):96-107 (in Japanese with English abstract).

Ohyama. K. 1998. Development of energy saving air conditioner- the reason why electric energy consumption rate of air conditioner had been half-. The Transactions of the Institute of Electrical Engineers of Japan 118-D (6):813 (in Japanese).

Šetlík, I. and Z. Šesták. 1971. Use of leaf tissue samples in ventilated chambers for long term measurements of photosynthesis. In: Šesták, Z., J. Èatský and P.G. Jarvis, (ed.),Plant photosynthetic production manual of methods, Dr W. Junk n. v., publishers. pp. 316-342.

Yamamoto, S. 1998. Fundamental knowledge of quality and quantity of lighting for energy saving. Railway & electrical engineering 9 (12):5-11 (in Japanese).

Yoshinaga, K., K. Ohyama and T. Kozai. (2000) Energy and mass balance of a closed type transplant production system (3) carbon dioxide balance. J. of High Technology in Agriculture. (in Japanese with English abstract) (Submitted).

# MICROPRECISION IRRIGATION SYSTEM FOR TRANSPLANT PRODUCTION

Haruhiko Murase
Osaka Prefecture University, College of Agriculture, 1-1, Gakuen, Sakai 599-8531 Japan.
E-mail: hmurase@bics.envi.osakafu-u.ac.jp

**Abstract.** While Low-Input Sustainable Agriculture (LISA) has been losing the support of the agricultural sector, since LISA decreases profits, the implementation of Precision Agriculture (PA) has been embraced. PA has become a promising way to handle a complex system and has received significant support from both agricultural and industrial sectors. Although a plant factory is a large-scale complex system, it is much simpler than an open field system. Most of the environmental factors in a fully controlled plant factory are observable and controllable; a plant factory can be optimized more easily than an open field. Limits and optimum design concepts must be applied in order to establish economically feasible, fully controlled plant growing factories. This paper proposes Microprecision Agriculture for a fully controlled plant factory. A practical example of microprecision technology in biomechatronics is introduced.

**Key index words.** closed system, mechatronics, plant factory, plug production, precision agriculture, variable rate technology.

## 1. Introduction

LISA (Low-Input Sustainable Agriculture) program was launched to reduce the use of off-farm inputs with the greatest potential to harm the environment or the health of farmers and consumers (Hess, 1991). This is one of the operational definitions of sustainable agriculture. However, LISA has not gained appreciable support from the agricultural sector, since LISA decreases profits. On the other hand, the implementation of Precision Agriculture (PA) has been embraced (Blackmore, 1999). The difference between PA and LISA is that PA requires technological innovation, whereas LISA always involves revising or improving traditional practices. The extensive application of information technology, including GPS and GIS, to agriculture, and the development of agricultural machinery for PA are major factors in the difference between PA and LISA. The machinery includes equipment to implement variable rate technologies, and vehicles specifically designed to gather the information required to generate yield and soil maps.

The open field agricultural system is a typical example of a large-scale complex system that has attracted the attention of researchers and scientists in various scientific and engineering fields. Now PA promises to handle such complex systems, and has received significant support from both agricultural and industrial sectors. Although a plant factory is also a large-scale complex system, it is much less complex than an open field system. There are quite a few plant factories operating commercially in Japan. PA is nothing but integrated technology, designed to optimize the cultivation process. The fully controlled environment of a plant factory can be considered an ideal cultivation system in terms of alternative agriculture. Most of the environmental factors in a fully controlled plant factory are observable and controllable; a plant factory can be optimized more easily than an open field. This paper proposes using microprecision agriculture in a fully controlled plant factory. Microprecision agriculture can be attained by using plant factories to realize profitable alternative agriculture. This article reviews the scientific and technological achievements of plant factories as alternative agriculture, and



*C. Kubota and C. Chun (eds.), Transplant Production in the 21st Century,* 33–37.

introduces a hardware system developed to implement microprecision agriculture in a plug seedling production factory.

## 2. Development of Plant Factories

In 1970, a plant growth system consisting of systematically integrated growth chambers was used to demonstrate that plant growth can be significantly improved by applying optimum growth conditions in terms of environmental factors such as temperature, humidity, light intensity, and $CO_2$ gas concentration. Those scientific achievements motivated the early development of a closed plant growing system with a controlled artificial environment. The research and development was extended to the development of plant factories that involve technologies such as process control for the plant growth environment, mechanization for material handling, system control for production, and computer applications. The advantages of a plant factory include production stabilization, higher production efficiency, and better quality management of products through a shortened growing period, better conditions, lower labor requirements, and easier application of industrial concepts (Takatsuji, 1987).

A precise definition of a plant factory has yet to be established. In a broad sense, a plant factory is defined as "a production system in which plants are under continuous production control throughout the growth period until harvest". A narrow definition is "a year-round plant cultivation system in a completely artificial environment. There are many commercially operating greenhouse-type plant factories, which are heavily equipped with sophisticated environment control systems, machines, instrumentation, and computers. Some use only natural light, but others occasionally use artificial light as a supplement during the seasons with low solar radiation. Ultimately, a greenhouse-type plant factory is not an ideal system because of unavoidable, unpredictable, and uncontrollable external disturbances. However, growers have more readily accepted greenhouse type plant factories, mainly because of current energy costs and the high initial investment required for a fully artificial plant factory.

## 3. Microprecision Technology

A closed fully controlled plant-growing factory is much more advantageous in terms of minimizing waste. Limits and the optimum design concept` have to be applied in order to establish an economically feasible, fully controlled, plant-growing factory. To achieve this objective, microprecision technologies must be developed. Microprecision does not mean a higher order of engineering precision. Microprecision in agriculture is the technological means to first identify what is needed and how much is needed as precisely as possible, and then to perform the job required to fulfill the identified quantitative and qualitative needs as precisely as possible. Microprecision technologies involve sensing, modeling, control, information, and mechatronics for plant production. Basic microprecision technologies are already available; they are SPA (Speaking Plant Approach to environmental control), AI (Artificial Intelligence: expert systems, neural networks, genetic and photosynthetic algorithms, etc.), bioinstrumentation, non-invasive measurement, biomechatronics, and biorobotics (Hashimoto and Nonami, 1992).

## 4. Microprecision Irrigation System

This section introduces an example of microprecision technology that has actually been implemented in a plug seedling production factory. A microprecision irrigation system was developed for a plug production system. It is a kind of variable rate technology for precision agriculture. The traditional irrigation method for plug production is overhead watering, which provides growing plug seedlings with excess nutrient solution. In the traditional system, some of the irrigated nutrient solution is absorbed by the substrate (soil) and then taken up by the plant, some stays on the leaf surface, and the rest falls to the ground and is wasted. This has been a major drawback of using the traditional irrigation method in plug production, both economically and environmentally.

### 4.1 Irrigation concept

The seedling should be irrigated with only the proper amount of water (nutrient solution) for the particular plant, delivered to the developing roots. This concept assures the minimization of waste irrigation water and does not leave a residual solution on the leaf surface. No recycling of the nutrient solution is considered, since there is no excess nutrient solution. Therefore, the water (nutrient solution) must be supplied from the bottom of the cell.

### 4.2 Design concept

The nutrient solution should be injected directly into the substrate (soil) where the roots develop, so that the leaves are not moistened by the irrigating solution. The nozzle for water injection should be inserted from the bottom of a plug cell, therefore the cells must have an appropriate-sized hole in the bottom. The injection process should be completed as quickly as possible, since a very large number of seedlings has to be irrigated. Leakage of solution from the cell should be avoided or at least minimized during and immediately after injection.

### 4.3 Irrigation device

A microprecision irrigation device was developed. The device was designed to fit a 300 x 600 mm cell tray containing 72 cells. The solution is discharged from 72 nozzles fixed on an aluminum plate, which can be moved up and down. It is actuated by a ball screw hooked to a servomotor, which controls the vertical position of the injection nozzle tip relative to the interior of a plug cell filled with a substrate (soil mass) and roots. The amount of solution discharged from each nozzle is regulated by the time that a solenoid valve connected to the nozzle remains open. Seventy-two solenoid valves are controlled individually, so that the amount of solution discharged from each nozzle can be varied as required. This can be considered as a variable rate technology such as those highlighted in PA. Figure 1 shows the irrigation device with a schematic diagram of its mechanism.

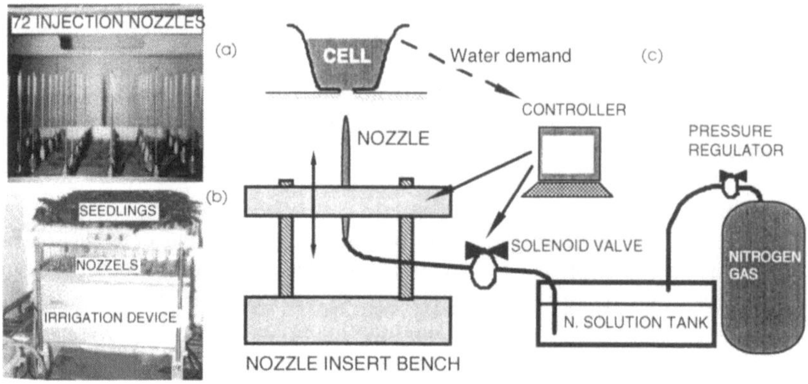


*Fig. 1* Pictures (a, b) of irrigation device and a schematic representation of the device mechanism (c).

## 4.4 Irrigation system

Figure 2 illustrates the irrigation system schematically. The irrigation device is mounted on a multi-functional shifter that can both position a cell tray on the irrigation device and weigh each cell tray before and after irrigation, to monitor the amount of water used or evaporated from the cell tray. Many cell trays are stored vertically in a two-dimensional array. The irrigation unit can move the irrigation device to the location of each cell tray requiring irrigation. Irrigation scheduling and operation should be controlled by computer. Fig. 2. Microprecision irrigation system. The irrigation unit is mounted on the transport equipment which can position the irrigation device precisely under the plug tray which is to be irrigated .

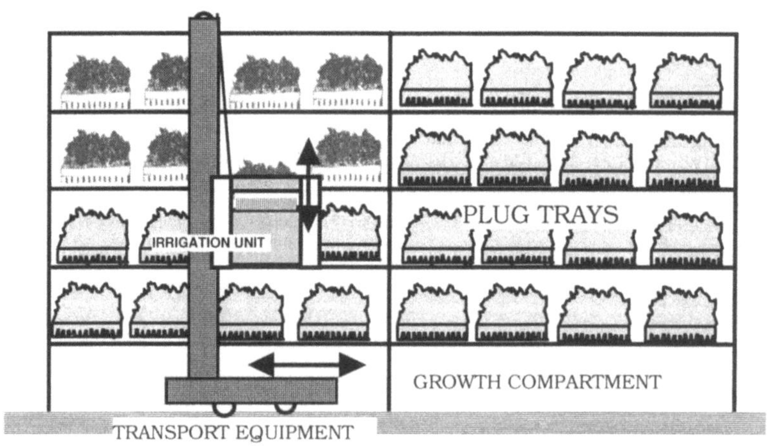


*Fig. 2* Microprecision irrigation system. The irrigation unit is mounted on the transport equipment which can position the irrigation device precisely under the plug tray which is to be irrigated.

## 5. Results and Discussion

Three microprecision irrigation devices were installed in a pilot plug seedling factory located at Chiba University. One unit was used to check the performance of the entire production system. Sweetpotato was used in this experiment. Currently, the daily weight change of each cell tray is used to schedule irrigation, but this only provides the average water requirement of the cells. When seedlings are missing from a cell tray due to poor initial growth, there is a severe solution leak from the cell with the missing plant. Since the new irrigation device is designed to irrigate cells independently, it is possible to avoid irrigating those cells, provided that the locations of empty cells are known. In the future, the microprecision concept should be applied to monitoring the growth of each seedling, so that the variable rate function of the irrigation device can be fully realized. Initially, leakage of a small amount of water from cells during irrigation was a minor problem, but this was solved by placing a thin synthetic sponge under the cell tray. Overall, the irrigation device is acceptable.

## 6. Concluding Remarks

A practical example of microprecision technology in biomechatronics was introduced. A fully controlled plant factory is an ideal system, where microprecision farming can be used to nearly eliminate energy and material waste and environmental impact. This approach should make agriculture profitable, while dealing with environmental concerns and feeding the world's growing population. Microprecision agriculture using plant factories is becoming economically feasible, but requires more research and development.

### References

Blackmore, B.S. 1999. Understanding Variability in Four Fields in the U.K. J. J. Soc. Agri. Machinery. 61(1):17-26 (in Japanese).

Hashimoto, Y. and H. Nonami. 1992. Measurement and Control in Transplant Production Systems. (In: Transplant Production Systems, eds. K. Kurata and T. Kozai), Kluwer Academic Publishers, Dordrecht, The Netherlands. pp. 117-136.

Hess, C.E. 1991. The U.S. Department of Agriculture Commitment to Sustainable Agriculture. (In: Sustainable Agriculture Research and Education in the Field, ed. B. J. Rice). National Academy Press, Washington D.C., USA. pp. 13-21.

Takatsuji, M. 1987. An Introduction to Plant Factories (Shokubutsu-kohjyo Nyumon). Ohmu-sha, Tokyo, Japan. pp. 156. (in Japanese).

# DESIGN CONCEPTS OF COMPUTERIZED SUPPORT SYSTEMS FOR LARGE-SCALE TRANSPLANT PRODUCTION

Takehiko Hoshi [1], Yasumasa Hayashi [2] and Toyoki Kozai[3]
[1]Department of Biological Science and Technology, School of High-Technology for Human Welfare, Tokai University, Nishino 317, Numazu, Shizuoka 410-0395, Japan. E-mail: hoshi@fb.u-tokai.ac.jp
[2]Development Division, ESD Co. Ltd. Koraku, Bunkyo-Ku, Tokyo 112-0004, Japan.
[3]Faculty of Horticulture, Chiba University, Matsudo, Chiba 271-8510, Japan.

**Abstract.** In large-scale transplant production systems, the adoption of production support systems by computers is indispensable. The design concept of two types of support systems, an autonomic decentralized production control system using the object-oriented programming technique and a production planning system that was based on each order were proposed. The results of computer simulation using the prototype systems were also reported. The control system developed on the autonomic decentralized concept could be scaled up to large-scale transplant production systems. The planning system could make a proper production plan for a transplant production system that consisted of many orders. Concepts of these support systems should be very effective to develop and manage large-scale transplant production systems.

**Key index words.** autonomic system, computer network, cutting, decentralized system, object-oriented system, planning, management system.

## 1. Introduction

In an attempt to find solutions to global food and global environmental problems, the scale and role of transplant production systems have entered a new phase (Kurata and Kozai, 1992, Kozai et al., 1999). Large-scale transplant production systems have extensive instrumentation and mechanization for automating and optimizing production. Additionally, many plants are continuously grown and stored in the systems for continuous mass production.

In such transplant production systems, the adoption of production support systems by computers is indispensable. However, past software techniques based on centralized control methods are not suitable for modern support systems, because the instruments, machines, and transplants that need to be monitored and controlled are too numerous to be controlled by a centralized software system. Additionally, if a centralized system is used, a single software problem can easily cause a general breakdown of the transplant production systems. A general stoppage would cause significant economic losses, because large numbers of plants in the transplant production systems would be damaged. In this case, it is desirable to adopt hierarchical distributed and autonomic decentralized software using the object-oriented programming technique (Gauther and Guay, 1990; Hoshi, 1992) to develop production support systems. Support systems based on this software design concept can also be applied to remodeling the transplant production system without software revision.

At present, it is a problem that a large-scale transplant production system is too big for only one grower (customer) to supply whole transplant products. Additionally, the desired transplant character will change every moment with a change in cultivation strategy and season, even if the transplant production system does business with only



*C. Kubota and C. Chun (eds.), Transplant Production in the 21st Century,* 38–43.

one grower. Therefore, we must always produce many kinds of transplants in the transplant production systems to correspond to various orders from many growers. We also need production planning systems to operate the transplant production systems efficiently.

In this article, the design concept of two types of support systems is discussed: An autonomic decentralized production control system (subsequently termed "control system") and a production planning system based on each order (subsequently termed "planning system"), for large-scale transplant production systems.

## 2. Materials and Methods

### 2.1 Target transplant production system

For this study, a typical transplant production system in the study was defined in order to design the support systems. The proposed system was a closed type transplant production facility with artificial lighting. The propagation method for the transplant production was cutting. The plants were planted in plug trays.

A logical hierarchy of the transplant production system was proposed as a basis for designing an autonomic decentralized system. Each hierarchical level was defined as follows. (1) Production Space: It was enclosed with walls to cut off the outflow and inflow of energy and substances. (2) Room: A room partitioned the production space for placing the modules. The rooms were used for either growing plants or plant storage. (3) Module (Basic Module): A shelf for accommodating the plug trays was designed. Artificial lighting, air conditioning, and medium (soil) control devices were also installed in the module. (4) Plug Tray: A plug tray was the production unit in the transplant production system. (5) Plant: A propagated plant by cutting.

### 2.2 Design concept

#### 2.2.1 Control system

Production management in the control system was defined as moving each plug tray either into the most suitable shelf of the modules or into the cultivating treatment area at the most suitable time. We did not describe all behaviors of the items in the transplant production system to the software as procedures. Each logical hierarchy in the transplant production system was represented as a different class in the software, and each class had action criteria for efficient production. The objects of the same number of items in the transplant production system were copied from each class. The objects autonomously acted in the control system, and the result reflected the actual items in the transplant production system. A currency was created that allowed resource, quality and time to be expressed in the same units. Then trading could be performed between the objects. Each object was programmed to gain the largest profit with the smallest payment. We chose the three typical transplant production characteristics that were (1) a resource conservation serious production, (2) a high quality serious production, and (3) a delivery date serious production. Balancing the transplant production characteristics could be done by changing the amount of funds in the objects and by changing the currency conversion rates of the objects.

In the design concept, all actions for producing transplants were defined as services in the control system. All classes were classified as either a service consumer or a service supplier. The service consumers searched for the cheapest service supplier, and they received the service from the service suppliers by paying the cost. Each service consumer had a required value that was the difference between the present condition and the ideal condition. If the consumer's condition was not good, the required value was changed to improve it. The service suppliers gained the currency by offering the service. If a conflict among the consumers occurred, the supplier chose the consumer that had the greatest required value. We have illustrated the relationship between the service supplier and the service consumer in Fig. 1.

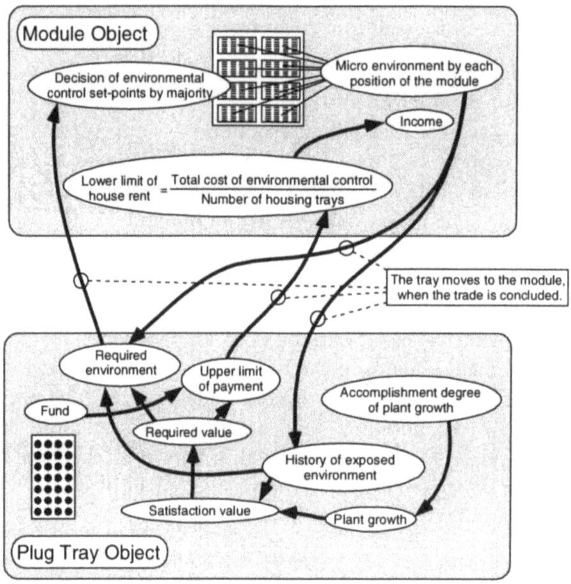


*Fig. 1* An example of the relationship between the objects. In this case, the relationship is limited to a between one tray object and one module object.

### 2.2.2 Planning system

The unit of the planning system was an order, *i.e.*, an order for a set of transplants having the same characteristics and delivery date. The goal of the planning system was to allocate time and space in the transplant production system to satisfy each order.

Figure 2 shows the production process in the supposed transplant production system. The source plants and the cuttings were mainly grown in the growing rooms until delivery. If an overflow of the growing rooms and/or a time lag of plant growth on the same order occurred, the storage rooms were used in order to adjust production.

The planning system estimated whether an order was acceptable using the backward tracing method from the time of shipping. Figure 3 illustrates the concept of the planning system by the backward method.

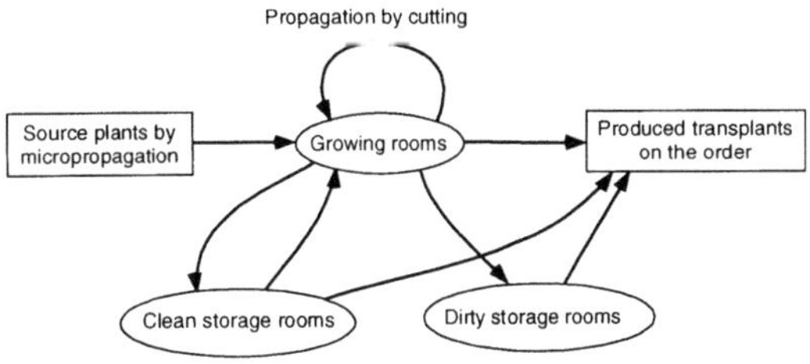


*Fig. 2* Production flows and constitution of the rooms.

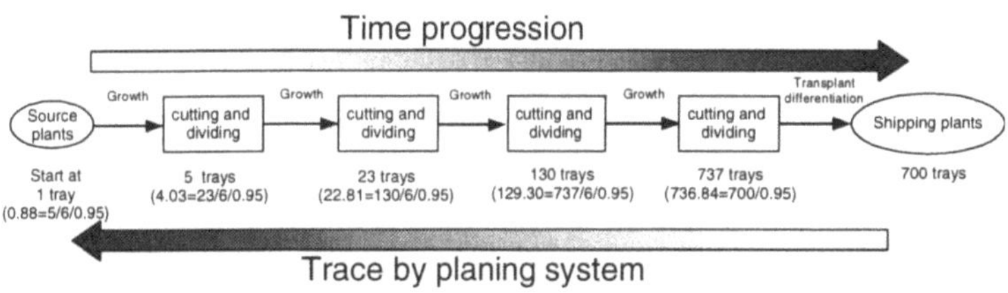


*Fig. 3* Concept of the planning system by the backward method.

The example shows a plan for producing 700 plug trays. The yield rate on propagation was 95%, and the rate of increase was 6 times per propagation cycle.

In the supposed transplant production system, not all production processes have been automated. Especially, the process of cutting and dividing needed hand work by humans. Therefore, a function for estimating daily working hours was also necessary for the planning system. Keeping the number of daily working hours relatively constant and avoiding the need for temporary employment were important for realizing low-cost transplant production.

## 2.3 Prototype systems

### 2.3.1 Control system

The prototype system was developed to examine characteristics of the autonomic decentralized production control system. In order to facilitate prototype development, the following function restrictions were adopted. (1) Each module accommodates one plug tray. (2) The modules can control air temperature, relative humidity, and photosynthetic photon flux (PPF). (3) The required environment in each plug tray is grouped by each order. (4) Transport of the plug trays for irrigation and propagating work is not considered. (5) The delivery date serious production is not considered.

### 2.3.2 Planning system

The prototype planning system was designed to plan for up to 30,000 plug trays in a growing room and to produce transplants in 300 days. The maximum capacity of the clean storage room and the dirty storage room was 500 plug trays and 1,500 plug trays, respectively. The longest storage period of transplants was 14 days.

The following restrictions were adopted in the development of the prototype planning system. (1) All the cuttings belonging to an order have the same plant growth rate. (2) The system does not arrange the average daily working hours. (3) The storage rooms are mostly used once per order. (4) The storage rooms are only used just before cutting or shipping of plants. (5) There is only one order for any given shipping date in the system.

### 2.3.3 Computer hardware and software

The source code of the prototype systems was written in the objective Pascal programming language of Borland Delphi version 3.0J. The prototype systems were evaluated on the PC with an Intel Pentium 133MHz, 48MB of memory, and Microsoft Windows 95 operating system.

## 3. Results and Discussion

### 3.1 Control system

To evaluate the prototype control system, three kinds of tray groups were assumed. There were 400 modules in the cultivating room, and 100 plug trays of each group were prepared respectively.

The frequency of movement of the plug trays was changed by the set point value of the transplant production characteristic. The plug tray could endure to move even in the bad environment by the decision of the transplant production characteristics.

As time passed, the plug trays formed communities in the room. The plug trays that belonged to the same group were gathered. Furthermore, the gap between communities increased with increasing difference between the environmental set points of each plug tray. The arrangement of the plug trays in the culture shelf is done so purposefully that is seems as if a human is doing it. The behavior of the control system autonomously appeared and was not programmed beforehand.

A control system based on this concept could perform only by changing the number of the copied object with an increase or decrease of the items in the transplant production system. It was unnecessary to change the software algorithms. The control system was properly able to control transplant production, even if the compositions of the modules and the plug trays changed. We conclude that the control system based on the concept can be applied to the real large-scale transplant production systems.

### 3.2 Planning system

In this test, we assumed that a transplant was divided into 6 cuttings having a node, the propagating cycle was 30 days, and the yield rate was 95 %. In the evaluation, the planning system could simulate the plan from February 18 to December 14 in 1999 (for 300 days).

The planning system was able to make a plan for transplant production combining many orders. For example, we made a plan for shipping 1,000 plug trays every weekday (Monday to Friday) from July 1 to December 14 using the system. The system pointed out that the storage rooms were unnecessary to produce 1,000 plug trays every weekday. The manager of the transplant production system could estimate manpower and source plug trays using the results of the system every day.

In another example, an irregular order occurred on July 24 at the transplant production system in which the production capacity was almost full. The order was for 1,400 plug trays. The planning system responded that the order was acceptable if the dirty storage room were used. The planning system suggested using the dirty storage room from July 21 to 24, because the growing room became full. The order required that 1 source plug tray is started on April 25 and that 1 source plug tray is also started on February 20. The early source plug tray was propagated to 1,457 plug trays on July 21, and they were stored for 4 days. Another source tray was propagated to 17 plug trays on July 24, and they were joined with the stored plug trays. The potential output of the production became 1,474 plug trays considering a yield rate (95%).

These results showed that the planning system was very effective at systematically planning transplant production. The planning system that we developed could make a proper production plan for a transplant production system that consisted of many orders. It is necessary to improve the planning system with respect to allocating daily man-hours in the future.

**References**

Gauther, L. and R. Guay. 1990. An Object-Oriented Design for a Greenhouse Climate Control System. *Trans. ASAE* 33:999-1004.

Hoshi, T. 1992. Object-Oriented Software Development Support System of Environmental controllers in Plant Growth Factories (in Japanese). *SHITA Journal* 3(2):129-136.

Kozai, T., K. Ohyama, F. Afreen, S. Zobayed, C. Kubota, T. Hoshi and C. Chun. 1999. Transplant Production in Closed Systems with Artificial Lighting for Solving Global Issues on Environment Conservation, Food, Resource and Energy. Proceedings of Transplant Production in Closed System for solving the Global Issues on Food, Rutgers University on July 23. 1-17.

Kurata, K. and T. Kozai (eds.). 1992. Transplant Production Systems. Kluwer Academic Publishers, Dordrecht, The Netherlands. pp. 335.

# 2. TECHNOLOGY IN TRANSPLANT PRODUCTION

# MODELING AND SIMULATION IN TRANSPLANT PRODUCTION UNDER CONTROLLED ENVIRONMENT

Chieri Kubota
Department of Bioproduction Science, Faculty of Horticulture, Chiba University, Matsudo, Chiba 271-8510, Japan. E-mail: ckubota@midori.h.chiba-u.ac.jp

**Abstract.** Modeling and simulation are discussed with regard to their roles in transplant production under controlled environment. Little research has been done toward analysis and optimization of transplant production system based on modeling and simulation. In this article, research on modeling and simulation of growth and development of transplants as well as production scheduling are reviewed. Some models and simulated results are introduced for use in different modes of transplant production (vegetative propagation, plug seedling production, and storage of transplants). Integration of the knowledge of environmental factors affecting transplant growth and development is required for using modeling/simulation techniques as strategic production tools of transplants.

**Key index words.** crop models, production planning, scheduling, transplant quality.

## 1. Introduction

Quality transplants are necessary to achieve high yields and quality in the final harvested products of horticultural, agricultural, and forestry-based production. During recent decades, the transplant production process has become separated from the total crop production process, as more growers have begun to buy quality transplants produced at specialized commercial operations. This organizational separation of transplant and final crop production has contributed to the development and standardization of methods and techniques used to produce transplants formerly produced in-house. Along with this trend, cultural methods and techniques for producing quality transplants have been investigated.

Modeling techniques are generally considered to be useful for integrating growth response with the environmental conditions, by applying mathematical equations. Modeling and simulation can be applied to: (1) environmental control, where environmental conditions (set points) will be selected for optimizing plant growth while minimizing production cost, (2) forecasting plant growth, such that the finishing dates of the final products will be predicted, and (3) production scheduling, where one can optimize labor, greenhouse space usage, and energy/resource input. However, a limited amount of research has been reported regarding modeling and simulation techniques for transplant production. In this article, the possible roles of modeling and simulation in transplant production under controlled environment are discussed and preliminary research examples are provided.

## 2. Modeling transplant growth and development

### 2.1 Vegetative propagation

Vegetative propagation is the production of genetically identical propagules (plants, organs, or tissue as starting materials for regeneration), and especially in floriculture, it is considered as another significant means of transplant production. Along the introduction of genetically manipulated new cultivars, more traditional seedling



*C. Kubota and C. Chun (eds.), Transplant Production in the 21$^{st}$ Century,* 47–52.

transplants are produced via vegetative propagation. Environmental control in vegetative propagation has been recognized as important to produce quality propagules. Research towards modeling and simulation has been done in micropropagation (vegetative propagation using micro-propagules under aseptic conditions). This is probably because micropropagation is usually conducted under controlled environment with artificial lighting.

In micropropagation, in vitro dry weight increase of plantlets under different light, $CO_2$, and photoperiod conditions were simulated successfully by Niu and her coworkers (Niu et al.,1996 and 1997; Niu and Kozai, 1997). Niu's models were based on $CO_2$ concentrations inside the vessel changing dynamically during the culture period, which were affected by environmental conditions and the plantlets' photosynthetic activities, This work on photoautotrophic micropropagation models was considered as an improvement of those models originally reported by Fujiwara and Kozai (1995) for the $CO_2$ exchange rate of plantlets in vitro. However, in vitro plantlets are conventionally cultured photomixotrophically (using both sugar in the medium, and $CO_2$ in the head space as their carbon sources). Modeling photomixotrophic growth has not yet been investigated.

Multiplication rate is number of propagules produced per plant (or plantlet) and is considered as more important than mass or volume increase of plants in vegetative propagation. For micropropagation, Walker (1991 and 1995) developed an empirical model relating the response of multiplication rate to the environmental conditions. Kubota et al. (unpublished) found that the number of sweetpotato harvestable propagules per stock plant ($P_S$), increased linearly with time (t). The linear equation ($P_S = 1 + (t - t_0) R_P$) was characterized with two parameters, days to first harvestable propagule ($t_0$) and propagule production rate ($R_P$), that were affected by different environmental variables (i.e., temperature, PPF (photosynthetic photon flux), and photoperiod) (Fig. 1).

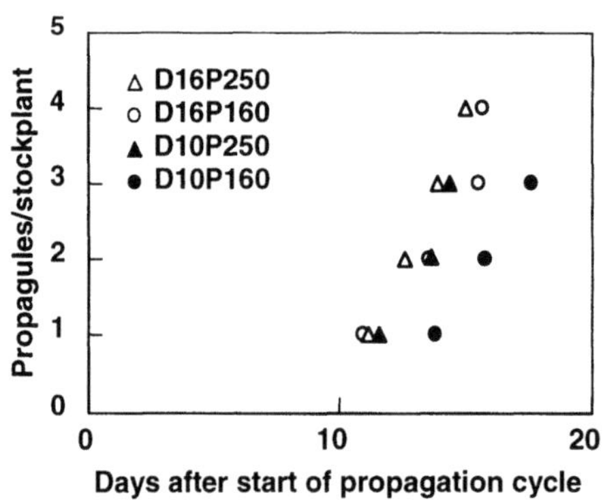


*Fig. 1* Number of propagules (leafy single node cuttings) harvestable per stock plant ($P_S$) on different days after start of propagation cycle, as affected by photoperiod (D = 10 and 16 h $d^{-1}$) or photosynthetic photon flux (P = 160 and 250 μmol $m^{-2}$ $s^{-1}$). For each set of conditions, data were fitted with the model $P_S = 1 + (t - t_0) R_P$. Parameter $t_0$ was 11.1, 10.7, 11.4, and 12.9 and parameter $R_P$ was 0.64, 0.45, 0.57, and 0.46 for treatments D16P250, D16P160, D10P250, and D10P160, respectively. (Kubota et al., unpublished).

## 2.2 Plug seedling production

Little research has been done on modeling growth and development of plug seedlings using environmental conditions as independent variables. One of the possible reasons is that plug transplants often create a dense canopy, and the differences of the environmental conditions inside and outside the canopy are often significant (e.g., Kim et al., 1996). The complexity of the plug seedling system results in greater difficulty in the development of theoretical models employing environmental factors as independent variables. Another possible reason is that unlike agronomic crop production models, quality must be considered as important as growth (i.e., increase in mass and volume). Although we often refer to quality transplants, the definition remains variable, being based on many factors of different value to different people. It likely consists of many factors including visual and physiological characteristics of the transplants (i.e., growth and developmental stages, color/chlorophyll concentration, leaf area index, photosynthetic capacity, shoot/root ratio, stem length, and uniformity). Development of a model incorporating multiple dependent variables may be necessary.

## 3. Modeling production scheduling

Transplant production planning is regulated by the seasonal market. The number of transplants and propagules produced should meet the constraints of narrow market windows. Use of modeling and simulation techniques for optimizing production scheduling is an important area where those techniques can be used effectively. Such techniques are also effective for decision making to select proper environmental conditions and production methods.

### 3.1 Vegetative propagation planning

Simonton and Thai (1988) and Walker (1991) reported modeling and simulation for production planning in micropropagation. The models were designed for scheduling where the objectives were matching production to demand (Simonton and Thai, 1988) and maximizing the profit (Walker, 1991). The simulations demonstrated the high potential usage of such techniques as strategic production tools. Conventional vegetative propagation (i.e., cutting production) is also recognized as an area where models can contribute to enhanced production efficiency. Kubota and Kozai (1999) have demonstrated that the transition matrix model, an ecological modeling technique, successfully simulated the discrete production of the propagules and the number of stock plants (plants grown for producing propagules) accommodated in the production areas, as affected by different propagation methods and environmental conditions. The model could estimate the numbers of each constituent (i.e., stock plants of different ages, cuttings produced, stock plant bases to be discarded, etc.) for each day after start of the propagation process. It was advantageous to quantitatively understand the propagation process and to simulate the effects of any unexpected modifications in propagation methods during the propagation period. Re-scheduling of production is therefore easily performed with such a day-by-day simulation. Fig. 2 shows the simulated results of number of cuttings, total number of stock plants growing in the production system for the successive cutting production, and stock plants bases to be discarded after repeated usage as a starting material (propagule). Matrix transition models with daily observation interval were applied (Kubota and Kozai, 1999). The propagation began with 500 cuttings on day 0.

### 3.2 Plug seedling production planning

In plug seedling production, crop cycling requires the projection of planting and cropping plans for the next several months or several years. Plug seedling production with diverse groups of plant species is logistically problematic. Although greenhouse conditions should be optimized to achieve the best quality of transplants for each species, the available combinations of specific environmental conditions and greenhouse plant growing area creating these conditions are limited. Modeling and simulation should optimize production scheduling by facilitating the identification of areas for germination, growing-on, and finishing for each production lot. The cost and profit for each possible production plan should be a consideration when scheduling the crop cycling.

A concept of "square foot week" (area multiplied by time) used by Healy (1995) for plug seedling production scheduling has potential for incorporation into mathematical models, and it could be used in both single-crop production or the more complex, simultaneous, multi-crop production. Pearson et al. (1998) also focused on the serious needs of modeling and simulation for plug transplant production and developed software for assisting growers to create production schedules. Kubota and Kozai (1999) have applied their transition matrix models to plug transplant production, since plug transplant production can be considered as a transition of stages from germination, growing-on, and finishing according to the growth stages defined in Koranski and Karlovich (1989).

Fig. 3a is an example result output, showing the number of trays finishing when five different crops are grown simultaneously and each crop has different total growth periods after sowing. Matrix transition models with weekly observation interval were applied (Kubota and Kozai, 1999). In this simulation, the first and last weeks are considered as germination and finishing stages, respectively, for all the crops, and the same number of trays of germination stage are replenished immediately after shipping. Fig. 3b shows the number of trays allocated to different growth stages. When starting with 100 trays for each crop on week zero, the number of trays in Stage 3 (the growing-on stage; Koranski and Karlovich, 1989) was either 500 or 300 except for week 21 (100 trays). The number of trays in this stage averaged for the 50 weeks was 373, and thus the average area use efficiency of the stage 3 greenhouse area was 74.5% assuming that the greenhouse had a capacity of 500 trays. Similar work for scheduling multi crop production was done by Fang (1989).

**4. Modeling transplant storage**

Storage of transplants is recognized to be an important technique since it allows adjustment of production scheduling, and can be considered as another production management tool. Modeling and simulation techniques can be applied to transplant storage by predicting storability (storage time period) based on potential deterioration of the transplants. With such techniques, transplant storage can be pre-scheduled and not considered only as an emergency technique to avoid the entire loss of the transplants. Preliminary research has been done by Kubota (1994) and Niu (1996) on predicting changes in dry weight of transplants during storage under different combinations of temperature and PPF. The storability of transplants may be correlated to the carbohydrate status (Rajapakse et al, 1996; Wilson et al., 1998), as well as, dry weight. It will be necessary to understand the environmental conditions affecting carbohydrate allocation of transplants during storage. Light environmental control in storage has been recognized as important for maintaining transplant quality for longer durations of period (Heins et al., 1994; Kubota and Kozai, 1995; Kubota et al., 1997).

## 5. Conclusion

Modeling and simulating transplant quality, as well as, growth, is challenging, but it is necessary for improving transplant production. Manipulation of growth and development by controlling environment and production planning are areas where the need for modeling and simulation have been recognized. These techniques will be used strategically and most effectively in transplant production systems using artificial lighting under controlled environment.

**References**

Fang, W. 1989. Strategic planning through modeling of greenhouse production systems. Ph.D. Dissertation. Rutgers University, NJ, USA.

Fujiwara, K. and T. Kozai. 1995. Control of environmental factors for plantlet production. –With some mathematical simulation- In: (F. Carre and P. Chagvardieff eds.) Ecophysiology and Photosynthetic In Vitro Cultures. CEA Cadarache, Cedex, France. pp. 109-120.

Healy, W. 1995. Scheduling bedding. In: (D. Hamrick ed.) Grower Talks. Plug II. Ball Publishing, Batavia, IL, 1-4.

Heins, R.D., N. Lange, T.F. Wallace Jr., and W. Carlson. 1994. Plug storage: cold storage of plug seedlings. Greenhouse Grower, Willoughby, OH, USA.

Kim, Y.H., T. Kozai, C. Kubota, and Y. Kitaya. 1996. Effects of air current speeds on the microclimate of plug stand under artificial lighting. Acta Hort. 440:354-359.

Koranski, D.S. and P. Karlovich. 1989. Plugs: problems, concerns and recommendations for the grower. Grower Talks. 53(8):28, 30, 32, 34.

Kubota, C. 1994. Growth regulation in plant micropropagation by controlling in-vitro physical environment. Ph.D. Dissertation. Chiba University, Japan. pp. 96-158.

Kubota, C. and T. Kozai. 1995. Low-temperature storage of transplants at the light compensation point: air temperature and light intensity for growth suppression and quality preservation. Sci. Hortic. 61:193-204.

Kubota, C. and T. Kozai. 1999. Use of transition matrix models for transplant production under controlled environments. ASAE Paper No. 995061. ASAE, St. Joseph, MI 49085-9659.

Kubota, C., N.C. Rajapakse, and R.E. Young. 1997. Carbohydrate status and transplant quality of micropropagated broccoli plantlets stored under different light environments. Postharvest Biol. and Tech. 12:165-173.

Niu, G. 1996. Simulation of the effects of physical environmental factors on the growth of in vitro plantlets in micropropagation. Ph.D. Dissertation. Chiba University, Japan. pp. 101-135.

Niu, G. and T. Kozai. 1997. Simulation of the growth of potato plantlets cultured photoautotrophically in vitro. Transactions of the ASAE. 40(1):255-260.

Niu, G. T. Kozai, M. Hayashi, and M. Tateno. 1997. Time course simulations of $CO_2$ concentration and net photosynthetic rates of potato plantlets cultured under different lighting cycles. Transactions of the ASAE. 40(6):1711-1718

Niu, G., T. Kozai, and Y. Kitaya. 1996. Simulation of the time courses of $CO_2$ concentration in the culture vessel and net photosynthetic rate of *Cymbidium* plantlets. Transactions of the ASAE. 39(4):1567-1573.

Pearson, S., P. Hadley and P. LeMiere. 1998. Crop forecasting in horticulture. The Horticulturist 7(2):2-8.

Rajapakse, N.C., W.B. Miller, and John W. Kelly. 1996. Low temperature storage of rooted chrysanthemum cuttings; relationship to carbohydrate status of cultivars. J. Amer. Soc. Hort. Sci. 121:740-745.

Simonton, W. and C.N. Thai. 1988. Plant tissue culture production planning using computer simulation/optimization. Transactions of the ASAE. 31(5):1616-1622.

Walker, P.N. 1991. Linear programming for a micropropagation enterprise. ASAE Paper No. 91-7533. ASAE, St. Joseph, MI 49085-9659.

Walker, P.N. 1995. System analysis and engineering. In: (J. Aitken-Christie, T. Kozai and M.A.L. Smith eds.) Automation and Environmental Control in Plant Tissue Culture. Kluwer Academic Publishers, Dordrecht, The Netherlands. pp. 65-85.

Wilson, S.B., K. Iwabuchi, N.C. Rajapakse, and R.E. Young. 1998. Responses of broccoli seedlings to light quality during low-temperature storage in vitro: II. Sugar content and photosynthetic efficiency. HortScience 33(7):1258-1261.

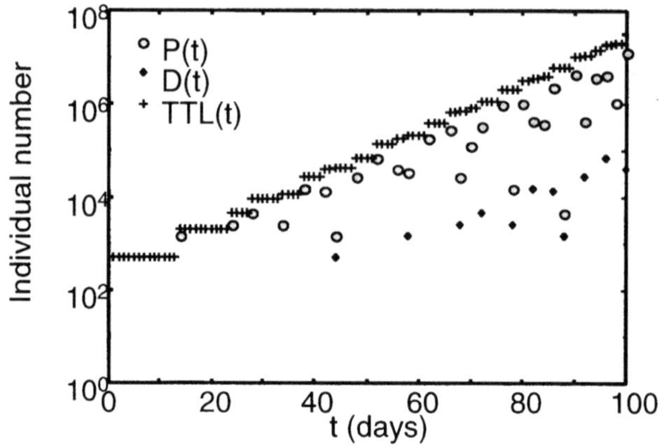


*Fig. 2* Simulated number of cuttings to be produced (P), stock plant bases to be discarded (D), and total number of stock plants maintained in the propagation system (TLL). Matrix transition models with daily observation interval were applied. The propagation began with 500 cuttings on day 0. (Kubota and Kozai, 1999)

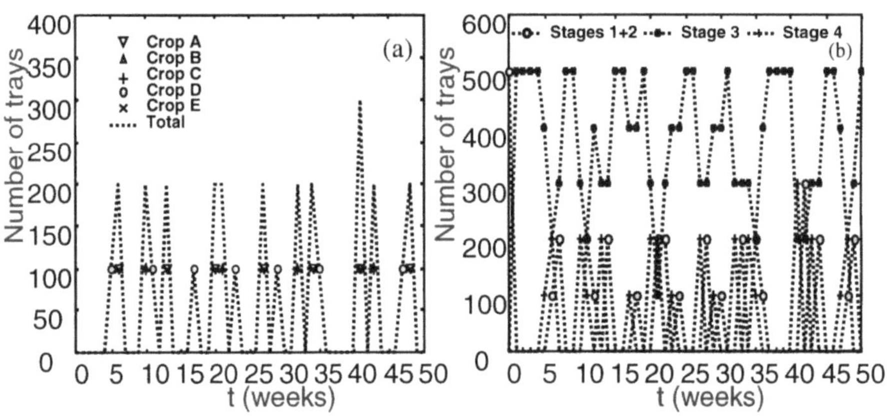


*Fig. 3* Simulated number of trays containing plug seedlings of 5 different crops ready for shipping (left, figure 3a). Simulated number of trays subjected to each growth stage of plug seedling production (right, figure 3b). Crops A, B, C, D, and E have 7, 11, 11, 6, and 7 weeks, respectively, of total growth period from sowing to finishing. Matrix transition models with weekly observation interval were applied. The production started with 100 trays of stage 1 (sowing and germination) for each crop on week 0. (Kubota and Kozai, 1999)

# OBJECT-ORIENTED ANALYSIS AND MODELING OF CLOSED PLANT PRODUCTION SYSTEMS

David H. Fleisher and K.C. Ting
Department of Bioresource Engineering, Rutgers, The State University of New Jersey, 20 Ag Extension Way, New Brunswick, New Jersey 08901-8500, USA. E-mail: ting@bioresource.rutgers.edu

**Abstract.** In order to obtain high level of control over plant production, systems of high degree of closure for growing plants have been developed. These "closed" systems frequently exhibit the integration of automation, plant cultural requirements, and environmental control (i.e. the concept of ACESYS). Examples include plant factories, biomass production units for space journeys, and transplant production facilities. An object-oriented approach was taken to analyze these plant production systems. The purpose was to develop a set of foundation classes that could be used to effectively describe the components of closed plant production systems. Eight foundation classes were developed as the result of the object-oriented analysis, namely: *Automation, Culture_Plant, Culture_Task, Culture_Facility, Environment_Rootzone, Environment_Aerial, Environment_Spatial*, and *Shell.* An object-oriented model representing closed plant production systems was subsequently developed. The first version of the model is a crop production model for systems study of biomass production units within an advanced life support system for long duration human exploration of space. This JAVA based computer model is capable of calculating crop yield, inedible plant material, transpiration water, power usage, labor requirement, etc. over time for various crop mixes and scheduling scenarios. This biomass production model can be modified for simulating other closed plant production systems.

**Key index words.** automation, culture, environment, JAVA, system studies, simulation.

## 1. Introduction

For many years, plants have been grown under modified/controlled environment to achieve results not obtainable in the natural environments. Frequently, controlled environmental factors that affect plant growth and development include temperature, light, $CO_2$, and humidity. The degree of system closure that plants are subject to depends on the practical need, technological feasibility, and economical viability. Historically, temperature has been the primary factor considered. Gradually, with the advances in technologies, it became worthwhile to also control the other environmental factors. In recent years, automation and innovative cultural devices have been added to the controlled environment plant production systems. Hence, the concepts of "Phytomation" and "ACESYS" were developed (Ting, 1999). Phytomation is a term describing any plant-based engineering system and ACESYS is a methodology for systematic analysis and integrated application of technologies related to automation, plant culture, and environmental control for implementing phytomation systems.

A plant factory is a commonly known form of closed plant production system. It resembles the systems normally seen in a well-organized manufacturing facility of industrial products. One major difference is that a plant factory handles products that rely on the environmental factors to change their shape and size. Furthermore, the shapes and sizes are relatively less uniform than most of the products in a manufacturing plant.



*C. Kubota and C. Chun (eds.), Transplant Production in the 21st Century*, 53–58.

Recently, the concept of closed plant production systems has been extended to grow plants for purposes other than delivering plant products (such as transplants, fruits, vegetables, and flowers) to the market. One example is a phytoremediation facility, employing rhizofiltration processes, where plants serve as a water processing device (Fleisher et al., 1997). Another example is a biomass production subsystem, within an advanced life support system (ALSS) for long-term human space journeys, where plants provide food, consume $CO_2$, generate $O_2$, and clean water to support human lives (Henniger et al., 1986). The complexity of closed plant production systems calls for an effective methodology and a functional tool for the purposes of analysis, planning, design, and management of various types of production systems.

The traditional approach for development of analysis tools is to decompose the chosen system's complexity with a sequential, algorithmic approach. The result is a set of step-by-step procedural events used to simulate the system. The object oriented approach, however, decomposes the system into a set of objects that identify and characterize the critical properties through a process called systems abstraction. Abstraction is a method that selects important components and ignores unimportant ones from a system to facilitate the study of the system (Booch, 1994). An abstraction is therefore a model representing the system under study. The level of detail included in the process of abstraction depends on the purpose of the study. The key is to have sufficient, but not overwhelming, amount of detail to effectively carry out the study. The result is the development of simpler, more robust mathematical models / programs (Chao et.al., 1997).

The extent of sophistication of an abstraction can be described using a top-down approach. A top level model (i.e. abstraction) has less detail than a lower level model. The concept of ACESYS provides a guideline for conducting the abstraction of phytomation systems. The highest (top-most) level of abstraction for a phytomation system based on the ACESYS approach has three components, namely automation, culture, and environment.

Automation deals with information processing and task execution related to a system's operation. The purpose of automation is to equip the system with the capabilities of perception, reasoning/learning, communication, and task planning/execution. Culture includes the factors and practices that can directly describe and/or modify the biological growth and development of plants. The cultural factors, such as morphological and physiological conditions of plants are important in monitoring plant growth and development. The cultural practices may include operations such as multiplication, rooting, transplanting, pruning, water and nutrient delivery, pesticide application, harvesting, post-harvest processing, and etc., which directly alter plant morphological and/or physiological conditions. Environment encompasses the surroundings of plants, which consists of climatic and nutritional, as well as structural/mechanical conditions. Each of the three components can be further divided into a number of sub-components and so on. Accompanied by the objected-oriented analysis technique, the components and sub-components can be described by a set of abstract foundation classes containing appropriate attributes (variables) (Chao et al., 1997). Objects are then formed, by assigning values to class attributes, and used to simulate the system.

The objectives of this study were (1) to develop a top-level computer model for closed plant production systems. (The process of model development was guided by the ACESYS concept accompanied by the object-oriented analysis, design, and programming techniques.), and (2) to demonstrate the utility of the top-level model by applying the model to simulating ALSS biomass production.

## 2. Object-Oriented Approach For Closed Plant Production Systems

Object-oriented analysis (OOA) was applied to a general closed plant production system diagrammed in figure 1. Crops are typically produced within several growth chambers or production facilities. Crop growth is associated with cultural, automation, and environment necessities, and these requirements affect system 'top-level' variables. Production of biomass, $O_2$, and transpired water, and consumption of $CO_2$, nutrients and water are involved in plant production. These interactions must be accounted for in a closed system since resource availability is limited. Eight object classes were identified in Table 1 to address these automation, cultural, and environment aspects (Fleisher et al., 1999).

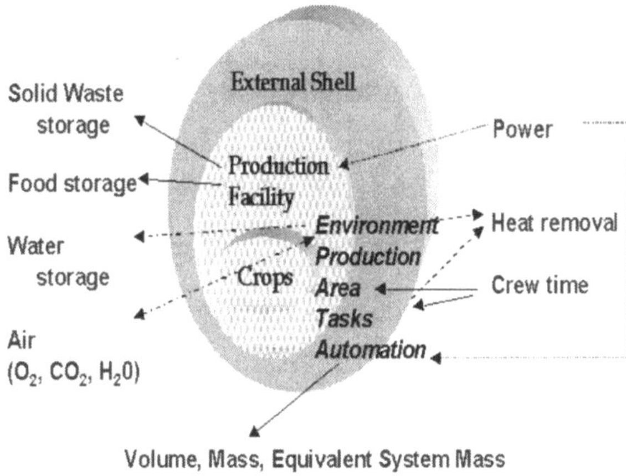


*Fig. 1* Automation, culture, and environment connections for a generic closed plant production system. Top level variables used to describe the system are shown outside the shaded areas.

The logical (classes and objects to be utilized) and physical (programming architecture) structure of the system described by the 8 classes was determined next. This step followed object-oriented design (OOD) principles. Static and dynamic interactions between classes were identified and quantified with OOD charts and diagrams.

Object-oriented programming was the final step and integrated results of OOA and OOD into a functional mathematical modeling tool for systems analysis. JAVA was selected for the programming language so that the simulation could be implemented on one commonly accessible platform, the Internet. Taking advantage of the OOD and world wide web platform, it was possible to develop a real-time data linkage with our model and a Sybase SQL designed database using Symantec's dbANYWHERE server.

## 3. An Application Example: The Biomass Production Model (BPM)

### 3.1 Development

Based on the general approach outlined above, a case specific model was developed for biomass production within an Advanced Life Support System. The problem domain was identical to the closed plant production system shown in figure 1, where the variables shown outside the grey areas in the figure are top level variables which describe system performance. Thus, the classes listed in Table 1 were used for the development of the biomass production model (BPM).

*Table 1* Classes utilized for BPM design. The number of static attributes is listed in parenthesis for each class. Attributes for Culture-Plant class, for example, include 'Final_edible_dry_mass' and 'Time_period'. All objects of each class share the same attributes, usually with different values.

| BPM Class Name | Description |
|---|---|
| Culture_Plant (15) | Tracks growth status of each crop. Predicts daily carbohydrate gain, $CO_2/O_2$ consumption/production, transpiration, and etc. |
| Culture_Facility (25) | Physical aspects of cultural system including equipment volume, mass, and power requirements. Details on lamp numbers, heat removal systems, hydroponic reservoir tanks, and irrigation pumps. |
| Culture_Task (9) | Determine task resource requirements for particular crop types in simulation |
| Environment_Aerial (9) | Aerial environment of crops including light intensity, temperature, air composition, water vapor, and etc. |
| Environment_Rootzone (6) | Track rootzone environment of crops |
| Environment_Spatial (4) | Production information including required production area and associated physical dimensions |
| Automation (14) | Determine automation resource requirements and availability of necessary equipment |
| Shell (6) | Maintains descriptive information on the biomass production system's external enclosure |

The OOD allowed maximum flexibility in the model for selection of multiple simulation scenarios. Classes were linked together so that several crops, each represented by a different Culture_Plant object, could be grown independently of one another within its own production facility, represented by a Culture_Facility object. Each crop was also associated with individualized objects for Environment_Aerial, Rootzone, and Spatial classes. The simulation thus allowed the environment and cultural system to be customized for each particular crop within the same simulation run. Culture_Task objects, seeding/transplanting, gardening/pruning, harvesting, etc., were assigned to each Culture_Plant object. Each task was accomplished either by manual labor or Automation objects. Methods were utilized to describe interactions between hydroponic growth of each individual Culture_Plant object, environment, and automation equipment. Top level variables were adjusted based on these interactions.

The biomass production model was coded in JAVA and implemented on the NJ-NSCORT ACESYS web site on the internet. A front-end graphical user interface was developed to allow the user to run the program and access the database (figure 2). Time-history plots of each top-level variable were accessed by double clicking on the appropriate variable name.

### 3.2 Results

The BPM was utilized to examine the impact of crop selection (e.g. wheat, soybean, white potato, rice), production areas, scheduling, automation equipment, and mission durations on top level variables including solid waste storage (non-edible biomass), food storage (edible biomass), water storage (transpired $H_20$), $CO_2$ and $O_2$ concentrations, power and heat removal requirements, crew time, volume, and system mass (Fleisher et al., 1999). Time-history graphs for crew time and food storage illustrate a typical simulation run (figure 3). The plot of food storage shows that the edible production of biomass for each crop continues to increase throughout the mission duration. This occurred because no food consumption or spoilage was considered. The model can thus be used to determine appropriate production schedules that ensure enough edible biomass is grown to satisfy astronaut dietary needs within an ALSS. The crew time plot shows daily manual labor requirements (in person-hours) needed to produce crops. This information is critical for determining potential bottlenecks in the system. Further analysis could be done to see if increasing the numbers of automation equipment in the simulation would reduce crew time without significantly affecting power, system mass, and volume requirements.

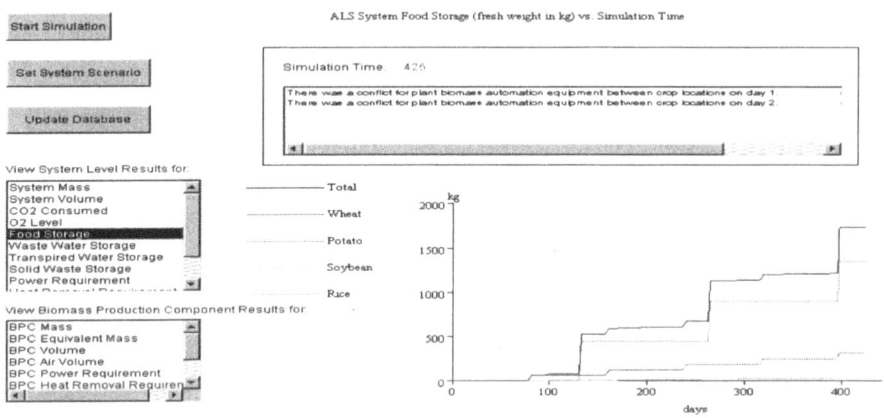


*Fig. 2* Main window for the BPM. Simulation results in the form of time history graphs can be obtained by double-clicking with the mouse on the appropriate variables. A plot of total food storage per crop is shown.

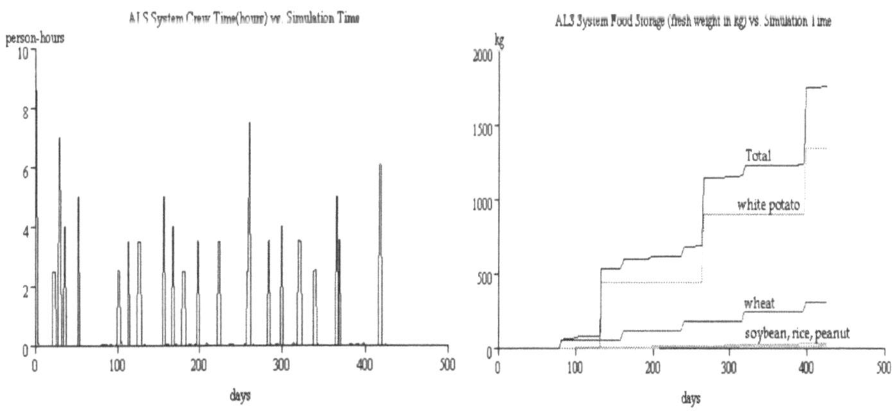


*Fig. 3* Simulation results for crew time (left) and food storage (right) over simulation time in days.

## 4. Conclusions

The object-oriented analysis, design, and programming approach was shown to be an excellent tool for studying the robustness and resource management of a closed plant production system. A flexible modeling tool was developed by decomposing the system into objects representing automation, culture, and environmental components. The approach was demonstrated for a particular biomass production system, but could easily be applied to nearly any closed plant production system since the object-oriented analysis and design would be similar.

Including a database linkage to the object-oriented based model enhanced the ability of the simulation. It has the added benefit of providing a single template for all researchers, regardless of scientific background, to describe components of a closed plant production system with the same 'language'. Users can specify virtually any scenario and change the values of all static attributes used in the model for each object in the simulation, thus tailoring the model to their particular research needs. This feature, plus the benefits of using object-oriented approach, yields a powerful method for systems studies and analysis.

**Acknowledgements.** NJAES Paper Number D-70501-02-00, sponsored by New Jersey-NASA Specialized Center of Research and Training (NJ-NSCORT) and NASA Graduate Student Research Program (GSRP) Fellowship.

**References**

Booch, G. 1994. Object Oriented Analysis and Design with Applications. 2nd edition. Addison-Wesley, Reading, MA.

Chao, K., K.C. Ting and G.A. Giacomelli. 1997. Foundation Class Library Design for Global BLSS Models. Paper No. 973115, ASAE Annual International Meeting, August 10-14, Minneapolis, Minnesota.

Fleisher, D.H., K.C. Ting, M. Hill and G. Eghbali. 1999. Top Level Modeling of Biomass Production Component of ALSS. The 29th International Conference on Environmental Systems, SAE Technical Paper No.1999-01-2041, Warendale, PA.

Fleisher, D.H., K.C. Ting and G.A. Giacomelli. 1997. Computer Model for Full-Scale Phytoremediation Systems Using Rhizofiltration Processes. Paper No. 973083, ASAE Annual International Meeting, August 10-14, Minneapolis, Minnesota.

Henninger, D., C. Lagle and D. Ming. 1986. A Lunar Derived "Soil" for the Growth of Higher Plants. In The First Lunar Development Symposium, G. Andrus (ed.), Lunar Development Council, Pitman NJ.

Ting, K.C. 1999. ACESYS and Phytomation. Proceedings of ACESYS III Forum: From Protected Cultivation to Phytomation, G.A. Giacomelli and K.C. Ting, Eds., The Center for Controlled Environment Agriculture, Cook College, Rutgers University, New Brunswick, NJ:105-110.

# ESTIMATING CUTICLE RESISTANCE OF SEEDLING SHOOT TIPS BASED ON THE PENMAN-MONTEITH MODEL

Hiroshi Shimizu[1] and Royal D. Heins[2]
[1] Department of Agricultural Engineering, Ibaraki University, Ibaraki, 300-0393 Japan.
E-mail: hshimizu@ipc.ibaraki.ac.jp
[2] Department of Horticulture, Michigan State University, 288 Plant and Soil Science Building Michigan State University East Lansing, MI 48824, USA.

**Abstract.** Morphological characteristics of seedlings such as height and size uniformity are important requirements in transplant production and are affected by temperature at the plant shoot-tip. Plant temperature is described by the energy budget equation based on the Penman-Monteith model. The energy budget equation of a plant shoot-tip consists of environmental factors and the shoot-tip cuticle resistance. The energy budget equation was algebraically modified to an equation that expresses the relationship between Vapor Pressure Deficit (VPD) and ambient air temperature ($T_{air}$) - plant shoot-tip temperature ($T_p$). This same relationship was also obtained by a simple experiment where VPD was regulated at constant $T_{air}$. Comparing the coefficient of these two equations, the cuticle resistance of plant shoot-tips was calculated. The experiments were conducted using *Campanula* 'Birch Hybrid', *Campanula carpatica* 'Blue Clips', and *Geranium dalmaticum*. The values of cuticle resistance of each plant are 524.9, 157.2 and 57.8 s/m, respectively.

**Key index words.** convection, cuticle resistance, energy balance, radiation, shoot tip, transpiration.

## 1. Introduction

One of the purposes of transplant production in closed system is to achieve effective production of high quality seedlings. Plant shoot-tip temperature influences plant morphogenesis, and correct morphogenesis is important for high quality plants. Plant shoot-tip temperature is determined by the energy balance of the shoot-tip where incoming and outgoing terms of the energy balance equation principally consists of shortwave radiation, longwave radiation, sensible heat convection, and transpiration. Shortwave and longwave radiation can be measured by sensors, and convective heat transfer of sensible heat is calculated using the techniques and the achievements developed in the field of thermodynamics. Transpiration is a function of boundary layer resistance, cuticle resistance, and the difference in vapor pressure of the ambient air and the shoot-tip surface. Boundary layer resistance and vapor pressure can be theoretically computed, however cuticle resistance is a species dependent factor and must be experimentally measured. The objective of this research was to derive a formula for calculating cuticle resistance and develop a method for estimating the cuticle resistance with a simple experiment.

## 2. Materials and Methods

### 2.1 Penman—Monteith model and its modification

The Penman-Monteith model was used to describe the energy balance of plant shoot tip. Jackson et al.(1981) discussed the energy balance considerations to show how the



*C. Kubota and C. Chun (eds.), Transplant Production in the 21st Century,* 59–62.

difference between leaf temperature and air temperature is related to vapor pressure deficit. Their equation expressed the relation between temperature difference and vapor pressure deficit included the net radiation, however the following modification gives the liner function of difference between shoot-tip and air temperature as outgoing longwave radiation.
The energy balance equation of plant shoot-tip is described as follows:

$$Radiation + Convection + Transpiration = 0 \quad [1]$$

The radiation term includes incoming shortwave radiation and incoming and outgoing longwave radiation:

$$Radiation = SW_{in} + LW_{in} - LW_{out} \quad [2]$$

where, $SW_{in}$: incoming shortwave radiation, $LW_{in}$: incoming longwave radiation, $LW_{out}$: outgoing longwave radiation.

$SW_{in}$ and $LW_{in}$ are directly measured and are not a function of plant shoot-tip temperature ($T_p$). On the other hand, $LW_{out}$ is a function of $T_p$, and given by Stepfan's law.

$$LW_{out} = \varepsilon_p \sigma (T_p + 273.15)^4 \quad [3]$$

where, $\varepsilon_p$: emissivity of plant shoot-tip, $\sigma$: Stefan-Boltzmann Constant.

It is empirically known that differences between the shoot-tip temperature ($T_p$) and ambient air temperature are not relatively big in absolute temperature units. Using the Taylor expansion, equation [3] is changed to:

$$LW_{out} = \varepsilon_p \sigma (T_{air} + 273.15)^4 - 4\varepsilon_p \sigma (T_{air} + 273.15)^3 (T_{air} - T_p) \quad [4]$$

Convective heat transfer is described as:

$$Convection = \frac{\rho Cp}{r_H}(T_p - T_{air}) \quad [5]$$

where, $\rho$: density of air, Cp: specific heat of air, $r_H$: resistance for convective heat transfer.

Transpiration is expressed by the Penman-Monteith model as follows:

$$Transpiration = \frac{\rho Cp\left(e_s(T_p) - e(T_{air})\right)}{\gamma r_V} \quad [6]$$

where, $e_s(T_p)$: saturated vapor pressure at shoot tip temperature, $e(T_{air})$: vapor pressure of air, $\gamma$: Psychrometer constant, $r_V$: resistance for transpiration.

Substituting equation [2], [4], [5] and [6] for [1], we have:

$$SW_{in} + LW_{in} - \varepsilon_p \sigma (T_{air} + 273.15)^4 + 4\varepsilon_p \sigma (T_{air} + 273.15)^3 (T_{air} - T_p) + \frac{\rho Cp(T_{air} - T_p)}{r_H} - \frac{\rho Cp\left(VPD - \Delta(T_{air} - T_p)\right)}{\gamma r_V} = 0 \quad [7]$$

where, VPD: vapor pressure deficit, Δ: slope of saturated vapor pressure curve. Combining like terms:

$$SW_{in} + LW_{in} - \varepsilon_p \sigma (T_{air} + 273.15)^4 - \frac{\rho Cp VPD}{\gamma r_V} + \left(4\varepsilon_p \sigma (T_{air} + 273.15)^3 + \rho Cp\left(\frac{1}{r_H} + \frac{\Delta}{\gamma r_V}\right)\right)(T_{air} - T_p) = 0 \quad [8]$$

Then rearranging equation [8] as follows:

$$VPD = \left(\frac{4\gamma r_V \varepsilon_p \sigma (T_{air} + 273.15)^3}{\rho Cp} + \gamma\left(\frac{r_V}{r_H}\right) + \Delta\right)(T_{air} - T_p) + \frac{\gamma r_V}{\rho Cp}\left(SW_{in} + LW_{in} - \varepsilon_p \sigma (T_{air} + 273.15)^4\right) \quad [9]$$

On the other hand, it is not difficult to obtain data that express the relationship between ($T_{air}$-$T_p$) and VPD experimentally.

$$VPD = b_1 (T_{air} - T_p) + b_0 \quad [10]$$

Comparing coefficients in equation [9] with [10]:

$$\frac{4\gamma r_V \varepsilon_p \sigma (T_{air} + 273.15)^3}{\rho Cp} + \gamma\left(\frac{r_V}{r_H}\right) + \Delta = b_1 \quad [11]$$

Arranging equation [11]:

$$r_V = \frac{b_1 - \Delta}{\frac{4\gamma \varepsilon_p \sigma (T_{air} + 273.15)^3}{\rho Cp} + \frac{\gamma}{r_H}} \quad [12]$$

The resistance for transpiration consists of boundary layer resistance and cuticle resistance. The boundary layer resistance is theoretically computed as 0.93 times $r_H$. Finally the following equation for calculating $r_C$ is derived:

$$r_C = \frac{b_1 - \Delta}{\frac{4\gamma\varepsilon_p\sigma(T_{air} + 273.15)^3}{\rho Cp} + \frac{\gamma}{r_H}} - 0.93 r_H \qquad [13]$$

**3. Results and Discussion**

The experiments were conducted in a glass greenhouse to obtain the relationship of equation [11] using seedlings of *Campanula* 'Birch Hybrid', *Campanula carpatica* 'Blue Clips', and *Geranium dalmaticum*. Drybulb temperature was kept at 25C and VPD was regulating by ON-OFF control of steam injection. Coefficient $b_1$ was 1.103, 0.51 and 0.354, respectively. These $b_1$ values were substituted into equation [14] and the cuticle resistance values for each species were calculated (Table 1).

The cuticle resistance is essential to calculate the energy budget of plant shoot-tips and estimate their temperature. However, there are few reports of the cuticle resistance in the literature. Bakker (1991) measured leaf conductance of egg plant, cucumber and tomato, and determined that the leaf conductance of tomato and cucumber was greater than that of eggplant. Hanan (1997) reported that the cuticle resistance of cucumber was 310 to 1350 (s $m^{-1}$). It is expected that the cuticle resistance is species dependent and probably is plant stage dependent as described above. This means it is dangerous to utilize published cuticle resistance values. We believe that the measurement method described here for determining the cuticle resistance of a plant shoot-tip can be used to predict the shoot tip temperature when combined with measured incoming shortwave and longwave radiation, air velocity, drybulb and wetbulb temperature.

*Table 1* Calculated cuticle resistance

| Plant species | Cuticle Resistance (s/m) |
|---|---|
| *Campanula* "Birch Hybrid" | 524.9 |
| *C. carpatica* "Blue Clips" | 157.2 |
| *Geranium dalmaticum* | 57.8 |

**References**

Bakker, J.C. 1991. Leaf conductance of four glasshouse vegetable crops as affected by air humidity. Agricultural and Forest Meteorology 55:23-36.

Hanan, J.J. 1997. Greenhouses –Advanced technology for protected horticulture. CRC Press. pp.361.

Jackson, R.D., S.B.Idso, R.J. Reginato and P.J. Pinter Jr. 1981. Canopy temperature as a crop water stress indicator. Water resources research 17(4):1133-1138.

Monteith, J.L and M. Unsworth. 1990. Principles of environmental physics (2nd edition). Anrnold. pp.180-187.

# MEASUREMENT OF pH IN GUARD CELLS USING A CONFOCAL LASER SCANNING MICROSCOPE

Masahiro Yabusaki, Yasuomi Ibaraki, Kenji Kurata and Keiko Iwabuchi
Department of Biological and Environmental Engineering, The University of Tokyo, Bunkyo-ku, Tokyo, 113-8657, Japan. E-mail: aa96160@mail.ecc.u-tokyo.ac.jp

**Abstract.** Two experiments were conducted to establish a method to measure pH in guard cells *in situ* for analysis of the relationship between pH changes in guard cells and stomatal aperture changes. To measure pH in guard cells of potato plants (*Solanum tuberosum* L. cv. Benimaru), ratio imaging was conducted using a confocal laser scanning microscope system with SNARF-1/AM as a pH fluorescent probe. Emitted fluorescence was divided into two ranges (580 and 640 nm) and the ratio of fluorescent intensities at 580 and 640 nm (F580/F640) was calculated. *In vivo* and *in vitro* calibrations showed an increase in the F580/F640 ratio with decrease in pH. In experiment 1, the relationship between stomatal aperture and pH in guard cells was investigated when the leaf disc was bathed in low pH solution (citric acid solution: pH 2.7). It was observed that stomata opened and F580 increased. Ratio imaging revealed that stomata with guard cells, which were initially at pH 7.0, opened as pH in the guard cells decreased. In experiment 2, to measure pH in guard cells for an attached leaf, 4.4 mM SNARF-1/AM solution was injected into the stem of a potato plant. However, the pH in the guard cells could not be measured because of low fluorescence intensity.

**Key index words.** confocal laser scanning microscope, fluorescence, guard cells, pH, ratio imaging, stomata.

## 1. Introduction

Stomatal aperture changes in response to the surrounding environment. Many factors are involved in the mechanism of the stomatal movement, and intracellular pH of guard cells is one of these factors. Previous studies have indicated that pH in guard cells plays an important role in stomatal opening. Gehring et al. (1998) reported that a reduction in cytoplasmic pH of the guard cells preceded stomatal opening induced by indole-3-acid (IAA) in epidermal strips of the *Paohiopedium tonsum* L. orchid. Also, pH changes in the apoplast during stomatal opening by lighting the epidermis of leaves of *Commelina communis* were reported (Edwards et al., 1988). These researches used epidermal strips from detached leaves. In order to clarify the role of pH in the stomatal response to the environmental factors, the relationship between pH in guard cells and stomatal opening should be examined for attached leaves. The objective of this study was to establish a method to measure pH in guard cells *in situ*.

A confocal laser-scanning microscope (CLSM) was used to measure pH in guard cells. A CLSM is advantageous for quantitative analysis of fluorescence intensity. As normal fluorescent microscopes take in fluorescence from parts of the sample other than the focal plane, the images are not clear. In contrast, a CLSM can minimize blur in the image because the CLSM does not take in fluorescence from points other than the focal plane owing to the confocal pinhole. Moreover, the CLSM can be used to observe a cross-section of the sample without special treatments such as slicing of the sample.



*C. Kubota and C. Chun (eds.), Transplant Production in the 21$^{st}$ Century*, 63–66.

## 2. Materials and Methods

### 2.1 Plant materials

Potato (*Solanum tuberosum* L. cv. Benimaru) nodal cuttings were cultured for 35 days in test tubes with Murashige and Skoog (1962) medium containing 2% sucrose. Fluorescent lamps provided lighting of 170 μmol $m^{-2} s^{-1}$ PPFD (14 h light, 10 h dark) and the temperature was maintained at 23℃. Then, the obtained plantlets were transferred to flasks containing nutrient solution 0.1% HYPONeX α (Hyponex Japan Co., Osaka) and covered with plastic film. The plantlets were acclimatized in the same environment for over 10 days. The plantlets were exposed to the outside air by gradually increasing the number of holes in the plastic film. After acclimatization, the potato plants had approximately 12 leaves, and a healthy leaf was selected for the sample of observation.

### 2.2 Systems for observing stomata

Two systems were developed. Fig. 1 shows the schematic layout of the systems for observing stomata, which included a CLSM system (TMD-300, Nikon, Tokyo, and MRC1000, Bio-Rad, Tokyo). In system 1, a leaf disc was fixed in a specially constructed plastic petri dish (hereafter, "glass bottom dish"). A circular hole was cut out from the bottom of the petri dish and a glass cover slip was positioned to cover the hole. The glass bottom enabled acquisition of clear images of the leaf. The glass bottom dish was set on the stage of the CLSM and nutrient solution was circulated through the dish (Fig. 1-a).

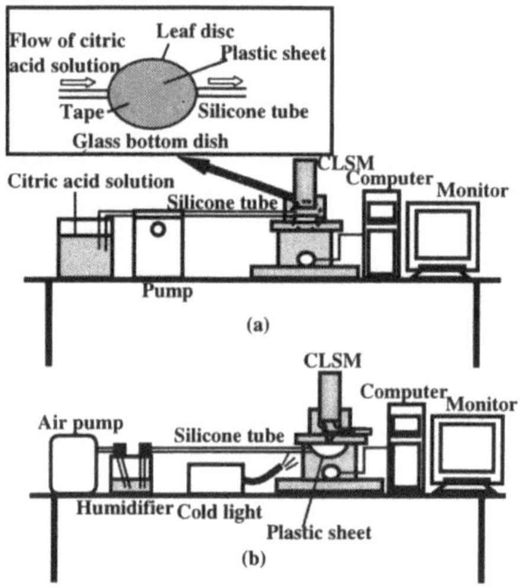


*Fig. 1* Schematic diagram of observation system for (a) leaf disc and for (b) attached leaf.

System 2 was developed for observation of attached leaves (Fig. 1-b). The attached leaf was fixed on the stage of the microscope without a cover slip. A plastic sheet covered the leaf and humidified air was supplied to regulate the environment surrounding the leaf. The leaf was illuminated by light through a fiber cable from a 100W-halogen lamp.

### 2.3 pH measurement

To measure pH in guard cells, ratio imaging was conducted using SNARF-1/AM (Molecular Probes, Eugene, Oregon) as a pH fluorescent probe. The emitted fluorescence was divided into two ranges (580 nm DF32 and 640 nm DF40) using an optical filter set (SA2R Bio-Rad, Tokyo), and the ratio of the fluorescence intensities F580/F640 was calculated. Bassenett et al (1990) described a method of measuring pH in cells by fluorescent ratios. Calculation of the ratio of the fluorescence intensities results in cancellation of artifactual variations in the fluorescence signal. Artifacts include photobleaching, leakage of the indicator, variable cell thickness, and nonuniform indicator distribution. *In vivo* and *in vitro* calibrations were conducted to determine the relationship between pH and the fluorescence ratio F580/F640. For *in vitro* calibration, filter papers were bathed in buffers (100 mM KCl + 1 mM $MgCl_2$ + 50 mM MES) of pH 4.8, 6.0 and 7.2. For *in vivo* calibration, potato leaf discs of which the epidermis was partially peeled were bathed in buffer of pH 6.0 containing 10 μM SNARF-1/AM, and then bathed in buffers of pH 4.88 and pH 7.0 containing 10 μM $H^+$ ionophore (nigericin) for 10 minutes. After this treatment for introducing SNARF-1/AM into guard cells, high intensity fluorescence was observed on the pore-side of the peripheral region of the guard cells. The fluorescence intensity in this part of the guard cells was used for pH measurement though it could not be determined whether the fluorescence originated from cytoplasm or apoplast.

## 2.4 Experimental procedures

*Experiment 1: Measurement of pH in guard cells of leaf discs*

The relationship between stomatal aperture and pH in guard cells was investigated when a leaf disc was exposed to low pH solution using system 1. The leaf disc (5 $mm^2$) was soaked in 10 μM SNARF-1/AM solution for 2 h to introduce the fluorescent probe into the guard cells. Then, the leaf disc was placed in the glass bottom dish and bathed in low pH solution (citric acid solution: pH 2.7).

*Experiment 2: Measurement of pH in guard cells for an attached leaf*

A solution of 4.4 mM SNARF-1/AM was injected into the stem of a potato plant. Then, an attached leaf of the plant was fixed on the stage of the microscope and fluorescence images were acquired using system 2.

## 3. Results and Discussion

In both systems, the stomata were successfully observed using the CLSM. The stomata had normal functions such as closing in dark (data not shown). Therefore, the stress due to fixing the leaf to thc microscope stage seemed to be minimal in the system.

*In vivo* and *in vitro* calibrations showed an increase in the F580/F640 ratio with a decrease in pH (Fig. 2). So, it seemed possible for fluorescence ratio to correspond to the pH. In this study, the results of *in vivo* calibration with potato leaf discs (solid line in Fig. 2) were used to measure pH in guard cells for experiments 1 and 2.

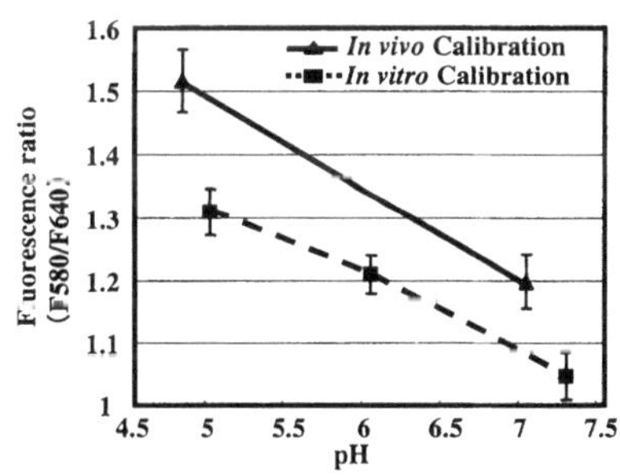


*Fig. 2* Relationship between fluorescence ratio and pH.

*Experiment 1*

For the leaf discs, tendencies were found that stomata opened and fluorescence intensity at 580 nm gradually increased following low pH solution treatment (Fig. 3). Stomata, initially closed, increased their aperture rapidly, while other stomata, which were initially open, increased their aperture more slowly. Ratio imaging revealed that stomata with guard cells initially at pH 7.0 opened as pH in the guard cells decreased during the low pH solution treatment (Fig. 4).

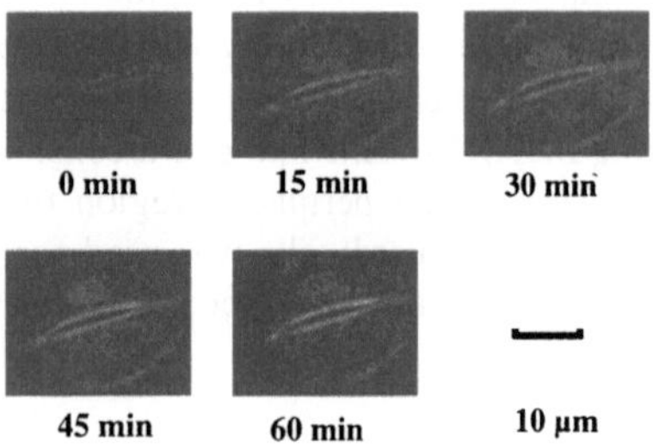


*Fig. 3* Increases of Stomatal aperture and intensity of fluorescence (580nm) with low pH solution treatment.

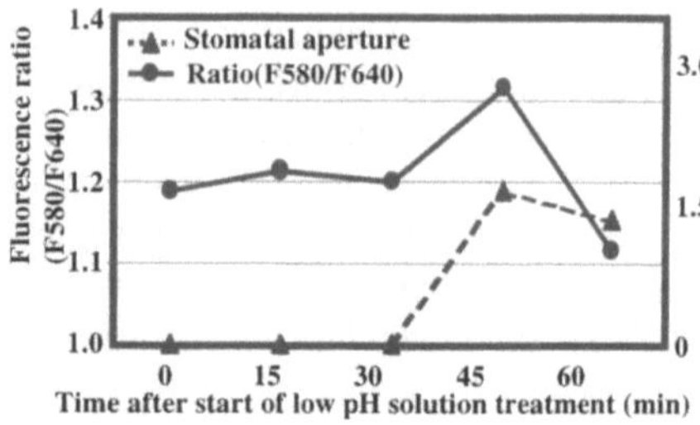


*Fig. 4* An example of relationship between fluorescence ratio and stomatal aperture.

*Experiment 2*

A significant change in the fluorescence intensity of the guard cells could not be detected in the attached leaf when SNARF-1/AM was injected into the stem. Therefore, pH in the guard cells could not be measured. Further research on the method of introducing of fluorescent probes into the attached leaf is necessary.

## 4. Conclusions

In this study, systems which could measure stomatal aperture and fluorescence intensity from guard cells using a CLSM were developed for leaf discs bathed in solution and for attached leaves. Using the system for leaf discs, increase in stomatal aperture was observed when the pH in the guard cells decreased. However, for attached leaves pH in the guard cells could not be measured because the fluorescent probe could not be successfully introduced into the guard cells. It should be possible to measure pH in guard cells *in situ* if the proper method of introducing fluorescent probes into guard cells is established for attached leaves.

**References**

Bassenett, S., L. Reinisch and D.C. Beebe. 1990. Intracellular pH measurement using excitation-dual emission fluorescence ratios. Am. J. Physiol. 258 (1):C171-C178.

Edwards, M.C., G.N. Smith and D.J.F. Bowling. 1988. Guard cells extrude protons prior to stomatal opening – a study using fluorescence microscopy and pH micro-electrodes. J. Exp. Bot. 39: 1541-1547.

Gehring, C.A., R.M. McConchie, M.A. Venis and R.W. Parish. 1998. Auxin-binding-protein antibodies and peptides influence stomatal opening and alter cytoplasmic pH. Planta 205:581-586.

Murashige, T., and F. Skoog. 1962. A revised medium for rapid growth and bioassays with tabacco tissue cultures. Physiol. Plant 15:473-497.

# DOES ELECTROLYZED-REDUCED WATER PROTECT PLANTS FROM PHOTOINHIBITION?

Keiko Iwabuchi, Kazuyuki Seyama, Kenji Kurata and Yukiko Hiruta
Department of Biological and Environmental Engineering, Graduate School of Agricultural and Life Sciences, University of Tokyo, Bunkyo-ku, Tokyo 113-8657, Japan. E-mail: aa96157@mail.ecc.u-tokyo.ac.jp

**Abstract.** Electrolysis of water produces reduced and oxidized water near the cathode and anode respectively. Electrolyzed-reduced water (hereafter referred to as "reduced water") exhibits high pH, and extremely negative oxidative redox potential (ORP). Reduced water was reported to scavenge active oxygen species that have a high potential to react with biological macromolecules destructively (Shirahata et al., 1997). In plants, photoinhibition of photosynthesis is a well known phenomenon resulting from oxidative damage of photosynthetic apparatus caused by active oxygen species generated in chloroplasts. Photoinhibition is often observed to occur under stressful conditions. Therefore, if the antioxidative function of reduced water can act also in plant cells, it may be possible to improve plant growth under stressful conditions by application of reduced water. In this study, growth of plants irrigated with deionized water or reduced water was studied under two types of stressful conditions: water stress in Experiment 1, and low temperature and high light intensity in Experiment 2. In both experiments, lettuce seeds (*Lactuca. sativa* L. var. capitata L.) were germinated in reduced water or deionized water in an environmentally controlled room at day/night air temperature of 23/18 ℃ and 190±20 μmol $m^{-2}$ $s^{-1}$ PPF. In Experiment 1, the plants were watered every 2 days for the first 15-day period, then watered every 4 days for the following 15 days to obtain mild drought conditions. In Experiment 2, 25-day-old plants were cultivated at low temperature and high light intensity (5 ℃ throughout a day and 500± 100 μmol $m^{-2}s^{-1}$ PPF) for following 20 days. Under the both photoinhibitory conditions, leaf yellowing was observed, especially for newly developed leaves. This implied that photoinhibition of photosynthesis occurred in both experiments. However, there were no significant differences in fresh and dry weights, unfolded leaf number, and chlorophyll concentration between reduced water and deionized water treatments in Experiment 1. In Experiment 2, fresh weight was lower in the reduced water treatment than in the deionized water, suggesting that very low pH of reduced water was likely to negatively affect plant growth. These results indicated that mere irrigation with reduced water could not enhance plant growth under photoinhibitory stressful conditions.

**Key index words.** Electrolyzed-reduced water, ORP, pH, photoinhibition of photosynthesis, stressful condition.

## 1. Introduction

Electrolyzed-reduced water is water that is produced near the cathode during electrolysis of water and exhibits high pH and extremely low oxidative redox potential (ORP). Recently, Shirahata et al. (1997) reported that strongly electrolyzed-reduced water could scavenge $O_2^{\cdot-}$ and $H_2O_2$ completely, and protect single-strand breakage of DNA from active oxygen species generated by the Cu(II)-catalyzed oxidation of ascorbic acid in vitro. These results indicate that reduced water may have potential to protect biological macromolecules from oxidative damage in living cells. In green plants,



*C. Kubota and C. Chun (eds.), Transplant Production in the 21st Century, 67–71.*

it is well known that photoinhibition of photosynthesis is a phenomenon resulting from photooxidative inactivation of the photosynthetic reaction center caused by reactive oxygen species generated in chloroplasts (Asada, 1994, Foyer et al., 1994). This phenomenon is commonly observed under stressful conditions, such as drought, low or high temperatures. The environmental stress limits photon utilization for photosynthesis, and excessive photon energy produces reactive oxygen species, leading to suppression of plant growth. Thus, it may be possible that if reactive oxygen species generated in plant cells can be scavenged by reduced water, photoinhibition of photosynthesis under stressful conditions would be suppressed. In commercial crop production, reduced water is occasionally utilized for irrigation, because it has been reported informally that application of reduced water improves plant growth or product quality. However, there has been few research on the effect of reduced water on plant growth and development. In addition, the physiological effects of reduced water on plant cells or tissues are entirely unknown.

In this study, plants were irrigated with reduced water, and their growth was examined to assess the antioxidative function of reduced water for plants under photoinhibitory conditions. Two typical photoinhibitory conditions, mild drought, and low temperature and high light intensity were tested for lettuce in a growth chamber. These environmental conditions are widely observed in the field, and are known to cause photoinhibition of photosynthesis. In this paper, the experiments were carried out as the first step to study the feasibility of reduced water utilization for plants.

## 2. Material and Methods

### 2.1 Reduced water production

Deionized water was produced with a water purifying system (Type ERN-W, Muromachi Chemical Industry Co. Tokyo), which was equipped with a carbon filter cartridge and ion exchange cartridges. Potassium chloride (1 g $L^{-1}$ of KCl ) was added to the deionized water to increase the electrical conductance enough to be electrolyzed. Then, the KCl solution was electrolyzed for 10 minutes at a maximum 15 V by using a water electrolysis system (SUPEROXYDOLABO JED-02, Aoi Engineering Co. Nagoya), which was equipped with a platinum-coated titanium electrode. Produced electrolyzed-reduced water has extremely high pH and negative ORP. However, the value of ORP increased gradually and reached to a similar level of ORP as deionized water in one week, while the pH was relatively stable (Fig. 1). Therefore, reduced water was used immediately after production.

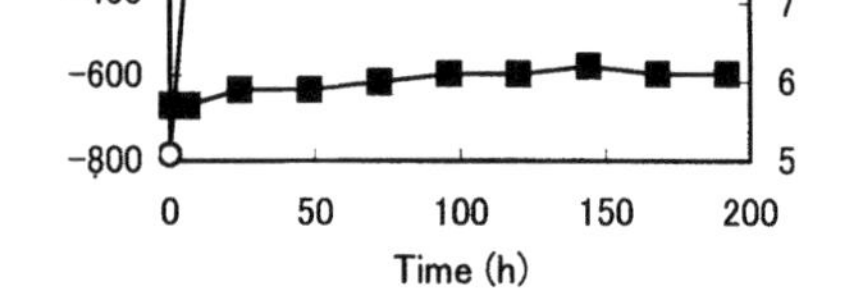


*Fig.1* Changes in pH (close symbols) and ORP (open symbols) of deionized water (squares) and electrolyzed-reduced water (circles).

## 2.2 Plant cultivation

Lettuce (*Lactuca.sativa* L. var. *capitata* L) seeds were sown and cultivated in plastic pots filled with vermiculite and culture soil mixure in an environmental controlled room. Hyponex fertilizer (Hyponex Japan) was added with every other watering in 1000 times dilution. Air temperature and PPF were 24/18℃ for light/dark periods and 190±20 μmol $m^{-2}$ $s^{-1}$, respectively. The environmental conditions are summarized in Table 1.

*Experiment 1: water stress*

In Experiment 1, the effect of reduced water on water stressed plants was examined. For the first 4 days after seeding, all plants were grown with deionized water, and then, reduced water or deionized water was used in 2 day intervals of watering. At 15 days after the germination, the watering interval was increased to 4 days for the plants in the water stress treatment. The volume of water at each watering was 50 ml per pot. Eight plants were grown for each treatment, and their fresh and dry weights were measured, and leaf number were recorded at 36 days after germination.

*Table 1* Environmental conditions for lettuce plants in Experiments 1 and 2.

| | Expt. 1 | Expt. 2 |
|---|---|---|
| Air temperature (day/night, (℃)) | 23/18 | 23/18 → 5/5* |
| Relative humidity(%) | 60±10 | 60±10 |
| PPF(μmol $m^{-2}$ $s^{-1}$) | 190 | 190 → 600* |
| Photoperiod (h $d^{-1}$) | 14 | 14 |

PPF: photosynthetic photon flux.
*Change at 25 days after seeding.

*Experiment 2: low temperature and high light*

In Experiment 2, plant growth under low temperature and high light intensity was examined. Plants were irrigated with deionized water for the first 5 days after seeding, and then cultivated for 25 days in four different types of water: deionized water, KCl solution (1 g $L^{-1}$), reduced water, and diluted reduced water (16 times). The plants were transferred and cultivated at 5℃ of air temperature throughout the day and 500±100 μmol $m^{-2}s^{-1}$ of PPF for the following 10 days. Five to six plants from each treatment were sampled at 25, 35 and 45 days after germination. Their fresh weight, dry weight, numbers of leaves, water content and chlorophyll content were measured. The treatment description of these experiments is summarized in Table 2.

*Table 2* Description of the treatments in Experiments 1 and 2.

| Expt. 1 (water stress) Treatment | Type of water | Watering Interval* (day) | Expt. 2 (High light and low temperature stress) Treatment | Type of water |
|---|---|---|---|---|
| Cont. | deionized water | 2 | Cont. | deionized water |
| ERW | electrolyzed-reduced water | 2 | ERW | electrolyzed-reduced water |
| | | | d-ERW | 1/16 strength of electrolyzed-reduced water |
| ws-Cont | deionized water | 4 | | |
| ws-ERW | electrolyzed-reduced water | 4 | KCl | 1 g/L KCl solution |

*Watering interval after day 15.

**3. Results and Discussion**

Table 3 shows shoot fresh and dry weights, dry matter content, and unfolded leaf number of lettuce in Experiment 1. Both fresh and dry weights of stressed plants were noticeable smaller than those of well-watered plants. There was no significant difference in growth rate between reduced water and deionized water irrigation treatment for both stressed and well-watered plants.

*Table 3* Shoot fresh and dry weights, dry matter content, unfolded leaf number of lettuce in Experiment 1. Means±SE are shown.

| Treatment | Fresh weight (g) | Dry weight (g) | Dry matter content (%) | Unfolded leaf number |
|---|---|---|---|---|
| Cont. | 9.8±1.4 [a] | 0.71±0.14 [a] | 7.3±0.5 [a] | 23.9±1.4 [a] |
| ERW | 9.3±1.1 [a] | 0.71±0.10 [a] | 7.7±0.6 [a] | 23.5±1.3 [a] |
| ws-Cont | 1.2±0.2 [b] | 0.25±0.04 [b] | 17.1±1.4 [b] | 12.6±0.5 [b] |
| ws-REW | 1.4±0.4 [b] | 0.24±0.06 [b] | 15.6±1.4 [b] | 13.6±1.3 [b] |

Mean within a row by different letters are significantly different at p=0.05 by Tukey test.

The time course change in fresh weight in Experiment 2 is shown in Fig. 2. Fresh weight at 45 days after the germination was similar in plants irrigated with deionized water and KCl solution, and was lowest in the reduced water treatment. This tendency was also observed in dry weight and leaf number (Table 4). These results indicated that the addition of KCl did not influence plant growth. However, irrigation with reduced water was likely to negatively affect plant growth.

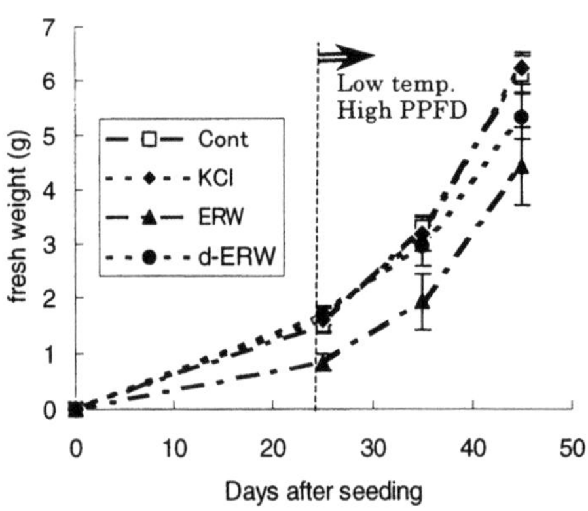


*Fig. 2* Effect of irrigated water on fresh weight of lettuce in Experiment 2. Vertical bars indicate standard errors.

For both stressful conditions, it was likely that photoinhibition of photosynthesis occurred because leaf yellowing was observed, especially for newly developed leaves. However, application of reduced water did not enhanced the plant growth under the either of the photoinhibitory conditions. This indicates that mere irrigation with reduced water dose not improve plant growth in these experimental conditions. Nutrient availability and root growth were strongly dependent on root zone pH. The vermiculite and soil mixture could be considered as the supply of nutrients and pH buffer to the roots. However, the pH of reduced water was extremely high, so it could be interpreted that plant growth was depressed due to the high pH. In addition, it is not clear whether reduced water can permeate into the plants while retaining the antioxidative function. To prove whether reduced water has an effect on photoinhibition of photosynthesis, direct assessment of photoinhibition of leaves is required. Furthermore, it is also necessary to test other methods to apply reduced water to plants.

*Table 4* Shoot dry weight and unfolded leaf number of lettuce on 25 and 45 days after seeding in Experiment 2. Means ± SE are shown.

| Treat. | days | Dry Weight (g) | Unfolded leaf number |
|---|---|---|---|
| Cont. | 25 | 0.14±0.02 | 13.6±0.6 |
| | 45 | 1.04±0.07* | 22.3±1.0 |
| KCl | 25 | 0.14±0.02 | 14.6±0.4 |
| | 45 | 0.90±0.05* | 22.7±0.3 |
| ERW | 25 | 0.08±0.02 | 12.4±0.7 |
| | 45 | 0.61±0.12 | 20.3±0.9 |
| d-REW | 25 | 0.15±0.01 | 14.8±0.6* |
| | 45 | 0.82±0.08* | 20.7±0.4 |

Mean with asterisk are significantly from those of ERW at p=0.05 by Tukey test.

**References**


Asada, K. 1994. Production and action of active oxygen species in photosynthetic tissues. In: C.H. Foyer and P.M. Mullineaux (eds.), Causes of Photooxidative Stress and Amelioration of Defence Systems in Plants. CRC Press, Boca Raton, FL. pp. 77-104.

Foyer, C. H., M. Lelandais and K. J. Kunert. 1994. Photooxidative stress in plants. Physiologia Plantarum 92:696-717.

Shirahata, S., S. Kabayama, M. Nakano and T. Miura. 1997. Electrolyzed-reduced water scavenges active oxygen species and protects DNA from oxidative damage, Biochemical and Biophysical research communications 234:269-274.

# ENVIRONMENTAL CONTROL FOR IMPROVED PLANT QUALITY WITHIN CONTROLLED ENVIRONMENT PLANT PRODUCTION SYSTEMS

Stephen T. Kania and Gene A. Giacomelli
Department of Bioresource Engineering, Rutgers, The State University of New Jersey
20 Ag Extension Way, New Brunswick, New Jersey 08901-8500, USA. E-mail: giacomel@bioresource.rutgers.edu

**Abstract.** Improved seedling quality and uniform growth within controlled environment plant production systems have become crucial, as automated production systems become more prevalent, and as specialized, highly intensive seedling production systems are developed. The environment greatly affects the seedling growth habit, final seedling quality, and ultimately the quality of the mature plant. There is an ever-increasing need to improve environmental control and uniform distribution of environmental parameters such that (1) the effect on the plant's microenvironment (plant temperature, PPF, and vapor pressure), and (2) the subsequent effect on plant processes such as metabolism, photosynthesis and gaseous transfer, are similar for all plants within the controlled environment system. This presentation will discuss the challenges of developing precise and uniform plant environments within relatively low mass, closed plant growth systems.

**Key index words.** machine vision, plant-based control, production system, sensors, uniformity.

## 1. Introduction

Improvement of production quality and uniform seedling growth within controlled environment plant production systems have become crucial, as automated production systems become more prevalent, and as specialized, highly intensive seedling production systems are developed. Ultimately the plant growth is dependent upon the inherent genetic potential of the plant specie, as well as, the environment in which it was grown. With the advent of microprocessor–based control systems and the electronic sensors for monitoring the fundamental macro-environmental parameters, the possibility for maintaining the desired level of environmental conditions and to capitalize on the genetic crop potential, has never been so promising. This 'duality' of genetics and environment is an important common factor among the professions of plant biology, horticulture and engineering. Future bio-engineering knowledge of plant processes, plant environmental needs, and environmental controls must be incorporated into the engineering science and technological developments related to monitoring and control systems.

The current process of environmental control for plant production begins with non-contact sensing technology. Sensors to monitor the levels of air temperature, atmospheric carbon dioxide concentration and moisture, and photosynthetic photon flux (PPF), transmit raw data which is then calibrated within hardware or software algorithms within the microprocessor, and then manipulated for comparison to predetermined environmental set points. If a measured parameter is different than desired, then environmental control hardware is activated to attempt to modify the parameter that is out of the desired range (Ling, et al., 1995; Giacomelli, et al, 1998).



*C. Kubota and C. Chun (eds.), Transplant Production in the 21st Century,* 72–77.

**1.1 Machine vision reads "sign language" of "speaking plant"**

The "speaking plant" approach (Takakura, et al, 1974; Udink Ten Cate, et al, 1978) to environmental control, which includes direct monitoring of plant processes, and not solely the measurement of the local environmental parameters, can provide opportunities for improved control of plant growth (Takakura, 1992; Giacomelli, 1996). Automated monitoring with programmable computers using real-time data from their sensors must be able to determine the true state or condition of the plant, while imposing a minimum of physical disturbance to the plant and its environment (Ling, et al, 1995). Machine vision is a non-invasive, non-contact sensing technology being developed for direct monitoring of plant responses to environmental conditions (Sauser et al., 1998). Machine vision which utilizes the spectral features (by reflectance), or morphological features (physical shape or dynamic growth response) of the plant can be used for correlating to the real-time plant condition. The plant uses 'sign language' to communicate information about its physical condition, and needs.

Faculty within the Controlled Environment Plant Production Systems (CEPPS) program at Rutgers University utilized machine vision technology to measure lettuce and tomato seedling growth responses, as part of NASA's interest in automated food production in space. Precise monitoring of the plant canopy provided: (1) non-destructive measurement of dry weight biomass of lettuce and tomato, (2) estimated maturation rates for lettuce, (3) correlation to timing of flowering in tomato, and (4) indication of nutrient stress in lettuce. For example, the rate of change of Top Projected Canopy Area (TPCA) (i.e. the horizontal projection of the plant canopy) of lettuce seedlings was determined with machine vision technology. Differences of TPCA between control and treatment plants were detectable within 48 hours from the onset of an imposed nutrient stress (Giacomelli, et al, 1998).

Tomato plants were also used as a model to study morphology, growth and plant diurnal movement resulting from air temperature changes and the light/dark cycle (Li, et al, 1998). The morphological features used to characterize the plant growth and movement were plant TPCA and plant side profile area. The plants were grown to first flower and destructive measurements were used to measure plant dry weight.

The plant TPCA was automatically measured every thirty minutes while growing under the three different air temperature treatments. As expected the warmest air temperature regime produced the largest TPCA (i.e. greatest projected area and biomass), and the coolest produced the least. However, there was a daily fluctuation in TPCA that showed an increase in the early part of the photoperiod, followed by a slight decrease prior to the dark period. This appearance of a 'loss' of TPCA was proven not to be an error of the camera measurement system but actually the result of the natural biorhythm of leaf movement common to some plant species. The TPCA measurements were thus influenced by both plant growth (i.e. leaf and canopy expansion) and by the natural movement of the plant and its leaves during a photoperiod.

Most environmental control sensing to date has used non-plant parameters (air temperature, etc). Future goals should include monitoring of plant processes, not environmental parameters adjacent to the plant, to directly obtain rates of photosynthesis, respiration, nutrient uptake, or intermediary biochemical molecules such as RuBP (ribulose bisphosphate) carboxylase. The activity of the enzyme RuBP carboxylase is typically measured in studies of photosynthesis, as it is the most abundant plant protein. The challenges ahead for developing systems for determining real-time plant conditions are to understand *what processes to measure*, *how to measure them*, and then, most importantly, *what set point control value to assign*, such that the desired plant response, morphology, or ultimate plant quality are obtained.

### 1.2 Genetic variability

Given that plant growth is based on both the genetic potential and the environment in which nurtured, then the objective of every environmental control system should be to provide a uniform environment throughout the entire plant growth system. This environmental uniformity will promote the most uniform growth that the natural genetic variability within one specie will allow. All monitoring systems are limited by the number of sensors, therefore it is even more imperative to obtain a uniformly controlled environment to ensure that "point source" measurements are truly representative of the majority of the plants within the overall system.

Manolkidis, 1987 demonstrated the natural genetic potential variability for tomato seedlings, as well as, techniques which could be employed to minimize natural variability. Plants were grown at close proximity within an environmentally controlled greenhouse, with every effort made to help ensure uniform environmental and plant cultural conditions, including air temperature, PPF, water, and nutrients. Plant growth differences that were measured, were primarily attributable to the genetic variability inherent with the variety of tomato. Separating very young seedlings into similar groups produced groups of mature plants that were morphologically and physiologically (days to reach first flower) more uniform within their group, than among groups of random plants. Plant parameters that were proven as good separation criteria included: 2.5 mm radicle length (as measured 7 days after sowing (DAS)), 2.5 cm first true leaf length (as measured 17 DAS), and stem length, leaf area and plant weight (measured at any time up until first flower). Table 1 includes the averages and standard deviations for stem dry weight (g) and total dry weight (g), leaf area ($cm^2$), and stem length (cm). Variability as indicated by standard deviation from the mean was always statistically significantly less (99% confidence) for each of the screening techniques compared to the control for all physical parameters listed in Table 1, for both DAS observations. The absolute variability increased for all treatments during the period from 25 to 66 DAS, but variability was always greater for the control.

## 2. Demonstrating improved environmental control

### 2.1 Environmental control for comparative plant responses used for plant modeling

Plant growth and development modeling is essential for predicting plant response within future environmentally controlled food production systems. Development of accurate dynamic crop models requires precisely controlled and repeatable environmental conditions that are uniform among all the plants located within each treatment. This is even more critical in plant environmental growth chambers because of the small number of plants that can be grown within each treatment. A small variance in plant growth response within a treatment can cause a significant effect on estimated plant productivity. Sauser, 1998 demonstrated that a typical commercial environmental plant growth chamber (EPGC) may provide environmental conditions that are highly

*Table 1* Average and standard deviation of stem length (cm), stem dry weight (g), total plant dry weight (g), and total leaf area ($cm^2$) for tomato seedlings measured at 25 and 66 days after seeding (DAS) for three treatments: an unscreened control, 2 mm radical screening at 3 DAS, and 2.5 cm first true leaf length screening at 17 DAS.

| | Unscreened Control | | 2 mm Radicle | | 2.5 cm First True Leaf | |
|---|---|---|---|---|---|---|
| | Avg. | Std. | Avg. | Std | Avg. | Std. |
| **Stem Length** | | | | | | |
| 25 DAS | 8.5 | 1.52 | 10.07 | 0.55 | 9.38 | 0.55 |
| **Stem Dry Weight** | | | | | | |
| 25 DAS | 0.041 | 0.017 | 0.055 | 0.007 | 0.055 | 0.005 |
| 66 DAS | 3.13 | 1.11 | 4.79 | 0.45 | 4.76 | 0.31 |
| **Total Dry Weight** | | | | | | |
| 25 DAS | 0.220 | 0.088 | 0.281 | 0.038 | 0.286 | 0.019 |
| 66 DAS | 15.3 | 4.48 | 20.61 | 1.86 | 20.86 | 1.60 |
| **Total Leaf Area** | | | | | | |
| 25 DAS | 83.2 | 26.5 | 111.9 | 13.7 | 110.7 | 6.5 |
| 66 DAS | 3310 | 610 | 3890 | 424 | 3873 | 375 |

different from the expected set point values. For example, the irradiance distribution was measured on the total plant production area within one EPGC and found to be 400-450 on 5%, 450-500 on 30%, 500-550 on 50%, and 550-600 $\mu mol\ m^{-2}s^{-1}$ on 15% of the entire area.

## 2.2 Whole plant canopy gas exchange monitoring device

To help verify plant growth models, four small environmentally controlled plant growth chambers (PGC) were designed and constructed to monitor whole plant processes for various advanced life support food crops (NASA-NJNSCORT Yearly Report, 2000). Net carbon dioxide consumed by the plants using a high-resolution gas analyzer (ADC-2250 IGA with an integral multiplexer) provided continuous, real time monitoring of photosynthetic rates of plants in each of four chambers. The 0.55 $m^2$ acrylic boxes have been design, constructed, installed, instrumented, and calibrated within a 9.3 $m^2$ environmental plant growth chamber for precise monitoring of carbon dioxide gas exchange during the 90 day lifecycle of soybean. The purpose was to determine plant response and generate computer model data for the short- and long-term effects of air temperature, $CO_2$ and irradiance on canopy net photosynthesis, and dark cycle respiration, and then to apply this information to developing and evaluating crop models for Advance Life Support research for NASA.

Design efforts focused on uniformity of distribution of cooling air and PPF irradiance. The unique ventilation air distribution system provided cooling air from openings within the base of the chamber. The air was forced up through the crop canopy and out the top perimeter of the boxes (Figure 1). The surrounding walk-in growth chamber provided "ambient" air (21 °C), which was blown through the canopy (1 to 5 volume air changes $min^{-1}$; 0.43 to 2.2 $m^3\ min^{-1}$). Canopy PPF was 320 $\mu mol\ m^{-2}\ s^{-1}$. Irrigation was by ebb and flood to the root zone. Completing a three dimensional air temperature distribution in the chambers, demonstrated that air temperature was maintained at desired values within the soybean plant canopy (35 plants $m^{-2}$), and above the canopy, for all air exchange rates. Air velocity of less than 1 $m\ s^{-1}$ was uniformly distributed among the plants. Representative values for the temperature distribution in one of the chambers are shown in Table 2. The top of the plant canopy was 51 cm.

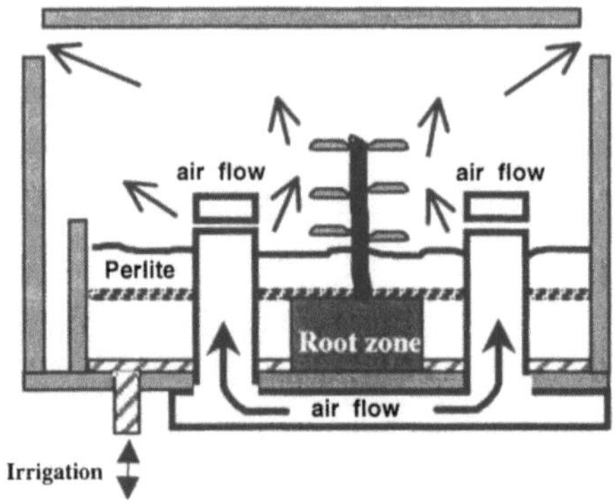


*Figure 1* Cooling air flow from openings within the base of the chamber, passing up through the crop canopy and out the top perimeter of the boxes.

*Table 2* Air temperature (°C) distribution (means and standard deviation) in environmentally controlled plant growth chambers (PGC) (Figure 1). Air exchange rate was 2 volume $min^{-1}$ (0.86 $m^3$ $min^{-1}$); cooling air temperature 21 °C; Canopy PPF was 320 μmol $m^{-2}$ $s^{-1}$; Point C is center point of the 67cm × 97 cm (x, y) PGC, Points A, B, D and E are corner points (12 cm) from edge.

| | Height from the bottom of the photosynthesis chamber | | | |
|---|---|---|---|---|
| | **15 cm** | **30 cm** | **51 cm** | **64 cm** |
| **A** | 21.5 ± 0.08 | 21.2 ± 0.10 | 22.6 ± 0.09 | 23.0 ± 0.08 |
| **B** | 21.9 ± 0.07 | 21.9 ± 0.09 | 22.7 ± 0.09 | 23.8 ± 0.11 |
| **C** | 21.7 ± 0.06 | 21.9 ± 0.08 | 22.8 ± 0.08 | 23.7 ± 0.09 |
| **D** | 21.4 ± 0.05 | 21.3 ± 0.08 | 22.8 ± 0.08 | 23.4 ± 0.08 |
| **E** | 21.9 ± 0.08 | 21.6 ± 0.09 | 22.1 ± 0.10 | 24.2 ± 0.09 |
| **Average** | 21.7 ± 0.06 | 21.6± 0.08 | 22.6± 0.09 | 23.6± 0.09 |

## 2.3 Microseedling production

Environmental control within tiny plant growth chambers (2.5 liter; 0.027 $m^2$) was demonstrated by Kania, 1989. The CEPA, controlled environment plant apparatus, was located within a walk-in environmentally controlled chamber (9.9$m^2$; 26 $m^3$) and was designed as an open system with respect to the chamber. Air temperatures were controlled to 0.5 °C above, and 1°C below set point temperature within the CEPA, by ventilating with the controlled air temperature of the walk-in chamber. Relative humidity was controlled within 5% of set point by modifying the ventilation air with desiccant dryers. Irradiance of 550 μmol $m^{-2}s^{-1}$ PAR was supplied by 2, 400W HPS lamps mounted above the CEPA. Monitoring and control were maintained with a commercially available process control software operating within a PC-based digital data acquisition and control hardware system. The uniform environment of the CEPA was shown by the maximum difference of spatial air temperature within the x-y planes, which were 1.1 °C and 1.5 °C at 3.8 and 6.4 cm (z), respectively, above the base of the seedlings. Air temperature variance over time was ±0.3°C.

**Acknowledgements.** NJAES Paper No. D-03130-04-00, supported by NASA Specialized Center of Research and Training (NJ-NSCORT) and CCEA, Center for Controlled Environment Agriculture.

**References**

Giacomelli, G.A.1996. Monitoring Plant Water Requirements with Integrated Crop Production systems. Proceedings of the International Symposium on Water Quality and Quantity in Greenhouse Horticulture. ACTA Horticulturae 458:21-27

Giacomelli, G.A., B.J. Sauser, P.P. Ling and Z. Li. 1998. Automated Machine Vision Monitoring and Feedback Control of Plant Growth. Proceedings of the 27th National Agricultural Plastics Congress, State College, PA. pp. 127-132.

Giacomelli, G.A., P.P. Ling and J. Kole. 1998. Determining Nutrient Stress in Lettuce Plants with Machine Vision Technology. HortTechnology 8(3):361-365.

Kania, S. T. 1989. A Controlled Environment Plant Apparatus for Seedling Production. M.S. Thesis, Bioresource Engineering, Graduate School—New Brunswick, Rutgers University.

Li, Z., Ling, P.P. and G. Giacomelli. 1998. "Machine Vision Monitoring of Plant Growth and Motion," *Life Support & Biosphere Science* 5(2):263-270.

Ling, P.P., Giacomelli, G.A. and Russell, T.P. 1995. Monitoring of Plant Development in Controlled Environment With Machine Vision. Advances in Space Research 18(4-5):101-112.

Manolkidis, A. K. 1987. Methods of Screening Tomato Seedlings to Improve Uniformity in Flowering. M.S. Thesis, Bioresource Engineering, Graduate School—New Brunswick, Rutgers University.

NASA NJ-NSCORT 1999 Yearly Report. Department of Plant Science, Foran Hall, Cook College, Rutgers University, New Brunswick, NJ. April 2000.

Sauser, B.J., Gene A. Giacomelli and Peter P. Ling. 1998. Development of the Basis for an Automated Plant-based Environmental Control System. SAE Technical Paper Series 981551, SAE, Int. Warrendale, PA.

Sauser, B.J. 1998. Modeling the Effects of Air Temperature Perturbations for Control of Tomato Plant Development. M.S. Thesis, Bioresource Engineering, Graduate School—New Brunswick, Rutgers University.

Takakura, T. 1992. Sensors in Controlled Environment Agriculture (CEA): Measuring Growth and Development. First International Workshop on Sensors in Horticulture. ACTA Horticulturae 304:99-102.

Takakura, T., Kozai, T., Tachibana, K. and Jordan, K.A. 1974. Direct Digital Control of Plant Growth. I. Design and Operation of the System. Transactions of the ASAE 17:1150-1154.

Udink Ten Cate, A.J., Bot, G.P.A., and van Dixhoorn, J.J. 1978. Computer Control of Greenhouse Climates. ACTA Horticulturae 87:265-272.

# ENVIRONMENTAL ENGINEERING FOR TRANSPLANT PRODUCTION

Chalermpol Kirdmanee and Kriengkrai Mosaleeyanon
National Center for Genetic Engineering and Biotechnology, National Science and Technology Development Agency, Rajdhevee, Bangkok 10400, Thailand. E-mail: ck@biotec.or.th

**Abstract.** Environmental engineering provides several tools and techniques needed to alter the environmental management *in vitro*. Development and application of methods *in vitro* for plant micropropagation, selection and improvement are encouraging. In past two decades, most of researchers focused on chemical environmental factors rather than physical environmental factors in order to promote the plantlet growth and to exhibit their hereditary characteristics. In this decade, the researches on physical environment under photoautotrophic condition (without an artificially supplied carbon source) have been extensively conducted. A system for photoautotrophic culture has been developed for various species and utilized to decrease the production cost. The photoautotrophic growth is utilized for *in vitro* plant micropropagation, selection, and improvement. The various applications of environmental engineering under photoautotrophic condition reflect the thrust and confidence in using such tools to improve transplant production.

**Key index words.** genetic modified plant, micropropagation, plant improvement, photoautotrophic condition, photosynthesis, plant selection.

## 1. Introduction

Growth and development of cells, tissues, and organs are determined by their genetic and environment. The gene expression of plantlets is limited by the environment *in vitro*. Therefore, control of the environmental factors is required to establish the procedures of plantlet production. Consequently, the cultures will exhibit their hereditary characteristics in a highly efficient and stable way (Fujiwara and Kozai. 1994). In past two decades, most of researchers focused on chemical environmental factors such as inorganic substances, plant growth regulators, sugar concentrations, etc., and/or chemical properties of the culture medium rather than physical environmental factors (gaseous components, light, humidity, temperature, etc.) in the culture vessel in order to promote the growth and to improve quality of plantlets. In this decade, the research on physical environmental factors has been extensively conducted. The controls of physical environment *in vitro* under photoautotrophic condition for plant micropropagation, selection, and improvement are discussed from the viewpoint of environmental engineering.

## 2. Environmental engineering under photoautotrophic condition

Explants, somatic embryos, and plantlets *in vitro* have been generally considered to have low photosynthetic ability. Therefore, they have been cultured under predominantly heterotrophic condition (on an artificially supplied carbon source such as sucrose) or photomixotrophic condition (on an artificially supplied and photosynthetically produced carbon source). However, it has been revealed that chlorophyllous plantlets have remarkable photosynthetic ability and grew better, in some cases, under photoautotrophic condition (without supply of carbon sources) than hetero- or photomixotrophic condition when the physical and chemical environment *in vitro*



*C. Kubota and C. Chun (eds.), Transplant Production in the 21st Century,* 78–82.

were properly controlled (Aitken-Christie et al., 1995). The $CO_2$ concentration in the culture vessel during the photoperiod was low (Fujiwara et al., 1987). Insufficient $CO_2$ supply into the vessel limits the photosynthesis during the photoperiod. $CO_2$ concentration in the culture vessel can be increased by using a gas-permeable microporous polypropylene film as a part of the vessel closure. Several reports have indicated positive effects of using a gas-permeable film as the closure under high photosynthetic photon flux density (PPFD) on the net photosynthetic rate and the growth rate of plantlets *in vitro* (Aitken-Christie et al., 1995; Solarova et al., 1996). Plantlets develop photoautotrophy and can grow well under photoautotrophic condition with high $CO_2$ concentration and high PPFD rather than under hetero- or photomixo-trophic condition.

## 3. Application of environmental engineering under photoautotrophic condition

### 3.1 Environmental engineering for micropropagation

Photoautotrophic growth of plantlets *in vitro* has been reported in many species (Kozai et al., 1992; Deng and Donnelly, 1993). An increase in PPFD promotes the net photosynthetic rate of plantlets *in vitro* (Kozai, 1991). However, the high PPFD may be the cause of the red discoloration on the leaves and stems (Kirdmanee et al., 1995; Dorcas and Marvin, 1994). The high $CO_2$ concentration and high PPFD (100-200 $\mu$mol $m^{-2}$ $s^{-1}$) are effective in promoting growth of potato (Kozai, 1991), tobacco (Mousseau, 1986) and *Eucalyptus* (Kirdmanee et al., 1995), and in decreasing hyperhydricity of carnation (Solarova et al., 1996). Photoautotrophic growth with an extensive root system and a high net photosynthetic rate of *Eucalyptus* plantlets *in vitro* is obtained in response to the high $CO_2$ concentration (1200 $\pm$ 100 $\mu$mol $mol^{-1}$) and use of vermiculite as the supporting material (Kirdmanee et al., 1995). The improved root system and rooting percentage are a consequence of an increase in net photosynthetic rate *in vitro* (Kirdmanee et al., 1995). The extensive root system of *in vitro* plantlets is necessary for their survival and growth *ex vitro*. Therefore, *in vitro* hardening is beneficial for the survival of plantlets *ex vitro*. Photoautotrophic growth can substantially increase efficiency of the production schemes due to the reduced biological contamination, limited processes of washing off sugar during transplanting, reduced *ex vitro* acclimatization processes, and increased survival percentage of plantlets *ex vitro*. An automated system for transplanting would be easy to develop (Kozai et al., 1997).

### 3.2 Environmental engineering for plant selection

There has been considerable interest in selection of the salinity-tolerant, and drought-tolerant species. However, the progress has been limited largely by inability to accurately identify those precise phenotypic characteristics due to environmental difference in selection. There are an increasing awareness of the potential and limitations of the techniques for *in vitro* selection (Sabbah and Moshe, 1990; McHughen and Swartz, 1984). Precise techniques of *in vitro* selection have been developed under photoautotrophic condition at different relative humidity (Fig. 1). *In vitro* relative humidity strongly affects the survival percentage of plantlets. In the medium containing NaCl, the survival percentage of plantlets is twice as high as those from the high relative humidity compared with the low relative humidity (Kirdmanee et al., 1997). Application of environmental control is necessary for *in vitro* selection to accurately identify those precise phenotypic characteristics. The photoautotrophic culture under low relative humidity is applied for selection of salt-tolerant trees. Net photosynthetic rate is used to identify the salt-tolerant ability at different NaCl concentrations (Fig. 2). The tree species with the high salt-tolerant ability show a high survival percentage after growing in the salinity land (Fig. 3). Therefore, the environmental control under photoautotrophic condition can substantially increase efficiency of the *in vitro* selection due to the less time consuming and high accuracy.

### 3.3 Environmental engineering for plant improvement

Recently, genetic engineering and mutation techniques have been widely used to improve the quality of plants. Genetic modified plants must be grown to observe the phenotype characteristics under closed system. Environmental engineering *in vitro* under photoautotrophic condition offers the bio-safety ways for testing genetic modified plants. Clonals of *Artemisia annua* cultured under photoautotrophic condition show the higher anti-malarial activity when compared with those cultured under photomixotrophic condition (Table 1). The anti-malarial activity of *Artemisia annua* under photoautotrophic condition is similar to those under field condition. It is possible to observe the precise phenotypic characteristics of genetic modified plants cultured *in vitro* under photoautotrophic condition. This system is feasible for genetic modified plant industry. Because it is bio-safety and low production cost.

## 4. Conclusion

Chlorophyllous explants, somatic embryos, and plantlets have photosynthetic ability. They grow well under photoautotrophic condition when the environment is properly controlled for promoting photosynthesis. Plantlets grown under photoautotrophic condition usually exhibit their precise phenotypic characteristics. It is possible that photoautotrophic growth can substantially increase efficiency of the plant micropropagation, selection, and improvement. These various applications of environmental engineering under photoautotrophic conditions reflect the thrust and confidence in using such tools to improve transplant production.

**Acknowledgments.** This work was supported by TRF/BIOTEC Special Program for Biodiversity Research and Training Grant BRT 639004 and BRT 139038.

**References**

Aitken - Christie J, Kozai T and Smith M.A.L. 1995. Automation and environmental control in plant tissue culture. Kluwer Academic Plublishers, Dordrecht. pp. 500.

Deng R and Donnelly D.J. 1993. *In-vitro* hardening of red raspberry through $CO_2$ enrichment and relative humidity reduction on sugar-free medium. Can. J. Plant Sci. 73:1105-1113.

Dorcas K.I and Marvin P.P. 1994. Rapid propagation of blueberry plants using *ex vitro* rooting and controlled acclimatization of micropropagules. HortScience 29:1124-1126.

Fujiwara K, Kozai T and Watanabe I. 1987. Measurements of carbondioxide gas concentration in closed vessel containing tissue cultured plantlets and estimates of net photosynthetic rates of the plantlets. J. Agric. Meteorol. 43:21-30.

Fujiwara K and Kozai T. 1994. The physical microenvironment and its effects. In: Aitken-Christie J. et al. (eds.) " Automation and environmental control in plant tissue culture". Kluwer Academic Publishers, Dordrecht, Netherlands. 319-369.

Kirdmanee C, Kitaya Y and Kozai T. 1995. Effects of $CO_2$ enrichment and supporting material on growth, photosynthesis, and water potential of *Eucalyptus* shoots/plantlets cultures photoautotrophically *in vitro*. In Vitro Cell. Dev. Biol.-Plant 31:144-149.

Kirdmanee C, Cha-um S and Wanussakul R. 1997. Morphological and physiological comparisons of plantlets *in vitro*: responses to salinity. Acta Hort. 457:181-186.

Kozai T. 1991. Micropropagation under photoautotrophic conditions. In: Debergh P.C., Zimmerman R.H. (eds.) "Micropropagation: technology and application" . Kluwer Academic Publishers, Dordrecht, Netherlands. pp. 447-469.

Kozai T, Fujiwara K, Hayashi M, et al. 1992. The *in vitro* environment and its control in micropropagation. In: Kurata T. (eds.) "Transplant production systems". Kluwer Academic Publishers, Dordrecht, Netherlands. pp. 247-282.

Kozai T, Kubota C and Jeong B.R. 1997. Environmental control for the large-scale pro-duction of plants through *in vitro* techniques. Plant Cell, Tiss. and Org. Cult. 51:49-56.

McHughen A and Swartz M. 1984. A tissue culture derived salt-tolerant line of flax (*Linum usitatissimum*). J. Plant Physiol. 117:109-117.

Mousseau M. 1986. $CO_2$ enrichment *in vitro.* Effect of autotrophic and heterotrophic cultures of *Nicotina tabacum* (var. *samsun*). Photosynthesis Research 8:187-191.

Sabbah S and Moshe T. 1990. Development of callus and suspension cultures of potato resistant to NaCl and mannitol and their response to stress. Plant Cell, Tiss. and Org. Cult. 21:119-128.

Solarova J, Souckova D, Ulmann J and Pospisilova J. 1996. *In vitro* culture: environmental conditions and plantlets growth as affected by vessel and stopper types. HortScience 23(2):51-58.

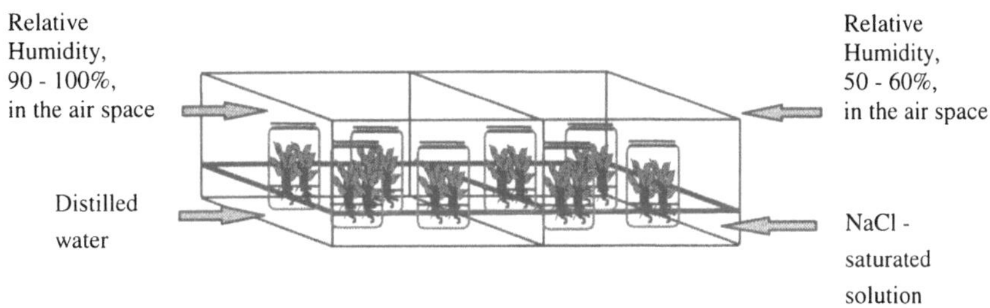


*Fig. 1* Schematic diagram of the relative humidity set up for salt-tolerant selection.

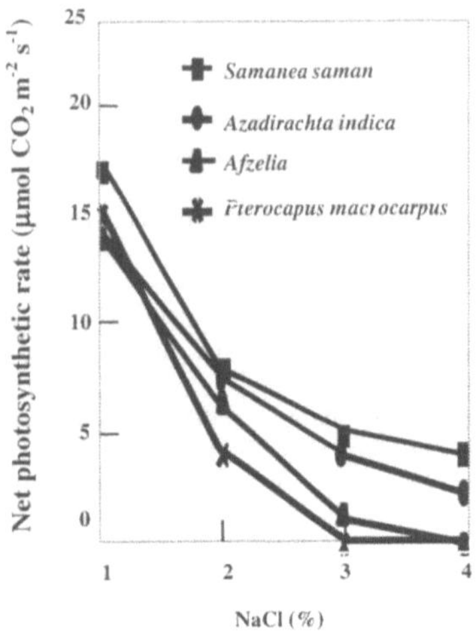


*Fig. 2* Effect of NaCl concentration on net photosynthetic rate of *Samanea saman*, *Azadirachta indica*, *Afzelia xylocarpa* and *Pterocapus macrocarpus* seedlings *in vitro*.

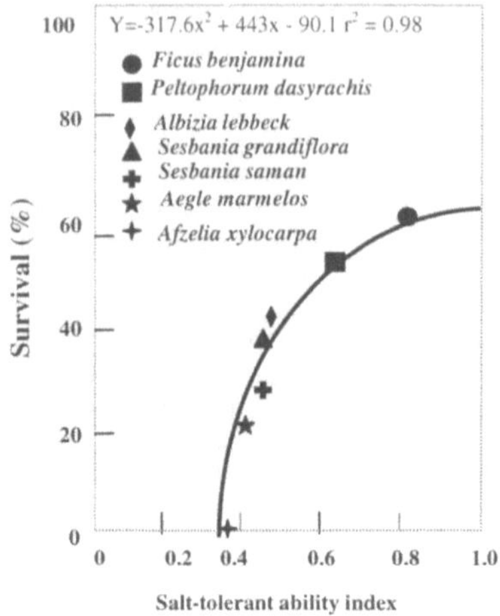


*Fig. 3* Relationship between the salt- tolerant ability index (net photosynthetic rate at 4% NaCl/net photosynthetic rate at 0% NaCl) and survival percentange of seedlings *in vitro* after growing in the salinity land.

*Table 1* Anti-malarial activity of *Artemisia annua* cultured under photomixotrophic, photoautotrophic and field (closed system) conditions.

| Condition | Anti-malarial activity ($EC_{50}$, μg $ml^{-1}$) | | |
|---|---|---|---|
| | A001 | A002 | A003 |
| Photomixotrophic | 24.5a | 24.5a | 23.7a |
| Photoautotrophic | 5.8b | 5.4b | 3.1b |
| Field | 5.6b | 4.8b | 1.0b |

Mean in a column followed by the same letters are not significantly different at *P=0.05*.

# EFFECTS OF AIR CURRENT ON TRANSPIRATION AND NET PHOTOSYNTHETIC RATES OF PLANTS IN A CLOSED PLANT PRODUCTION SYSTEM

Yoshiaki Kitaya, Johshin Tsuruyama, Masayuki Kawai, Toshio Shibuya and Makoto Kiyota
College of Agriculture, Osaka Prefecture University, Gakuen-cho 1-1, Sakai, Osaka 599-8531, Japan. E-mail: kitaya@envi.osakafu-u.ac.jp

**Abstract.** The effects of the air current speed less than 1 m $s^{-1}$ on transpiration (Tr) and net photosynthetic rates (Pn) of sweetpotato leaves and on a canopy of rice plants were determined using a chamber method and a combination of micro-meteorological and weighing methods, respectively. The effects of vertically downward moving and horizontal air currents on the Tr were also compared using a model plant canopy made from wet papers. The Tr and Pn of sweetpotato leaves were doubled as the air current speed increased from 0.01 to 0.3 m $s^{-1}$ and was almost constant at air current speeds 0.3-1.0 m $s^{-1}$. The Tr and Pn of the rice plant canopy increased linearly by 2.5 and 2 times, respectively, as the horizontal air current speed increased from 0.01 to 0.8 m $s^{-1}$. Horizontal air current speeds above 1 m $s^{-1}$ are necessary to obtain maximal Tr and Pn of the canopy. The Tr of the model plant canopy was 2-3 times greater with the vertically downward air current than in the horizontal air current in an air current speed of 0.15-0.3 m $s^{-1}$. A vertically downward air current at 0.3 m $s^{-1}$ around leaves would be adequate for promoting Tr and Pn and thus plant growth in a closed plant production system.

## 1. Introduction

The air current in a closed chamber without any adequate air circulation systems will be slower than that under greenhouse and field conditions because of low convection. Insufficient air movement around plants generally limits their growth by suppressing the gas diffusion in the leaf boundary-layer and thus by decreasing transpiration and photosynthetic rates (Yabuki and Miyagawa, 1970; Monteith and Unsworth, 1990). Control of air movement in a closed plant production system is thus essential to enhance the gas exchange between plants and the ambient air, and consequently promote growth of plants. In addition, low air current was shown to induce spatial variations in air temperature, $CO_2$ concentration and humidity inside the plant canopy (Kitaya et al., 1998). Poor uniformity of plant growth may thus be attributed to variability in the environme components around the plants. Precise control of the environment for growing pl uniformly in a closed chamber is therefore partly dependent on the control of air curre

The present research was initiated to establish an adequate air circulation system for enhancing plant growth in a closed plant production system with a large number of plants at a high density in a limited space. In the present study, the effects of the air current speed less than 1 m $s^{-1}$ on (1) transpiration and net photosynthetic rates of plant leaves and (2) on evapotranspiration and net photosynthetic rates of a plant canopy were investigated, and (3) the optimum air current direction was evaluated.



*C. Kubota and C. Chun (eds.), Transplant Production in the 21st Century,* 83–90.

## 2. Materials and Methods

### 2.1 Measurement of transpiration and net photosynthetic rates of plant leaves as affected by air current speeds

Measurements were made with leaves of sweetpotato (*Ipomoea batatas* (L.) Lam., cv. Beniazuma) in a leaf chamber as shown in Fig. 1. The measurement system was a modification of a commercial system for measuring the transpiration and photosynthetic rates of a single leaf (CIRAS-SC, PP-System Co. UK).

The air current speeds examined were varied from 0.01 to 1.0 m $s^{-1}$ by using controllable air circulation fans inside the leaf chamber. The air current speed was measured 5 mm above the leaf surface with an anemometer (Model 6071, Nihon Kanomax, Japan). The light source was a metal halide lamp and the photosynthetic photon flux was 1000 μmol $m^{-2}$ $s^{-1}$ at the leaf surface. The atmospheric conditions inside the leaf chamber were maintained at an air temperature of 28°C, a relative humidity of 65% and a $CO_2$ concentration of 380 μmol $mol^{-1}$. Leaf temperature was controled precisely at 25 °C.

### 2.2 Measurement of evapotranspiration and net photosynthetic rates of a plant canopy as affected by air current speeds

A dwarf mutant of rice (*Oriza sativa* L.) was grown in a plastic tray (265 mm in width and 530 mm in length) having 70 cells (50 x 50 $mm^2$ and 50 mm in depth each) under natural light conditions in a greenhouse. The plants on the 45th day after germination were used in the experiment. The planting density was 400 stocks per square meter. The plant height was 90 mm and the LAI was 1.4. Four sides of the plant canopy were covered with meshed cloth (60 mm in height) in order to reduce a possible effects of a horizontal air current (advection) inside the canopy during measurement (Fig.2).

The evapotranspiration and net photosynthetic rates of the plant canopy were determined following the method described by Shibuya et al. (1997). Absolute humidities and $CO_2$ concentrations were measured at two heights above the plant canopy using an infrared $H_2O$/ $CO_2$ analyser (CIRAS-1, PP-System Co., UK) to determine $H_2O$ and $CO_2$ gradients above the plant canopy. The evapotranspiration rate was determined by the time course of weight reduction of the plant canopy measured with an electronic balance (FY - 300, Kensei Kogyo Co., Japan). The gas diffusion coefficient was then determined ba on the evapotranspiration rate and the $H_2O$ gradient above the plant canopy. The photosynthetic rate was determined from the CO2 gradient above the plant canopy multiplied by the gas diffusion coefficient.

Air current speeds were controlled with a fan in a wind-tunnel-type growth chamber and measured 60 mm above the canopy with the anemometer. The PPF was 580 μmol $m^{-2}$ $s^{-1}$. The light source was a mixture of high-pressure sodium lamps and metal halide lamps. The atmospheric condition inside the growth chamber was maintained at an air temperature of 28 °C, a relative humidity of 65% and a $CO_2$ concentration of 380 μmol $mol^{-1}$.

### 2.3 Measurement of evaporation rates of a model plant canopy as affected by air current direction with various air current speeds

The effects of vertically downward and horizontal air currents on the transpiration rate were compared using a model plant canopy (150 mm of the canopy height and 2 of LAI) made from wet papers.

The model plants were settled in cell trays having small holes (3 mm in diameter) between the cells (21 mm in diameter and 25 mm in depth) in a wind-tunnel-type growth chamber (Fig. 3). The planting density was 500 model plants per square meter. The air inside the model plant canopy was forcibly ventilated through the small holes in the cell tray by suction made with a controllable air circulation fan. The speed of the downward air current was controlled in the range of 0.1-0.3 m $s^{-1}$. The evaporation rate was determined by the time course of weight reduction of the model plant canopy in 15-20 minutes interval measured with the electronic balance. The atmospheric condition inside the growth chamber was maintained at an air temperature of 25 °C and a relative humidity of 65% in darkness.

## 3. Results

The transpiration and net photosynthetic rates of leaves increased to 2.3 and 1.8 times, respectively, as the air current speed increased from 0.01 to 0.3 m $s^{-1}$and were almost constant at air current speeds 0.3-1.0 m $s^{-1}$ (Fig. 4). The increase in the transpiration rate with increasing air current speeds was more noticeable than that in the net photosynthetic rate. The leaf boundary-layer resistance measured with a wet paper replica of a leaf placed in the leaf chamber was 6.3 $m^2$ s $mol^{-1}$ at the air current speed of 0.01 m $s^{-1}$, which was double that at 0.3-1.0 m $s^{-1}$. The restricted air current speed limited the transpiration and net photosynthesis of the leaf through an increase in the leaf boundary-layer resistance.

The evapotranspiration and the net photosynthetic rates of the dwarf-rice plant canopy increased linearly by 2.5 and 2 times, respectively, as the horizontal air current speed increased from 0.01 to 0.8 m $s^{-1}$ (Fig. 5). The increase in the transpiration rate of the plant canopy with increasing air current speeds was more noticeable than that in the net photosynthetic rate as well as the single leaf. Horizontal air current speeds above 1 m $s^{-1}$ are necessary to obtain maximal transpiration and net photosynthetic rates of the plant canopy.

The evaporation rate of the model plant canopy increased with increasing air current speeds in the vertically downward and the horizontal air current treatments (Fig. 6). The evaporation rates were 2 and 2.7 times greater in the vertically downward air current than in the horizontal air current at air current speeds of 0.15 and 0.25 m $s^{-1}$, respectively. The enhancement of evaporation with increasing air current speeds was greater with the downward air current than with the horizontal air current.

## 4. Discussion

Forced air movement is more significant for a plant canopy than for a single leaf because of significant reduction of the air current speeds inside the canopy (eg., Kim et al., 1996). Shibuya and Kozai (1998) reported that the evapotranspiration and net photosynthetic rates of a canopy of tomato seedlings under an air current speed of 0.6 m $s^{-1}$ were, respectively, 1.9 and 1.4 times those under 0.1 m $s^{-1}$. The air current speed around the canopy of plants must be controlled properly because gas exchange between the plants and atmosphere is affected considerably by the air current speed. The retardation of gas exchanges with insufficient air movement would be mainly due to the increased leaf boundary-layer resistance and partly due to considerable differences in the levels of environmental variables between the inside and the outside of the plant canopy (Kim et al., 1996; Kitaya et al., 1998). Precise control of environmental variables inside the plant canopy with a sufficient air movement will allow the gas exchanges of leaves and thus the growth of plants to be more controllable.

## 5. Conclusion

The air current significantly affected the transpiration and net photosynthesis of plants at relatively low air current speeds. The appropriate air current speeds for enhancing gas exchanges of leaves were more than 0.3 m $s^{-1}$.in the vicinity of the leaves. Forced air movement with vertically downward air currents is essential in the closed plant culture system with a large number of plants at a high density and the air current speed around plants should be ensured to be at least 0.3 m $s^{-1}$.

### References

Kim, Y.H., T. Kozai, C. Kubota and Y. Kitaya. 1996. Effects of air current speeds on the microclimate of plug stand under artificial lighting, Acta. Hort. 440:354.

Kitaya Y., T. Shibuya, T. Kozai and C. Kubota. 1998. Effects of light intensity and air velocity on air temperature, water vapor pressure and $CO_2$ concentration inside a plants stand under an artificial lighting condition. Life Support & Biosphere Science 5:199-203.

Monteith J.L. and M.H. Unsworth. 1990. Principles of environmental physics, Edwward and Arnold Publishing Co., London. pp.101-120.

Shibuya T. and Kozai T. 1998. Effects of air current speed on net photosynthetic and evapotranspiration rates of a tomato plug sheet under artificial light, Environ. Control in Biol. 36(3):131-136.

Shibuya T., Y. Kitaya and T. Kozai. 1997. Dynamic Measurements of Net Photosynthetic and Evapotranspiration Rates, and Sensible and Latent Heat Transfer Rates of Plug Sheets Based on Micrometeorological and Weighing Methods. Environmental Control in Biology 35(1):71-76.

Yabuki, K. and H. Miyagawa. 1970. Studies on the effect of wind speed on photosynthesis, Japan. J. Agric. Met. 26:137-142.

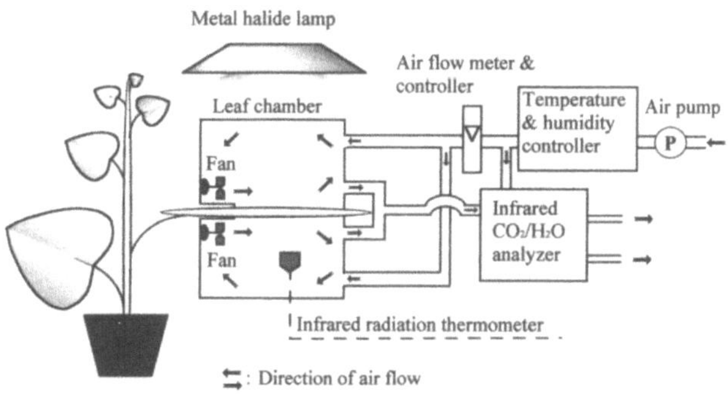


Fig.1. The apparatus for measuring the transpiration and net photosynthetic rates of a leaf.

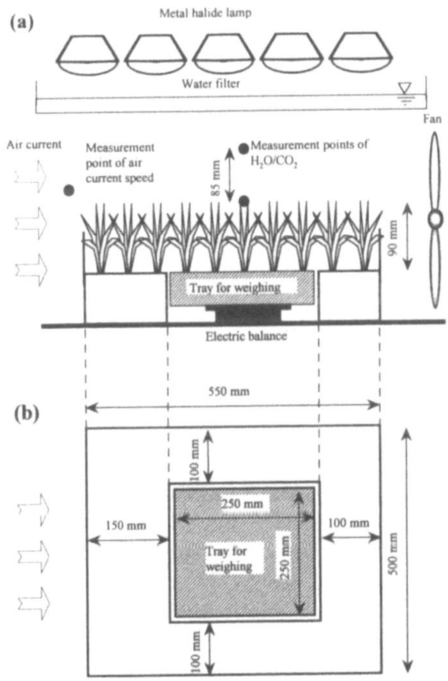


Fig.2. (a) Cross-sectional view and (b) top view of the apparatus for measuring the evapotranspiration and net photosynthetic rates of dwarf rice plant canopy.

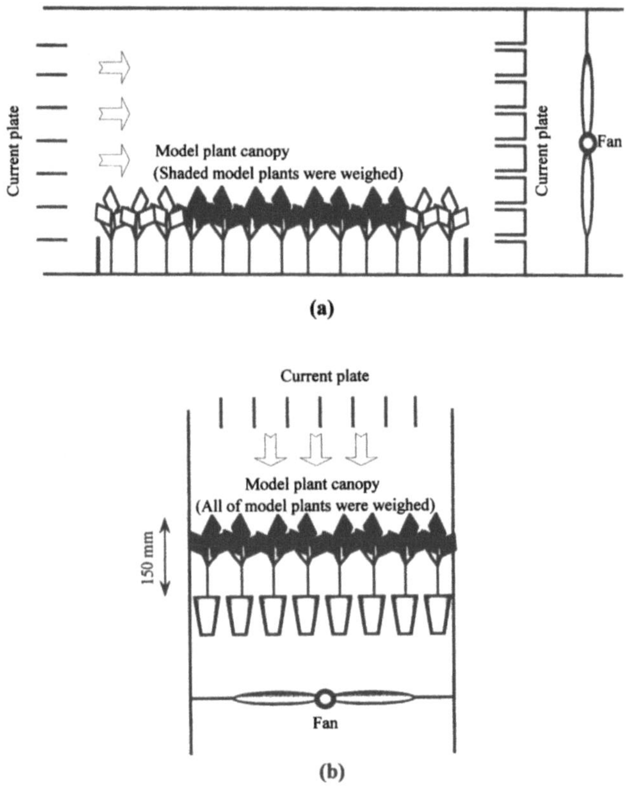


Fig.3. Cross-sectional views of the apparatus for measuring evaporation rates of model plant canopies with (a) horizontal and (b) vertically downward air currents.

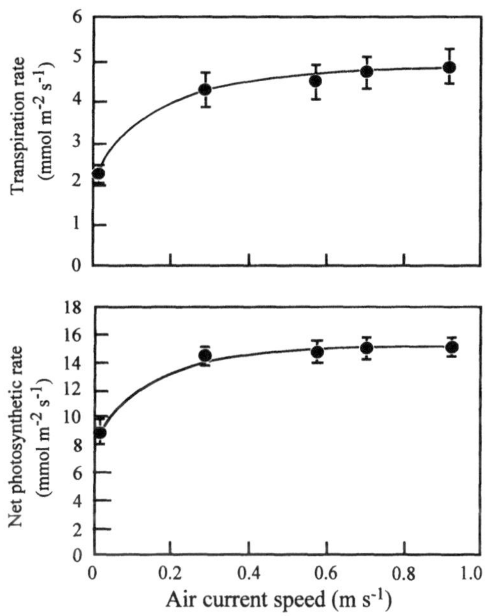


Fig.4. Effects of the air current speed on the transpiration rate and net photosynthetic rates of a single leaf of sweet potato.

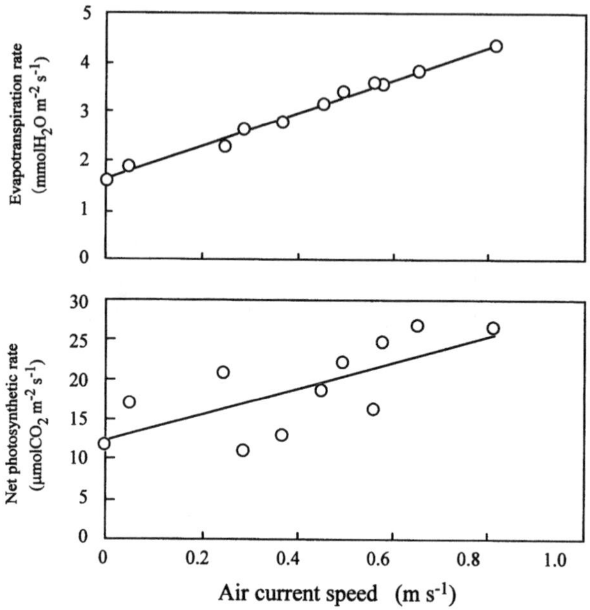


Fig.5.Effects of the air current speed on the evapotranspiration and net photosynthetic rates of the dwarf rice plant canopy.

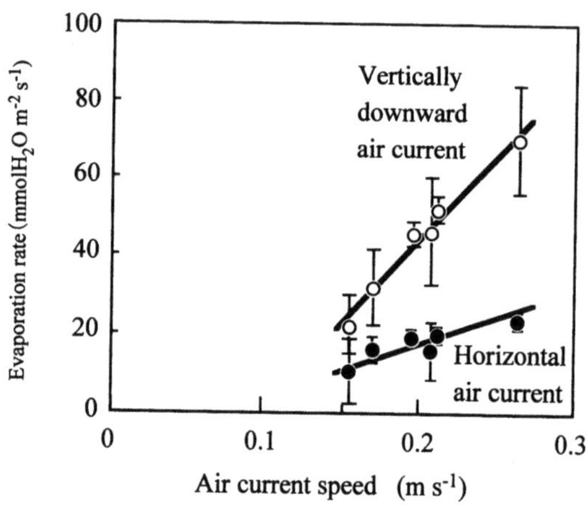


Fig.6. Effects of the speed and direction of the air current on the evaporation rates of the model plant canopies made from wet paper.

# EFFECTS OF AIR TEMPERATURE, RELATIVE HUMIDITY AND PHOTOSYNTHETIC PHOTON FLUX ON THE EVAPOTRANSPIRATION RATE OF GRAFTED SEEDLINGS UNDER ARTIFICIAL LIGHTING

Yong Hyeon Kim
Bioresource Mechanical Engineering Major, Division of Bioresource Systems Engineering, Chonbuk National University, Chonju 561-756, Korea, Institute of Agricultural Science and Technology, Chonbuk National University, Chonju 561-756, Korea. E-mail: yhkim@moak.chonbuk.ac.kr

**Abstract.** Four air temperature levels of 23, 25, 27 and 29C, three humidity levels of 85, 90 and 95%R.H. and two photosynthetic photon flux (PPF) levels of 30 and 50 μmol $m^{-2}$ $s^{-1}$ were provided to investigate the effects of air temperature, relative humidity and light intensity on the evapotranspiration rate (EVTR) of watermelon grafted seedlings. EVTR of grafted seedlings increased with increasing air temperature and the passage of time after grafting. Also EVTR increased with decreasing relative humidity. EVTR of grafted seedlings at dark period reduced by half of those at photoperiod. Effect of relative humidity on the EVTR of grafted seedlings was distinctly shown at relatively high PPF. As the vapor pressure deficit decreased, the graft-taking of grafted seedlings increased. Under the high humidity more than 90%, EVTR was gradually increased with days after grafting and thus the scion and rootstock were also smoothly joined together. It is required to control optimally the environment for decreasing the vapor pressure deficit and preventing the wilting of grafted seedlings under artificial lighting during first 2-3 days after grafting. And then it is suggested to lower the relative humidity and raise the PPF by steps for the robust joining of grafted seedlings.

**Key index words.** artificial lighting source, graft-taking, light intensity, vapor pressure deficit, watermelon.

## 1. Introduction

The production area of fruit-bearing vegetables using grafted seedlings has increased in Korea. Grafting of fruit-bearing vegetables has been widely used to increase the resistance to soil-borne diseases, to increase the tolerance to low temperature or to soil salinity, to increase the plant vigor, and to extend the duration of economic harvest time. However, grafting requires time, space and materials. Also a high expertise is required for grafting, healing and acclimation (Oda, 1995).

After grafting, it is important to control the environments around grafted seedlings for the robust joining of a scion and rootstock. Usually the shading materials and plastic film are used to keep the high relative humidity and low light intensity around grafted seedlings in greenhouse or tunnel. It is quite difficult to optimally control the environment for healing and acclimation of grafted seedlings under natural light. Therefore the farmers or managers rely on their experience for the production of grafted seedlings with high quality.

If artificial light is used as a lighting source for graft-taking of grafted seedlings, the light intensity and photoperiod can be easily controlled. Although Nobuoka et al. (1996, 1997) have studied the suitable conditions for healing of graft unions on tomato scions in a growth chamber, there was no report for the graft-taking enhancement of whole grafted



*C. Kubota and C. Chun (eds.), Transplant Production in the 21st Century,* 91–97.

seedlings under artificial lighting.

The purpose of this study was to investigate the effects of air temperature, relative humidity and light intensity on the evapotranspiration rate (EVTR) of watermelon grafted seedlings under artificial lighting.

## 2. Materials and Methods

### 2.1 Grafting and graft-taking

Watermelons having green rind with stripe and round shape (*Citrullus vulgaris* cv. Sweetdew, Hungnong Seed Co.) and bottle gourd (*Lagenaria siceraria* cv. FR-King, Hungnong Seed Co.) were raised on plug tray of 50 cells. The mixture of sphagnum peat moss and perlite (80:20, v/v) was used as medium.

One cotyledon and the apical meristem of rootstock was removed and a hole about 2mm in diameter was made at the top of the rootstock hypocotyl with a stick. The hypocotyl of scion was slantly cut at 45°. The cut hypocotyl of scion was then inserted into th hole of rootstock. Grafted seedlings were healed and joined for 5 days under cool-white fluorescent lamps (FL20SEX-d/18, Keumho Electric Co.) with photoperiod of 12 d $h^{-1}$ except dark period for one day after grafting in a closed graft-taking enhancement system developed by Kim (1999).

### 2.2 Measurement system of EVTR of grafted seedlings

EVTR was calculated as follows from data for the weight change of grafted seedlings, plug tray, and medium measured by a load cell (MLP-25, Transducer Techniques) shown in Fig. 1.

$$E = \Delta W/S \quad (1)$$

where E is EVTR of grafted seedlings ($gH_2O\ m^{-2}\ h^{-1}$), $\Delta W$ is the change of weight($gH_2O\ h^{-1}$), and S is the surface area of plug tray ($m^2$).

Load cell with aluminum plate was in a wind tunnel developed by Kim et al.(1996). The output from a load cell was recorded by a data logger (CR23X, Campbell Scientific Co.) at an interval of 1 hour. Fig. 2 represents the regressional relationship between load and output from load cell used in this experiment. Load cell has a good linearity at loading or unloading.

Air temperatures were measured with copper-constantan thermocouples ($\phi$ 0.3 mm) at the inlet and outlet of a wind tunnel. At the same point, humidity sensors (BEAM 2000N, Japan Beam Electronics Co., Ltd.) were used for the measurement of relative humidity at the same points..

Four air temperature levels of 23, 25, 27 and 29C, three humidity levels of 85, 90 and 95%R.H. and two photosynthetic photon flux (PPF) levels of 30 and 50 $\mu mol\ m^{-2}\ s^{-1}$ were provided to investigate the effects of air temperature, relative humidity and light intensity on EVTR of watermelon grafted seedlings. Table 1 represents the experimental treatments. Air current speed around the grafted seedlings was controlled to 0.1 $m\ s^{-1}$ in all treatments.

## 3. Results and Discussion

### 3.1 Effect of air temperature on EVTR

Effect of air temperature on the EVTR of grafted seedlings at relative humidity of

95%, PPF of 30 μmol $m^{-2}$ $s^{-1}$, and air current speed of 0.1 m $s^{-1}$ was shown in Fig. 3. At photoperiod, EVTR of grafted seedlings increased with increasing air temperature and the passage of time after grafting. EVTR of grafted seedlings at dark period, which increased slightly, reduced by half of those at photoperiod. The variation for EVTR of grafted seedlings at PPF of 50 μmol $m^{-2}$ $s^{-1}$ as shown in Fig. 4 was similar to those at PPF of 30 μmol $m^{-2}$ $s^{-1}$. Although PPF had no effect on the tendency of variation in EVTR at relative humidity of 95%, PPF seemed to affect absolute value of EVTR.

### 3.2 Effect of relative humidity on EVTR

Fig. 5 represents the effect of relative humidity on EVTR of grafted seedlings at air temperature of 27C, PPF of 30 μmol $m^{-2}$ $s^{-1}$, and air current speed of 0.1 m $s^{-1}$. At relative humidity of 85%, EVTR of grafted seedlings during first 1-2 days after grafting rapidly increased. But EVTR when 3 days after grafting elapsed decreased. Thus wilting was observed in some grafted seedlings grafted-taken at relative humidity of 85%. EVTR of grafted seedlings continually increased at relative humidity higher than 90%. Difference in EVTR between at relative humidity of 90% and at relative humidity of 95% was slightly observed.

Effect of relative humidity on the EVTR of grafted seedlings was distinctly shown at relatively high PPF. As shown in Fig. 6, EVTR of grafted seedlings at PPF of 50 μmol $m^{-2}$ $s^{-1}$ was highly observed during first 1-2 days.

EVTR increased with decreasing relative humidity. As relative humidity decreased and air temperature increased, vapor pressure deficit increased as shown Table 1. Therefore it is required to maintain a low level vapor pressure deficit for suppressing EVTR of grafted seedlings during first 1-2 days after grafting.

### 3.3 Graft-taking of grafted seedlings

The graft-taking of grafted seedlings was defined as the percentage of number of surviving seedlings to the number of grafted seedlings at different treatment. Effect of vapor pressure deficit on the graft-taking of grafted seedlings was shown in Fig. 7. As the vapor pressure deficit decreased, the graft-taking of grafted seedlings increased. Graft-taking higher than 90% was observed at the vapor pressure deficit less than 0.4kPa.

Fig. 8 and Fig. 9 represent the graft-taking of grafted seedlings according to different treatment at PPF of 30 and 50 μmol $m^{-2}$ $s^{-1}$. Graft-taking higher than 90% was observed at the high humidity of 90 and 95%. In these treatments, EVTR was relatively low during first 1-2 days after grafting and then gradually increased with days after grafting. Thus the scion and rootstock were also smoothly joined together. Grafted seedlings graft-taken under artificial light should move into a greenhouse under natural light for acclimation. Light intensity is very high and relative humidity is moderate under natural light

From the above results, it is required to control optimally the environment for decreasing the vapor pressure deficit and preventing the wilting of grafted seedlings under artificial lighting during first 2-3 days after grafting. And then it is suggested to lower the relative humidity and raise the PPF by steps for the robust joining of grafted seedlings.

**Acknowledgements.** This research was funded by the MAF-SGRP(Ministry of Agriculture & Forestry-Special Grants Research Program) in Korea.

**References**


Kim, Y.H and H.S. Park. 2000. Measurement of evapotranspiration rate of grafted seedlings under artificial lighting. Proceedings of the Korean Society for Agricultural Machinery Summer Conference 5(1):228-233.

Kim, Y.H. 1999. Design of a prototype room for graft-taking enhancement of grafted seedlings using artificial lighting. Proceedings of the Korean Society for Agricultural Machinery Summer Conference 4(2):112-117.

Kim, Y.H. and C.H. Lee. 1998. Light intensity and spectral characteristics of fluorescent lamps as artificial light source for close illumination in transplant production factory. J. of the Korean Society for Agricultural Machinery 23(6):591-598.

Kim, Y.H.. T. Kozai, C. Kubota and Y. Kitaya. 1996. Design of a wind tunnel for plug seedlings production under artificial lighting. Acta Hort. 440:153-158.

Nobuoka, T., M. Oda and H. Sasaki. 1996. Effects of relative humidity, light intensity and leaf temperature on transpiration of tomato scions. J. Japan. Soc. Hort. Sci. 64(4):859-865.

Nobuoka, T., M. Oda and H. Sasaki. 1997. Effects of wind and vapor pressure deficit on transpiration of tomato scions. J. Japan. Soc. Hort. Sci. 66(1):105-112.

Oda, M. 1995. New grafting methods for fruit-bearing vegetables in Japan. JARQ 29:187-194.


*Table 1* Description of experimental treatments.

| Treatments | Air temperature (C) | Relative humidity (%) | Vapor pressure deficit (kPa) |
|---|---|---|---|
| L*(H)11 | 23 | 85 | 0.42 |
| L(H**)12 | 23 | 90 | 0.28 |
| L(H)13 | 23 | 95 | 0.14 |
| L(H)21 | 25 | 85 | 0.48 |
| L(H)22 | 25 | 90 | 0.32 |
| L(H)23 | 25 | 95 | 0.16 |
| L(H)31 | 27 | 85 | 0.54 |
| L(H)32 | 27 | 90 | 0.36 |
| L(H)33 | 27 | 95 | 0.18 |
| L(H)41 | 29 | 85 | 0.60 |
| L(H)42 | 29 | 90 | 0.40 |
| L(H)43 | 29 | 95 | 0.20 |

* L stands for lower photosynthetic photon flux (30 μmol $m^{-2}$ $s^{-1}$).

** H stands for higher photosynthetic photon flux (50 μmol $m^{-2}$ $s^{-1}$).

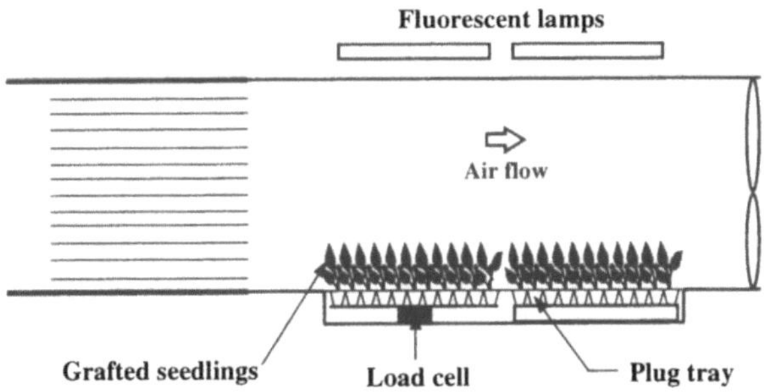


*Fig. 1* Measurement system of evapotranspiration rate of grafted seedlings using a wind tunnel under artificial lighting.

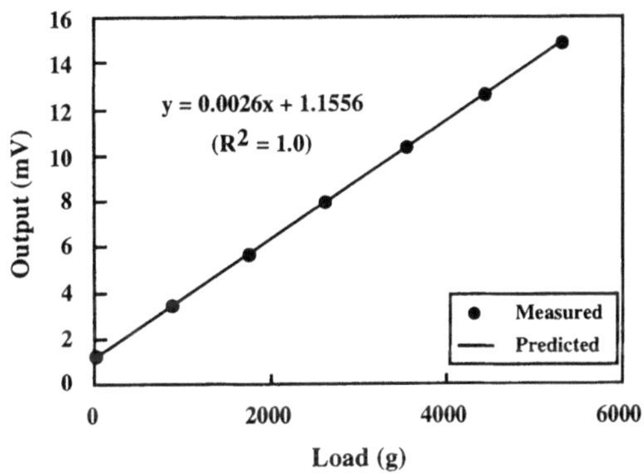


*Fig. 2* Regressional relation between load and the output from load cell.

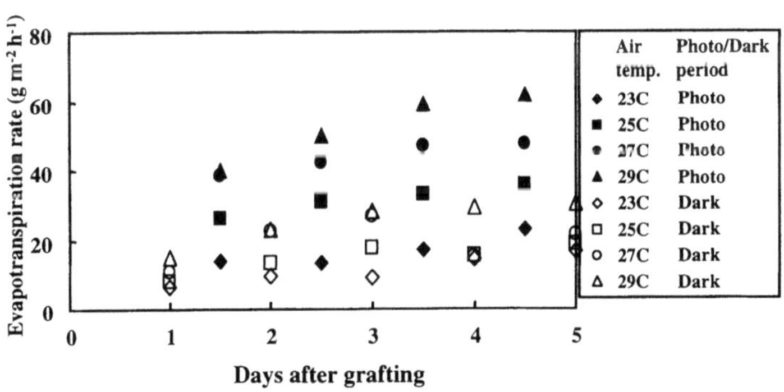


*Fig. 3* Evapotranspiration rate of grafted seedlings affected by air temperature at the relative humidity of 95% and PPF of 30 μmol $m^{-2}$ $s^{-1}$.

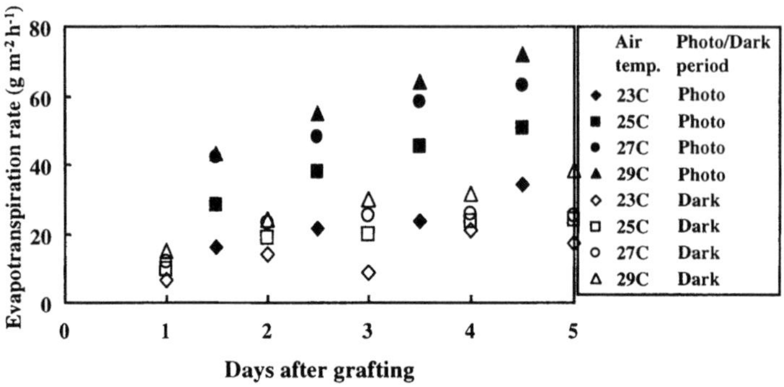


*Fig. 4* Evapotranspiration rate of grafted seedlings affected by air temperature at the relative humidity of 95% and PPF of 50 μmol $m^{-2}$ $s^{-1}$.

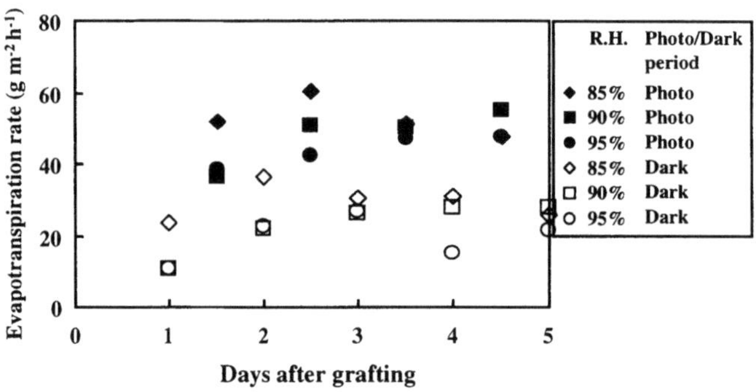


*Fig. 5* Evapotranspiration rate of grafted seedlings affected by relative humidity at the air temperature of 27C and PPF of 30 μmol $m^{-2}$ $s^{-1}$.

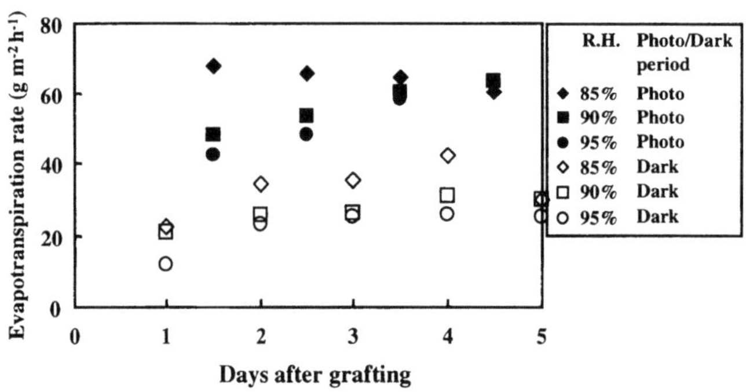


*Fig. 6* Evapotranspiration rate of grafted seedlings affected by relative humidity at the air temperature of 27C and PPF of 50 μmol $m^{-2}$ $s^{-1}$.

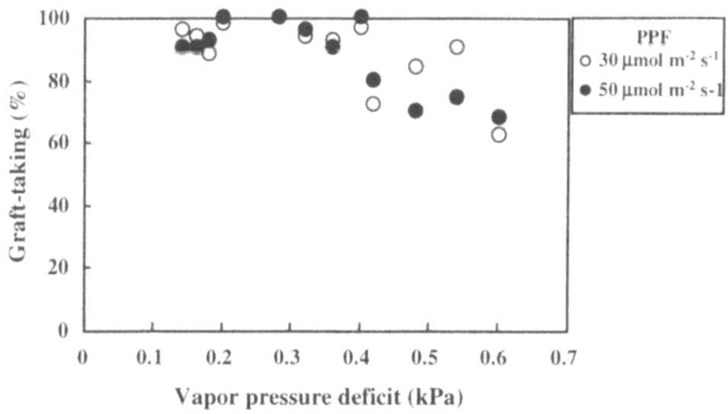


*Fig. 7* Graft-taking affected by vapor pressure deficit and photosynthetic photon flux.

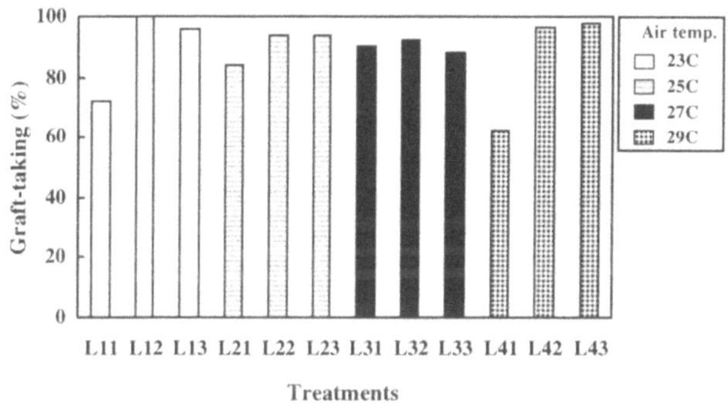


*Fig. 8* Graft-taking of grafted seedlings by the different treatments at PPF of 30 μmol m$^{-2}$ s$^{-1}$.

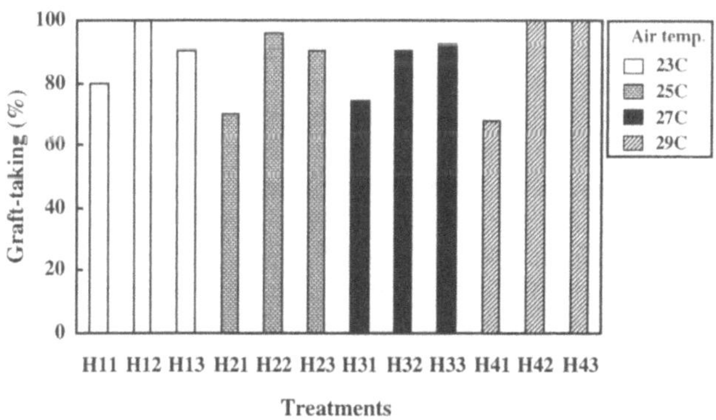


*Fig. 9* Graft-taking of grafted seedlings by the different treatments at PPF of 50 μmol m$^{-2}$ s$^{-1}$.

# GROWTH OF TOMATO (*LYCOPERSICON ESCULENTUM* MILL.) PLUG TRANSPLANTS IN A CLOSED SYSTEM AT RELATIVELY HIGH AIR CURRENT SPEEDS – A PRELIMINARY STUDY -

Watcharra Chintakovid and Toyoki Kozai
Faculty of Horticulture, Chiba University, Matsudo, Chiba 271-8510, Japan. E-mail: watchara@green.h.chiba-u.ac.jp

**Abstract.** Growth and its uniformity of tomato (*Lycopersicon esculentum* Mill., cv. Momotaro) plug transplants, with fully expanded cotyledon, in 128-cell plug trays were investigated in the wind tunnel at air current speeds of 0.5 and 1.0 m $s^{-1}$. The plug transplants were grown at PPF of 375 µmol $m^{-2}$ $s^{-1}$ as measured at the canopy surface and provided by fluorescent lamps, with air temperature of 26℃, and $CO_2$ concentration of 1000 µmol $mol^{-1}$. The results showed that dry mass of the plug transplants on Day 6 at air current speed of 0.5 m $s^{-1}$ was similar to that at 1.0 m $s^{-1}$ (119±35 and 120±39 mg per transplant, respectively). Net photosynthetic rates of plug transplants on Days 2, 4 and 6, respectively, were 10, 16 and 24 µmol $m^{-2}$ $s^{-1}$ at 0.5 m $s^{-1}$; and 7, 15 and 23 µmol $m^{-2}$ $s^{-1}$ at 1.0 m $s^{-1}$. The differences in $CO_2$ concentration between windward and leeward end of the plug tray at air current speeds of 0.5 and 1.0 m $s^{-1}$ were about 2 and 1 µmol $mol^{-1}$, respectively. The differences in water vapor concentration were 0.7 and 0.4 mmol $mol^{-1}$, respectively. Dry mass of plug transplants was significantly lower at rear half of the plug tray than at front half of the plug tray. The reason for this significant difference requires further investigation.

**Key index words.** artificial lighting, carbon dioxide concentration, *Lycopersicon esculentum*, tomato, water vapor concentration, wind tunnel.

## 1. Introduction

Transplant production in the closed system with artificial lighting will become commercially feasible in the near future, and thus has recently become an important research theme (Kozai, 1998). In the closed system, air current speed is one of the environmental factors affecting the transplant growth; and can be controlled easier than in greenhouses or in the field.

At low air current speeds, large concentration gradients of $CO_2$ and water vapor develop in the air flow stream over the plug tray, resulting in uneven growth of plug transplants. Increasing air current speeds is considered one possible way to decrease the large concentration gradients of $CO_2$ and water vapor over the plug tray, thereby increasing the uniformity of plug transplants. Net photosynthetic and evapotranspiration rates of plug transplants, increase with increasing air current speed from 0.1 to 0.6 m $s^{-1}$ under artificial light and ambient $CO_2$ concentration (Shibuya and Kozai, 1998). However, the effect of air current speed over 0.6 m $s^{-1}$, and the consistency of those effects on the growth of plug transplants in the closed system, is still unclear.

The objective of this experiment was to measure the growth and its uniformity over the plug tray in simulated condition of the closed system for transplant production at air current speed of 0.5 and 1.0 m $s^{-1}$. Plant dry mass and net photosynthetic rate were used as growth variables of the plug transplants.



*C. Kubota and C. Chun (eds.), Transplant Production in the 21st Century,* 98–101.

## 2. Materials and Methods

The experiment was conducted using two identical wind tunnels (Fig. 1), which were originally developed by Kim et al. (1996b). Air current speed in the wind tunnels was controlled by a suction fan connected with an inverter. Styrofoam blocks and a punched panel board were placed inside each wind tunnel to simulate the condition of the closed system for transplant production. The sole light source of twin fluorescent lamps (FPL55EX-N, Matsushita Electric Co. Ltd., Japan) were set above each wind tunnel.

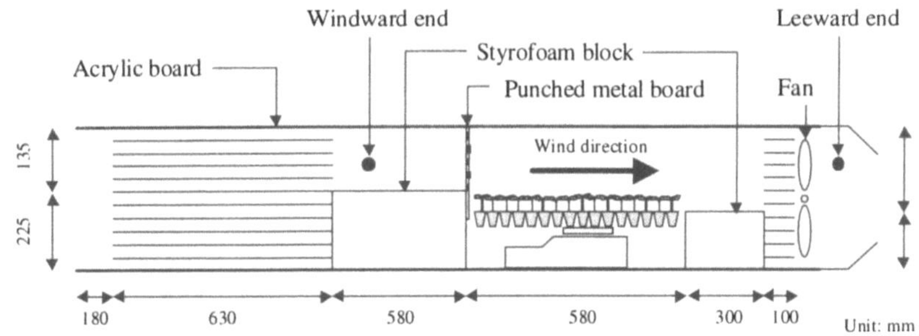


*Fig. 1* Schematic diagram of the wind tunnel, originally developed by Kim et al. (1996), used in this experiment. The wind tunnel was modified to simulate the condition of the closed transplant production system.

Tomato (*Lycopersicon esculentum* Mill., cv. Momotaro) plug transplants were grown from seed in 128-cell plug trays (270 mm × 600 mm, 25 ml volume) filled with commercial soil mixture (Yanmar Co. Ltd., Japan). After cotyledons were fully expanded, the plug transplants were placed in the wind tunnel at the air current speed of 0.5 or 1.0 m $s^{-1}$. Photosynthetic photon flux (PPF) on the canopy surface was 375±10 μmol $m^{-2}$ $s^{-1}$. $CO_2$ concentration was 1,000±200 μmol $mol^{-1}$. Air temperature was 26±3°C. Commercial nutrient solution (1/2 strength Enshi solution, Japan Tobacco Inc., Japan) was supplied with a fixed volume to each plug transplant prior to each photoperiod.

Plug transplants from the front and rear half of the plug tray, i.e., windward and leeward ends, respectively, were separately harvested for dry mass measurements on Day 6. Air temperature, $CO_2$ concentration and dewpoint temperature of the air at the windward and leeward ends of the plug tray were measured, respectively, by "T" thermocouples (copper-constantan, 0.1 mm in diameter), an infrared $CO_2$ gas analyzer (LI-6252, LI-COR, Inc., Nebraska) and a dewpoint analyzer (Model 911, Shimatsu Co. Ltd., Japan). Water vapor concentration was estimated from the dewpoint temperature. An anemometer (Climomaster 6521, Kanomax Japan Inc., Japan) measured air current speed at windward end of the plug tray. Net photosynthetic rate of plug transplants (Pn) was estimated using the methods described by Shibuya and Kozai (1998).

## 3. Results and Discussion

### 3.1 Growth at air current speeds of 0.5 and 1.0 m $s^{-1}$

Dry mass and net photosynthetic rate of plug transplants, at air current speed of 1.0

m $s^{-1}$, were similar to those at the air current speed of 0.5 m $s^{-1}$ (Table 1 and 2). Apparently, the air current speed of 0.5 m $s^{-1}$ was sufficient to promote the growth of plug transplants. However, the growth of plug transplants at air current speeds between 0.5 and 1.0 m $s^{-1}$ should be further investigated to determine if the growth of plug transplants is improved at air current speeds greater than 0.5 m $s^{-1}$ within the closed system. Slightly lower Pn on Day 2, at air current speed of 1.0 m $s^{-1}$, might be attributed to water stress of plug transplants at the early growth-stage. Consequently, a stepwise increase in air current speed, as dictated by transplant growth, is an important consideration for the further investigations.

*Table 1* Dry mass (average ± standard deviation) of tomato plug transplants on Day 6 at air current speeds of 0.5 and 1.0 m $s^{-1}$.

| Air current speed (m $s^{-1}$) | Location | Dry mass (mg) |
|---|---|---|
| 0.5 | Front half | 126±36[x] |
| | Rear half | 112±34 |
| 1.0 | Front half | 130±39 |
| | Rear half | 111±32 |
| Air current speed (A) | | NS |
| Location (B) | | ** |
| A × B | | NS |

*Table 2* Net photosynthetic rate of tomato plug transplants at air current speeds of 0.5 and 1.0 m $s^{-1}$.

| Air current speed | Net photosynthetic rate (μmol $m^{-2}$ $s^{-1}$) | | |
|---|---|---|---|
| (m $s^{-1}$) | Day 2 | Day 4 | Day 6 |
| 0.5 | 10 | 16 | 24 |
| 1.0 | 7 | 15 | 23 |

### 3.2 Uniformity of growth over the plug tray

Dry mass of plug transplants at the rear half of the plug tray or leeward end, was significantly lower than that at the front half or windward end (Table 1). Differences in $CO_2$ and water vapor concentrations, between windward and leeward end of the plug tray (Dc and Dw, respectively), at air current speed of 1.0 m $s^{-1}$ were almost 40-50% lower than those at air current speed of 0.5 m $s^{-1}$ (Table 3). There is little evidence in the literature to predict a difference in dry mass production as a consequence of a Dc of 1-2 μmol $mol^{-1}$ and Dw of 0.4-0.7 mmol $mol^{-1}$ (relative humidity difference of 1-2%).

In moving air, the water vapor boundary layer near the stomata are removed or diminished and vapor loss is more linearly related to the stomata aperture (Willmer and Fricker, 1996). The closing of stomatal aperture is ascribed mainly to the leaf temperature and water status of plants and soil, both of which are affected by air current speed. However, in this experiment, the data was insufficient to explain the lower dry mass at rear half of the plug tray.

*Table 3* Concentrations of $CO_2$ and water vapor of air at the windward end (W) and leeward end (L) of the plug tray on Day 6 at air current speeds of 0.5 and 1.0 m $s^{-1}$.

| Air current speed (m $s^{-1}$) | $CO_2$ concentration ($\mu$mol $mol^{-1}$) | | Water vapor concentration (mmol $mol^{-1}$) | |
|---|---|---|---|---|
| | W | L | W | L |
| 0.5 | 1189 | 1187 | 17.0 | 17.7 |
| 1.0 | 1061 | 1060 | 18.7 | 19.0 |

## 4. Summary

Dry mass and net photosynthetic rate of tomato plug transplants at an air current speed of 1.0 m $s^{-1}$ were not significantly different to those at 0.5 m $s^{-1}$. At either air current speed, of 0.5 or 1.0 m $s^{-1}$, dry mass at the rear half of plug tray was lower than that at front half of the plug tray.

**Acknowledgement.** Special thanks to Dr. Y. H. Kim of Chonbuk National Univerisity, Korea, and Dr. T. Shibuya of Osaka Prefecture University, Japan, for their valuable advice.

**References**

Kim, Y. H., T. Kozai, C. Kubota and Y. Kitaya. 1996a. Effects of Air Current Speeds on the Microclimate of Plug Stand under Artificial Lighting. *In* Proc. Int. Sym. Plant Production in Closed Ecosystems. Acta Hort. 440:354-359.

Kim, Y. H., T. Kozai, C. Kubota and Y. Kitaya. 1996b. Design of a Wind Tunnel for Plug Seedlings Production under Artificial Lighting. *In* Proc. Int. Sym. Plant Production in Closed Ecosystems. Acta Hort. 440:153-158.

Kozai T. 1998. Transplant Production under Artificial Light in Closed Systems - Quality Improvement of Seedlings and Plantlets by Environmental Control -. *In* Proc. 3rd Asian Crop Sci. Conf. 296-308.

Shibuya, T. and T. Kozai. 1998. Effects of Air Current Speed on Net Photosynthetic and Evapotranspiration Rates of a Tomato Plug Sheet under Artificial Light. Environ. Control in Biol. 36:131-136. (Japanese text with English abstract)

Willmer, C. and M. Fricker. 1996. Stomata; 2nd edition. Chapman & Hall. UK. pp. 375.

# ADVANCES AND CURRENT LIMITATIONS OF PLUG TRANSPLANT TECHNOLOGY IN KOREA

Byoung Ryong Jeong
Division of Applied Life Science, Graduate School of Gyeongsang National University, Chinju, Korea 660-701. E-mail: brjeong@nongae.gsnu.ac.kr

**Abstract.** Plug transplant technology was developed for efficient production of quality transplants using greenhouses. The technology, first introduced in Korea in 1992, has been established as an important industry. The technology was first adapted with the desire and emphasis on mass production of high quality seedlings of vegetable crops, but recently is rapidly widening to flower and other crops. However, some limitations of the technology are noticed, especially in terms of softwares. Regardless of large capital investment for construction of greenhouse systems with the intention for precise environment control, quality of transplants is hard to be precisely controlled for grower's satisfaction. Excessive stem stretch, uneven growth and development, and unpredictable performance of plants after transplanting are some of the problems, which are hard to correct despite of producers effort. A major factor, which causes undesirable transplant quality, is culture environment, which can not be precisely controlled in reality due to severe climatic conditions, such as extreme temperatures. Considering potential demands of transplants in Korea currently being estimated to be several billions, further refinement of the technology is needed.

**Key index words.** mechanization, nursery, plug production, propagation, transplant.

## 1. Introduction

Transplants in Korea, once produced in cold frames or hot beds by individual growers in an effort to widen growing seasons, are now being produced in plug greenhouses to meet many objectives. The production method and quality of transplants changed greatly during the developmental process. The advent and wide application of plug technology to meet these requirements are fortunate. In this paper the process of establishment of the plug industry in Korea will be discussed in relation to the importance of the industry, the needs of new technologies, current status of the technology, composition and development of system, related hardwares and softwares, and factors which currently limit transplant quality.

## 2. Concept of plug transplant and introduction of plug technology in Korea

### 2.1 Concept of plug transplant

Plug transplants are seedlings or small propagation plants raised in uniform individual cells called plugs, which are filled with a cohesive medium, and to be transplanted to other growing systems. In plug system usually seeds are sown to plug trays by an automated seeder, and with a few exceptions, only one plant per cell are raised. Plug transplants as compared to traditional transplants have advantages and disadvantages.

### 2.2 Gradation of transplant technology

Transplant production technology has changed gradually with different objectives. At the early stage of development, transplants were raised during late winter in cold



*C. Kubota and C. Chun (eds.), Transplant Production in the 21st Century,* 102–107.

frames or in hotbeds to avoid damages from late frost and to headstart. The protection of young plants from frost and cold was the prime objective at this stage. The growth and development of transplants were far advanced than those just sown on the field after the danger of final frost disappeared. Therefore, growing period was extended, early harvesting was possible, and farmland was more efficiently utilized.

However, as crop cultivation area under protected structures increased, changes in social environment such as necessity for stable production of transplants in cold seasons, and shortage and aging of farm labor force, other objectives were added. These include more active objectives such as stable and year-round supply of uniform and high quality transplants and increased efficiency in labor use and in environment manipulation.

Currently transplant production technology is developing into a stage of mass and commercial productions, which are not affected by season. In plug system overall operation is systematized, integrated and mechanized. Transplants with high uniformity in size and quality are produced year-round according to planned production schedules. Currently, transplant production is separated from crop cultivation, and is managed by specialized producers called plug growers. Crop growers just buy transplants of their choice from plug growers and devote their effort on growing of the crop according to a schedule desired.

### 2.3 Why plug system had to be introduced to Korea?

1) Need to emerge from the limitations of traditional technology: Traditional transplant production technology has several limitations and problems. Firstly, long cultivation time, use of a large amount of media, not uniform medium components, and not sterile media required large space and high management costs, and yet transplants are still exposed to various dangers. These led to low transplant quality and high production costs.

Secondly, large growing containers with a large amount of medium are heavy, and transportation and mechanization are difficult. Roots are severely damaged in the process of transplanting, and hence taking roots is delayed due to root damages.

Thirdly, production of own transplants by individual growers has limitations. Due to low technology level and small production size, materials and production environment are poor. This requires more labor input, but still transplants with poor quality are produced. As a result, year-round production is unstable, and hence diversification and more efficient utilization of diverse cropping systems are not possible. Those factors mentioned above lead to increase in costs and reduction in productivity (Table 1).

*Table 1* Efficiency of traditional vs. plug transplant production systems (Kim, 1998).

| | Traditional | Plug | Plug advantage |
|---|---|---|---|
| Labor (hour person per 10a) | 201 | 46 | 76% saving |
| Cost ($ per plant): Cucumber | 0.175 | 0.083~0.100 | 43~53% saving |
| Pepper | 0.083 | 0.050~0.058 | 30~40% saving |
| Medium (ml per plant) | 300~500 | 15~32 | 89~97% saving |
| Seed consumption (relative) | 100 | 70 | 30% saving |
| Production efficiency (plants $m^{-2}$) | 1,000 | 10,000 | 10x increase |

2) With worldwide market opening, production of quality commodities at low costs urgent: Demands for high quality horticultural commodities are steadily increasing,

and are high year-round regardless of season. This led to steady increase of greenhouse cultivation area. However, because of shortage and aging of farm labor force, greenhouse operation need to be more mechanized and automated. Also acquiring competitiveness of domestic commodities in price and quality against imported commodities from other countries is urgently needed. Plug transplants are high in quality and uniform enabling mechanization of transplanting, thus lowering production costs much lower than in traditional method.

3) Grower benefits with plug transplants: Mental and physical freedom from pressure imposed by difficult and complex processes of transplant production is realized, and more time and effort can be focused on growing crops with flexibility of adopting advanced technologies such as mechanical transplanting. Supply of professionally grown high quality transplants are easy, stable, and can be timed, even during the time when transplants are hard to be produced by individual growers. Handling and transplanting are easy, and time and cost saved. Efficiency of greenhouse space utilization is improved, market trends are noticeable, and proper adjustment to market demands are possible.

## 3. Demand and potential of plug transplants in Korea

### 3.1 Demand of vegetable and flower transplants in Korea

Cultivation areas of vegetables by group in Korea in 1996 were root (radish, carrot, etc.), 45,038 ha; leaf (chinese cabbage, lettuce, spinach, cabbage, edible chrysanthemum, etc.) 70,882 ha; fruit (watermelon, melon, cucumber, squash, tomato, strawberry, eggplants, green pepper, and muskmelon), 81,485 ha; condiment (hot pepper, garlic, onion, green onion, and ginger), and others, 24,773 ha (Ministry of Agriculture and Forestry, 1997a). Flower cultivation area was 5,342 ha including 3,274 ha of greenhouse area (Ministry of Agriculture and Forestry, 1997b).

Based on the estimation that overall about 16 billion vegetable and flower plants were grown and about a half of these were supplied as transplants, about 8 billion transplants were needed in 1996. If 50% of these are to be supplied as plugs, then the demand is about 4 billion transplants. Therefore, the potential for Korean plug industry is quite big.

In reality, maximum number of plugs that can be produced in a 4,860 $m^2$ greenhouse in a year varies by species (Table 2). Currently it is estimated that about sixty 4,860 $m^2$ plug greenhouses, including 14.3 ha greenhouses that were subsidized by government, are in production nationwide. Assuming each greenhouse produces 20 million plugs a year, then to meet the demand of 4 billion plug transplants minimum of about two hundreds 4,860 $m^2$ plug greenhouses are required. In addition, the demand will increase with the steady increase of plug demand and of cultivation areas, especially of greenhouses.

### 3.2 Current status of plug technology, problems, and perspectives in Korea

Currently plug technology in Korea is very advanced and is quite comparable to that in advanced countries. However, plug technology in Korea has following characteristics. The production of diverse group of species in small quantities presents numerous problems. The assortment of species demands diverse greenhouse environments for optimum plant quality, a demand difficult to obtain under one holding. Also the species and variety demands change with season and growing regions.

*Table 2* Numbers of plug transplants by crop that can be produced in a 4,860m$^2$ plug greenhouse a year (Jeong, 1998).

| Crop | Time required per crop (day) | Tray size (cell) | Cropping time per year | Yearly production (plants per greenhouse) |
|---|---|---|---|---|
| Pepper | 60 | 128 | 5 | 13,440,000 |
| Cucumber | 30 | 72 | 8 | 12,099,600 |
| Watermelon | 35 | 72 | 7 | 10,587,271 |
| Tomato | 30 | 128 | 8 | 21,504,000 |

Because of the small medium volume, precise control of heating, irrigation and fertilization is critical. Failure to proper control of the environment causes various disorders. Many plug growers go back to larger cell trays. For uniform growth, high quality seeds should be used, because not uniform germination leads to not uniform growth.

Development of proper technology, materials and equipment related to plug production is necessary, and this has to be supported by proper cropping systems and cultural management techniques. Plug transplants, because of small rooting medium, have very narrow windows of proper transplanting time, and relatively weak tolerance to cold or heat.

## 4. Facilities and machinery developed for the plug production system

### 4.1 Development of standard greenhouse models for the plug production systems

Considering major factors involved in each step of plug transplant production, two standard greenhouse models, an even-span steel frame glass house and an even-span greenhouse with rigid covers were developed. Both models were designed either as one gutter-connected greenhouse or as two separate gutter-connected greenhouses. The latter one was designed for the ease of environment control necessary for raising plug transplants of different crops at the same time.

### 4.2 Standardization of facility, equipment and materials for the plug production

Equipment and facilities, such as beds, an opening and closing system of thermal curtains, a travelling overhead fertigation system, a fertilizer injector, an automated chemical fogger, an integrated environment control system, a germination chamber, a graft union chamber, and an automatic seeder, were developed.

### 4.3 Development of automated systems

1) Development of automated sowing systems: The sowing systems developed are automatically capable step by step of medium mixing, tray supplying, medium filling and packing, sowing, covering, irrigation and moving to the germination chamber. The work process needs two persons and has the capacity of one hundred fifty 200-cell trays per hour.

Automated seeders of drum and nozzle types were designed and developed. The capacity of the nozzle seeder is slower with 180 trays per hour compared to the drum seeder, which can handle up to 300 trays per hour. However, the nozzle seeder has 98% sowing efficiency, compared to 90% of the drum seeder, and is more effective in terms of handling seeds of irregular shapes and sizes. Only the nozzle seeder was commercialized.

2) Development of plug tray moving lines: The movable greenhouse beds were

manufactured in sizes of 1,900 mm wide and less than 40 m long, so that 4 bed lines fit in each 9 m wide greenhouse. On top of each bed 7 rows of standard 560 mm x 280 mm trays fit longitudinally. This arrangement gives effective culture area of 76-79%. Two tray-moving systems, movable beds on top of bed rails and movable tray racks, were developed. Movable beds had 5 times higher efficiency than the tray racks. However, since movable beds cost 10 times more than the tray racks to manufacture and since the bed moving on rails is not commonly used, only the tray rack system was commercialized.

3) Germination room: 200 mm thick pressed styrofoam panels are used on walls. The germination room was estimated to need a cooling load of 36 kcal $h^{-1}$ $m^{-2}$, and cooling and heating capacity of 166.4 kcal $h^{-1}$ $m^{-2}$. Floor heating with 25 mm XL pipes arranged at 15 cm intervals as radiators gives sufficient heat. Fogging system was designed considering the maximum removal of 0.8 g $min^{-1}$ $m^{-2}$ of water during cooling. With fog particle size of 15 $\mu$ m and with fogging capacity of 5 g $min^{-1}$ $m^{-2}$, 95% RH was achieved in 10 minutes.

4) Development of a culture management system: Among the facilities and equipment, which have been improved by making prototype models, are a travelling fertigation system, a fertilizer injector, a cooling system, an ultra-fine fogger system, and an application of vertical curtain for the separation of a greenhouse area into sections.

5) Integrated environment control system: Hardware with an emphasis on the precision of different sensors, safety of the interface cards and controllers, optimization of the circuit composition of the local control panel, and prevention of surge, noise and shock were developed. Software development was focused on grower-friendly menus and control programs. Factors such as temperature, humidity, air circulation, $CO_2$ supply, insulation, water curtain, abnormal climate, and fertilizer supply are controlled.

6) Graft chamber: The major objectives of grafting are suppression of soil-born diseases, increasing resistance to adverse environment, and avoiding physiological disorders caused by successive cultivation. Grafting techniques are currently applied to majority of Cucurbitaceous species, and are further widening to tomatoes, eggplants and pepper. Commonly used methods include inarching, insertion grafting, cutting grafting, and pin grafting. A cabinet type with shelves for tray stacking, and a tunnel type graft chamber was developed and commercialized. The chamber is manipulatable to adjust temperature, RH, and air movement.

7) Others: Soil mixer, tray filler, medium compactor or auto dibbler, soil covering device, travelling boom sprayer irrigation system, fertilizer injector, grafting devices, and transplanting machines were also developed and commercialized.

**5. Development of plug culture techniques**

These areas are not as advanced as that of the hardware. Although these techniques need further refinement for production of high quality transplants, obvious limitations exist.

1) Growing medium and nutrient solution: Several media are produced or imported. Most are mixtures of peatmoss, coir, perlite, vermiculite, and such materials as soil and rice hulls. Most media currently blended in Korea have raw materials imported. Currently research is conducted to develop media using domestic resource materials such as pine and chestnut wood chips. Fertilizers are injected through an irrigation system. Not many growers use $CO_2$ fertilization yet.

2) Trays: Diverse sizes of plug trays, ranging from 32 to 288 cells, are currently in

trade. Most commonly plug trays have a dimensions of 56cm x 28cm x 5cm, and have squared cells. Trays are made of rigid or soft plastic materials, with ventilation holes between cells.

3) Seed treatment and germination: Seed techniques, such as priming and coating, are current in development and seed companies have interest. Germination chamber with high RH supplied by fog nozzles and optimum germination temperature is used at the stage of seed germination by most growers to enhance germination percentage and rate.

4) Growth regulation: The need for growth control of plug transplants is well realized. Although diverse measures, such as water and nutrition, growth retardants, mechanical stimulation, DIF, and light, are used in the advanced countries to avoid stem stretch of plug plants, only water restriction method is used now in Korea. Many producers often face with the problems of overstretching, especially when growers demand aged transplants. Not many growers try to hold transplants. Precooling of transplants to be shipped to growers is not practiced commonly yet. Plug transplants are boxed and shipped in trucks.

5) Cuttings and vernalized transplants: Plug technology, commonly applied to seedlings, is applied to such flower species as mums, roses and carnations, which are propagated asexually. Cuttings treated with rooting hormones and planted in clean medium are put under mist for rapid rooting. Perennial species such as strawberries and some flower species, which need low temperature treatment to induce reproductive phase, are propagated first in plug and then are put in a cold chamber.

## 6. Conclusions

Plug technology introduced to Korea is now well established as an important industry, dramatically changing horticultural production methods. Most of the equipment, machinery and facilities related to plug transplant production and use are already developed and commercialized. However, such techniques as growth control still need quite some improvement, quite probably through the use of closed production systems.

### References

Jeong, B.R. 1998. Plug transplant production. In Advanced Greenhouse Management and Cultural Techniques. Rural Development Corporation. pp. 571-662. (in Korean).

Kim, G.Y. 1998. Energy saving technology for greenhouse crops. In Symposium on cost saving technology for greenhouse production. Korean Research Society for Protected Horticulture. pp. 1-24. (in Korean).

Ministry of Agriculture and Forestry. 1997a. Vegetable Productions in 1996. (in Korean).

Ministry of Agriculture and Forestry. 1997b. Flower Productions in 1996. (in Korean).

# A REVIEW ON ARTIFICIAL LIGHTING OF TISSUE CULTURES AND TRANSPLANTS

Wei Fang and R.C. Jao
Department of Agricultural Machinery Engineering, National Taiwan University, 136 Chou-Shan Rd., Taipei, Taiwan. E-mail: weifang@ccms.ntu.edu.tw

**Abstract.** Tubular fluorescent lamps (TFLs) are principally used in multiple-layer tissue cultures (TC) and transplants production. HF electronic fluorescent lighting system with light regulation provides not only energy saving but also the capability of adapting the quantity of light to the stages in the development of the TC plants. Direction, uniformity, quality of light and various efficient light sources were studied worldwide. Super bright red, blue and far-red light-emitting diodes (LEDs) have many advantages over conventional light source for photosynthesis and morphology research. The characteristics of such LEDs were reviewed. Simulation models of spatial distribution of intensity, conversion factors among photometric, radiometric, and quantum units, and the electrical energy efficiency of TFLs and LEDs were discussed. The cost effectiveness of using LEDs in commercial TC and transplants production was discussed. A growth chamber using LEDs as light source, capable of adjusting light quantity and quality, can be a great tool for research and the development of such a system can be cost effective.

**Key index words.** artificial lighting, LEDs, light-emitting diodes, supplemental lighting.

## 1. Introduction

Various types of light source can be used in horticulture including incandescent bulb, tubular fluorescent, compact gas-discharge, high-pressure mercury, metal halide and high-pressure sodium lamps. Incandescent bulb is popular in photoperiodism control and compact gas-discharge lamps are normally used for decorative and display purposes. Besides these two types, others can be used for photosynthesis (Philips lighting, 1992). Tubular fluorescent lamps (TFLs) are principally used in multiple-layer tissue cultures (TC) and transplants production (Ikeda et al., 1992; Fang and Jao, 1996). HF electronic fluorescent lighting system with light regulation provides not only energy saving but also the capability of adapting the quantity of light to the stages in the development of the TC plants. The radiant efficiency (mW/W) and luminous efficacy (lm/W) of various light sources are listed in Table 1.

Energy cost is one of the major concerns for commercial applications when using artificial lights. Various efficient light sources were investigated worldwide. A microwave-powered lamp, developed by an American company (MacLennan et al., 1995), has many advantages over conventional lamps for use in artificial lighting of plants (Kozai et al., 1995; Kozai and Kubota, 1997).

Recent developments have resulted in greatly increased light output for red and blue light-emitting diodes (LEDs). LEDs with the characteristics of high energy-conversion efficiency and low thermal energy production, thus, making it a promising light source for plant growth in confined environment (Bula et al., 1991; Hoenecke et al., 1992).



*C. Kubota and C. Chun (eds.), Transplant Production in the 21st Century,* 108–113.

*Table 1* Radiant efficiency (mW/W) and luminous efficacy (lm/W) of various light source used in horticulture.

| Lamp type | Radiant efficiency (mW/W)* | Luminous efficacy (lm/W) |
|---|---|---|
| Incandescent | 62 | 14.8 |
| Fluorescent | 220-270 | 64-93 |
| Compact gas-discharge | 138-170 | 50-67 |
| High-pressure mercury | 124-166 | 40-57 |
| Metal halide | 227 | 78 |
| High-pressure sodium | 313-316 | 125-137 |

* Values adapted from Philips lighting (1992).

The objectives of this study were to review the means in applying tubular fluorescent lamps and LEDs - investigate on the promising new light source.

## 2. Lighting cycle, direction and quality of light

### 2.1 Lighting cycle

Morini et al. (1990) reported that the shorter lighting cycle promote the plant growth. Hayashi et al. (1993) examined the effects of the 24-, 6-, 1.5-, and 0.375-hour lighting cycles on growth of potato plantlets cultured photoautotrophically. The ratio of light/dark period is kept at 2:1 in all treatments. The result was consistent with Morini's conclusion. A continuing research of Hayashi et al. (1995) found that shorter lighting cycles resulted in a higher average $CO_2$ concentration during the photoperiod and a higher $CO_2$ exchange rate, thus, promoting the growth of plantlets.

### 2.2 Lighting direction

Traditionally, artificial light was provided from the top of plants. Hayashi et al., (1992) examined the lighting from the side of plantlets using TFL and showed a number of advantages including a reduction of shoot length, an increase in dry weight and a more efficient use of culture space. Kozai et al., (1995) used diffusive optical fibers to provide sideward lighting in a growth chamber. High quality transplants with short and thick stems can be obtained. Fang and Jao (1996) developed a movable TFL-mounting fixture attached to a multi-layer TC bench and investigated on downward, sideward and downward plus sideward lighting. The result showed that the movable downward lighting provide the most uniform distribution of light on bench. Advantages of using such a TFL-mounting fixture in addition to the uniform distribution of light including efficient use of space (lamp to plant distance can be reduced), less electricity cost and less number of lamps required.

### 2.3 Light quality

Light quantity and spectral quality have effects on plants in both photosynthesis and photomorphogenesis. TFL was the most popular artificial light source in tissue culture and growth room. Various types of TFL with different spectral quality were commercially available. Thimijan and Heins (1983) conducted a thorough investigation on the conversion factors among photometric, radiometric and quantum units of various types of artificial light. Among which, 4 types of TFL was included. Fang and Jao (1996) added 18 more commercially available TFL to the list of conversion factors.

Specially designed light source with different light quality were also under investigation. Sase and Ling (1996) used HID lamps providing white, blue, green and yellow light to investigate the growth of lettuce. Bula et al. (1991) showed that growing lettuce with red LEDs in combination with blue TFL is possible. Hoenecke et al. (1992) verified the necessity of blue photons for lettuce seedlings production using red LEDs with blue TFL. Super bright blue LED was not available until 1993. Nichia chemical industries of Japan succeeded in producing high intensity blue LEDs. Since then, companies such as Hewlett Packard of U.S., Panasonic and Toshiba of Japan and Everlight, Excellence of Taiwan started to produce super bright blue LEDs.

Yanagi et al. (1996) used super bright blue and red LEDs as the light source to investigate the effects of the quality and quantity of light to the growth and morphogensis of lettuce. Okamoto et al. (1996) used super bright blue and red LEDs as the light source to develop LED PACK, BIOLED, UNIPACK, and COMPACK with respect to their structure, function, circuit design and characteristics.

By changing the photon flux ratio in red (R, 600-700nm)/far red (FR, 700-800nm) radiation of artificial lights or daylight, the stem elongation can be controlled. R/FR and Blue(B, 400-500nm)/R ratios of 18 types of TFL and the combination of each TFL out of 18 types with red TFL or Agro-lite (Philips Lighting,1992) were investigated (Fang and Jao, 1996). Schuerger et al. (1997) showed that the effects of spectral quality on anatomical changes in stem and leaf tissues of peppers were correlated to the amount of blue light present in the primary light source. R/FR and B/R ratios of sunlight transmitted through various colored solid and liquid transparent filters were also investigated (Fang et al., 1999).

## 3. Light-Emitting Diodes

### 3.1 Characteristics

Super bright red, blue and far-red LEDs have many advantages over conventional light source for photosynthesis and morphology research (Bula et al., 1991; Miyashita et al., 1995). The characteristics of such LEDs available locally were measured and compared with the data sheet provided by the manufacturers. Table 2 shows 3 types of super bright red LEDs and 4 types of super bright blue LEDs. Table 3 shows the conversion factors between quantum and photometric units of super bright LEDs. Characteristics of LEDs used in the literature were also compiled as listed in Table 4.

*Table 2* Super bright red and blue LEDs measured in this study.

| Manufacturers | Model | mcd (no. of measured) | mcd from Data sheet | Peak wavelength |
|---|---|---|---|---|
| Everlight, Taiwan | 383URC-3 | 2372 (5) | 2000-3000 | 660 nm |
| Excellence, Taiwan | 5ERBCCW-DG | 8579 (30) | 6617 | 622 nm |
| Hewlett Packard | HLMP-EG08-VY000 | 3757 (30) | 3600-13800 | 626 nm |
| Nichia, Japan | NSPB500S | 3533 (4) | 3700 | 470 nm |
| Everlight, Taiwan | 333-UBC | 219 (14) | 630-1000 | 430 nm |
| Excellence, Taiwan | 5GBCCCT-EG | 1631 (30) | 2500 | 467 nm |
| Hewlett Packard,USA | HLMP-CB16 | 1900 (30) | 1575 | 472 nm |

*Table 3* Conversion factors of super bright LEDs measured in this study.

| Manufacturer (Color) | mcd[*1] | mW[*2] | Quantum[*3], µmole/m²/s | Photometric[*4] Lux | µmole/m²/s per lux | mW per µmole/m²/s |
|---|---|---|---|---|---|---|
| Hewlett Packard (Red) | 3230 | 37 | 11.03 | 374.4 | 0.029 | 3.35 |
| Excellence, Taiwan (Red) | 5878 | 45.8 | 11.08 | 479.6 | 0.023 | 4.13 |
| Everlight, Taiwan (Red) | 2207 | 37 | 10.27 | 135.1 | 0.076 | 3.60 |
| Hewlett Packard (Blue) | 1899 | 64.2 | 10.41 | 195.3 | 0.053 | 6.2 |
| Excellence, Taiwan (Blue) | 1670 | 70.4 | 4.83 | 99.6 | 0.048 | 14.57 |
| Everlight, Taiwan (Blue) | 205.6 | 80 | 1.09 | 14.9 | 0.073 | 73.39 |
| Nichia, Japan (Blue) | 3460 | 68 | 6.27 | 188.3 | 0.033 | 10.84 |

*1. Measured using photometer (J17) with J1805 LED head (TekLumaColor, Inc.).
*2. Forward current at 20mA.
*3. Measured 10 cm away using LICOR 190SB quantum sensor.
*4. Measured 10 cm away using photometer (J17) with J1811 Luminance head (TekLumaColor, Inc.).

*Table 4* The characteristics of LEDs listed in the literature.

| Company (Model) | Peak wavelength | Power consumed (standard current) | quantum yield (Luminous Intensity) | Source |
|---|---|---|---|---|
| N/A | 660 nm | N/A | N/A | Bula et al., 1991 |
| Stanley Electric Co. Ltd. (H1000) | 660 nm | 40 mW | N/A | Miyashita et al., 1995 |
| Stanley Electric Co. Ltd. | 730 nm | N/A | N/A | Miyashita et al., 1995 |
| Nichia (NLPB520) | 450 nm | 72 mW (20mA) | N/A | Okamodo et al., 1996 |
| Toshiba (TLRA120) | 660 nm | 36 mW (20mA) | N/A | Okamodo et al., 1996 |
| Shinko Denshi | 730 nm | N/A | N/A | Okamodo et al., 1996 |
| Panasonic (LNG992CF9) | N/A Blue | 68 mW (20mA) | 145 µmole/m²/s (1400 mcd) | Ono et al., 1997 |
| Panasonic (LNG901CF9) | N/A Blue | N/A | (500 mcd) | Ono et al., 1997 |
| Toshiba (TLSH180P) | 623 nm | 42 mW | 180 µmole/m²/s (7000 mcd) | Ono et al., 1997 |
| Panasonic (LN261CAL,UR) | 665 nm | N/A | (2000 mcd) | Ono et al., 1997 |
| Rohm (SLA570JT3) | 660 nm | N/A | (1000 mcd) | Ono et al., 1997 |
| Rohm (SLA570UM13) | 660 nm | N/A | (1000 mcd) | Ono et al., 1997 |
| Quantum Devices, Inc. (3009A001) | 660 nm | N/A | N/A | Schuerger et al., 1997 |
| Quantum Devices, Inc. (3009A002) | 735 nm | N/A | N/A | Schuerger et al., 1997 |

### 3.2 Cost effectiveness

LEDs have been proposed as a primary light source for space-base plant research chamber or bioregenerative life support systems (Bula et al., 1991; Barta et al., 1992). At that time, it was not ready for commercial operation. Recently, the price of both blue and red LEDs have reduced and the brightness have increased a lot. The cost effectiveness of using LEDs in a commercial TC production is still in question. The UNIPACK (Okamoto et al., 1996) consists of a cultural vessel (11 cm x 11cm x 14 cm) and a light source named LEDCAP containing 9 blue LEDs and 36 red LEDs. The price for blue and red LEDs are 27 and 3.3 NT$ (Price in Taiwan at August of 1999), respectively. LEDs in LEDCAP will cost 361.8 NT$ (11.67 US$ with exchange rate 31:1) which is not acceptable to most of the TC plantlets producers.

## 4. Simulation

Simulation models of spatial distribution of intensity of TFLs and LEDs were developed (Fang and Jao,1996; Takita et al.,1996). TFL and LED are considered as line and point light source, respectively. The TFL model is flexible in defining the arrangement of TFLs on top of a bench and the LED model can display a perspective view of the PPF distribution and a contour map of the B/R ratio with fixed arrangement of blue and red LEDs. Both models were validated using measured spatial data.

## 5. Conclusion

Various types of artificial light source were available in horticulture. FL was the most popular and cost effective light source used for TC and transplants production and still is. Searching for efficient artificial light source and better means to apply the light is a continuing task. At present, using super bright LEDs as primary light source looks promising but not cost effective for the design such as LEDCAP in a commercial scale operation. A growth chamber using LEDs as light source, capable of adjusting light quantity and quality, can be a great research tool. Development of such systems can be cost effective.

**Acknowledgment.** The authors are thankful to Splendor Green Co., Ltd. of Taiwan for the support of funds.

**References**

Barta, D.J., T.W. Tibbitts, R.J. Bula and R.C. Morrow. 1992. Evaluation of light emitting diode characteristics for a space-based plant irradiation source. Advances in Space Research 12:141-149.

Bula R.J., R.C. Morrow, T.W. Tibbitts, and D.J. Barta. 1991. Light-emitting diods as a radiation source for plants. HortScience 26(2):203-205.

Fang, W. and R.C. Jao. 1996. Simulation of light environment with fluorescent lamps and design of a movable light-mounting fixture in a growing room. Acta Hort. 440:181-186.

Fang, W., K.H. Lee, and R.C. Jao. 1999. Using colored solid and liquid filters to adjust light quality. Journal of Agricultural Machinery 8(3):23-33. (in Chinese)

Hayashi, M., N. Fujita, N. Kitaya and T. Kozai. 1992. Effect of sideward lighting on the growth of potato plantlets in vitro. Acta Hort. 319:163-167.

Hayashi M et al. 1993. Effects of lighting cycle on the growth and morphology of potato plantlets in vitro under photomixotrophic culture conditions. Environ. Control Biol. 31(3):169-175. (in Japanese)

Hayashi M et al. 1995. Effects of lighting cycle on daily $CO_2$ exchange and dry weight increase of potato plantlets cultured in vitro photoautotrophically. Acta Hort. 393:213-218.

Hoenecke, M.E., R.J. Bula, and T.W. Tibbitts. 1992. Importance of 'Blue' photon levels for lettuce seedlings grown under red-light-emitting diodes. HortScience 27(5):427-430.

Ikeda, A., Y. Tanimura, K. Esaki, Y. Kawaai and S. Nakayama. 1992. Lighting design of plant cultivation system suing fluorescent lamps. Acta Hort. 319:463-468.

Kubota, C., K. Fujiwara, Y. Kitaya, and T. Kozai. 1997. Recent advances in environment control in micropropagation. Goto et al. (eds.), Plant production in closed ecosystems. Kluwer Academic Publishers, Netherland. pp. 153-169.

Kozai, T., Y. Kitaya, and Y.S. Oh. 1995. Microwave-powered lamps as a high intensity light source for plant growth. Acta Hort. 399:107-112.

Kozai, T. and C. Kubota. 1997. Greenhouse technology for saving the earth in the 21st century. Goto et al. (eds.), Plant production in closed ecosystems. Kluwer Academic Publishers, Netherland. pp. 139-152.

MacLennan, D.A., et al. 1995. Efficient, full-spectrum, long-lived, non-toxic microwave lamp for plant growth. Proc. of International lighting in controlled environments workshop, Madison, Wisconsin, USA. pp. 243-254.

Morini S. et al. 1990. Effect of different light-dark cycles on growth of fruit tree shoots cultured *in vitro*. Advances Hort. Sci. 4:163-166.

Miyashita, Y., K. Kitaya, T. Kozai and T. Kimura. 1995. Effects of red and far red light on the growth and morphology of potato plantlets in vitro: using light-emitting diodes as a light source for micropropagation. Acta Hort. 393:189-194.

Okamoto, K., T. Yanagi, and S. Takita. 1996. Development of plant growth apparatus using blue and red LED as artificial light source. Acta Hort. 440:111-116.

Ono, E., J.L. Cuello and K.A. Jordan. 1997. Evaluation of high intensity light-emitting diodes as light source for plant growth. ASAE paper 974028.

Philips lighting. 1992. Artificial lighting in horticulture. Philips Lighting Application Information. pp. 21-22.

Sase, S. and P.P. Ling. 1996. Quantification of lighting spectral quality effect on lettuce development using machine vision. Acta Hort. 440:434-439.

Schuerger, A.C., C.S. Brown, and E.C. Stryjewsk. 1997. Anatomical features of pepper plants (Capsicum annuum L.) grown under red light-emitting diodes supplemented with blue or far-red light. Annals of Botany 79:273-282.

Takita, S., K. Okamoto, and T. Yanagi. Computer simulation of PPF distribution under blue and red LED light source for plant growth. Acta Hort. 440:286-291.

Thimijan, R.W. and R.D. Heins. 1983. Photometric, radiometric, and quantum light units of measure: a review of procedures for interconversion. Hortscience 18(6):818-822.

Yanagi, T., K. Okamoto, and S. Takita. 1996. Effects of blue, red and blue/red lights of two different PPF levels on growth and morphogensis of lettuce plants. Acta Hort. 440:117-122.

# LIGHT EMITTING DIODES (LEDs) AS A RADIATION SOURCE FOR MICROPROPAGATION OF STRAWBERRY

Duong Tan Nhut[1], Takejiro Takamura[1], Hiroyuki Watanabe[2] and Michio Tanaka[1]
[1]Faculty of Agriculture, Kagawa University, Miki-cho, Kagawa 761-0795, Japan.
E-mail: nhutduong@hotmail.com
[2]Mitsubishi Chemical Corp., Yokohama Research Center, Yokohama 227-8502, Japan.

**Abstract.** Strawberry 'Akihime' shoot explants with three leaves were cultured in three different culture systems; (1) 500 ml bottle fitted with MilliSeal with sugar-free half-strength Murashige and Skoog agar medium, (2) the "Culture Pack"-rockwool (4 by 4 blocks) system, and (3) the "Miracle Pack"-rockwool (5 by 5 blocks) system with sugar-free half-strength MS liquid medium. These culture systems were placed in the "LED PACK3", in which the red to blue light emitting-diodes (LED) ratio and the irradiation level were adjusted to 70% red +30% blue LED and 45 $\mu mol.m^{-2}.s^{-1}$, respectively. For comparison, they were also placed on the shelf under plant growth fluorescent lamps (PGF) in the culture room. In the "Culture Pack"-rockwool system, the number of leaves of plantlets under LEDs was higher than that of PGF. Shoot and root fresh weight of plantlets under LEDs were higher than that of PGF and the values were comparable to that of conventional culture systems. In the bottle culture and "Miracle Pack"-rockwool systems, the number of leaves and the total of shoot and root fresh weight of plantlets under LEDs was equal to that of PGF. Subsequent growth of plantlets cultured in the "Miracle Pack"-rockwool system under LEDs was examined after transferring to soil. The LED light source for *in vitro* culture of plantlets was found to contribute to an improved growth of the plants in acclimatization.

**Key index words**. acclimatization, *Fragaria*, plant growth fluorescent lamp.

## 1. Introduction

The use of light emitting diodes (LEDs) as a radiation source for plants has attracted considerable interest in recent years because of its vast potential for commercial application. The most attractive features of LEDs are small mass, volume, and long life (Bula et al., 1991; Brown et al., 1995). Because of these unique characteristics, there is suggestion that LEDs may be suitable for the culture of plants in a tightly controlled environment such as a space based plant culture system (Bula et al., 1991; Barta et al., 1992). Several plant species have been reported to grow successfully under LEDs. These include seedlings of lettuce, pepper, cucumber, wheat, spinach (Bula et al., 1991; Hoenecke et al., 1992; Brown and Schuerger, 1993; Scheurger and Brown, 1994; Yanagi and Okamoto, 1994; Okamoto and Yanagi, 1994; Tripathy and Brown, 1995) and *in vitro* potato plantlets (Miyashita et al., 1995). In an earlier paper, we reported enhanced growth of *Cymbidium* plantlets cultured *in vitro* under superbright red and blue LEDs the highest shoot and root fresh weight were obtained in strawberry plantlets cultured under 70% red + 30% blue LED ratio (Tanaka et al., 1998). Therefore, the current objectives were to investigate the *in vitro* shoot development of strawberry in various sugar-free culture systems under LEDs and plant growth fluorescent lamps under $CO_2$-enriched condition (3000 $\mu mol\ mols^{-1}$). Attempts were made to examine whether the light source during *in vitro* culture affects the subsequent growth of plantlets under different light source. The study presented here demonstrates the effectiveness of a



*C. Kubota and C. Chun (eds.), Transplant Production in the 21st Century*, 114–118.

total radiation system by using LEDs for micropropagation of strawberry and subsequent growth of plantlets after transferring to soil.

## 2. Materials and Methods

Unrooted strawberry 'Akihime' shoots, 30-35 mm in length, having three leaves were used as explants in all treatments. These shoots were excised from shoot masses derived from shoot-tip culture on sugar-containing half-strength Murashige and Skoog (MS) (1962) agar medium supplemented with 3 % sucrose and 0.2 mg/l 6-benzylaminopurine. The basal medium used for this research was sugar-free half-strength MS (1962) liquid medium. The pH of the medium was adjusted to 5.7 before autoclaving. Three types of culture vessels were used: (1) glass bottle (500 ml) with polycarbonate screw caps having a 3 mm diameter hole fitted with a circular self-adhesive gas permeable membrane (Milliseal™, pore size 0.5 μm; Millipore Ltd., Japan) of diameter 18 mm (aerated bottle), (2) the "Culture Pack" (CP), and (3) the "Miracle Pack" (MP). The CP (7.5 x 7.5 x 10.5 cm, the outer size of stainless frame) was made of fluorocarbon polymer film (Neoflon® PFA films, 25 μm in thickness, Daikin Industries, Japan) as described by Tanaka et al. (1988) and Tanaka (1991). The MP is the practical model of CP using Neoflon® PFA film (25 μm in thickness) (Tanaka et al., 1996). The medium substrates were agar (0.8 %, Wako Pure Chemical Industries Ltd., Japan) for bottle and rockwool (sixteen joined-blocks, 4 by 4 for CP and 5 by 5 for MP, of Grodan® Rockwool Multiblock™ AO 18/30, Grodania A/S, Denmark) with liquid medium for CP and MP. The rockwool was previously sterilized in a dry sterilizer (150°C, 1h), and placed in the CP and MP. The vessels were first autoclaved (35 min at 121°C) into which 100 ml or 160 ml of sterile liquid medium (autoclaved for 17 min at 121°C) was poured, respectively. The bottle with agar medium was autoclaved at 121°C for 17 min. These culture vessels were placed in the LED PACK 3 (Ryusho Industrial Co., Japan) in which the red to blue LED ratio and irradiation level were adjusted to 70% red + 30% blue LED and 45 $\mu mol.m^{-2}.s^{-1}$, respectively. For comparison, they were also placed on the shelf under plant growth fluorescent lamps (PGF) (45 $\mu mol.m^{-2}.s^{-1}$, Homo-Lux, National Electric Co., Tokyo, Japan) in the same culture room; and for another comparison conventional culture systems (sugar-containing agar medium in bottle under PGF and $CO_2$-nonenriched condition) were used.

To examine the effects of culture vessel and medium substrates on the shoot growth under LED lighting system and PGF, 16 shoots were placed on the agar-solidified medium. Sixteen and twenty five shoots were inserted in a hole (3 mm diameter and 10 mm depth) made on each block of the rockwool for CP and MP, respectively. Three vessels were used for each treatment. The cultures were placed in a temperature-controlled culture room (1.8 m x 1.8 m x 2.2 m) at 25°C and a 16 h photoperiod (45$\mu mol.m^{-2}.s^{-1}$, Homo-Lux, National, Japan).

For acclimatization, twenty-five plantlets cultured in MP for four weeks, were transferred to a soilless mixture (Metro-Mix® 350, Scottsco, Manisville, Ohio), and placed in an environment controlled-chamber at 25°C under halogen lamps (75 $\mu mol.m^{-2}.s^{-1}$).

The number of leaves, plant height, shoot fresh weight, number of roots, root length and root fresh weight of plantlets were recorded four weeks after planting. The acclimatized plants were analyzed after three months.

**3. Results and Discussion**

Photographs of *in vitro* strawberry plantlets on day 30 are shown in Fig. 1. The plant height of plantlets cultured in both bottle (BO) sugar-containing and sugar-free medium was higher than those of plantlets cultured in CP- and MP-rockwool systems (Table 1). The plant height was lowest in the CP-rockwool, regardless of light source. In BO and MP-rockwool systems, the number of leaves and shoot and root fresh weight of plantlets under LEDs was equal to that of PGF. In CP-rockwool system, the number of leaves and shoot and root fresh weight of plantlets under LEDs was higher than that of PGF (Table 1). The number of leaves of plantlets cultured in CP and MP under LEDs was equal to that of plantlets cultured in the conventional culture system (BO) under PGF. The shoot fresh weight of plantlets cultured in three types of culture vessel under LEDs was similar to these in the conventional culture systems. The results obtained here demonstrate the effectiveness of a total radiation system by using LEDs for *in vitro* growth of strawberry plantlets in three types of culture vessels. The use of film culture system can be applied to micropropagation of strawberry.

*Fig. 1* A comparison of *in vitro* growth of strawberry plantlets cultured under light-emitting diodes (LEDs) and plant growth fluorescent lamps (PGF) with $CO_2$ enrichment condition. *Left* two, bottle culture system with agar; *Right* two, "Culture Pack"-rockwool system.

*Table 1* *In vitro* growth of strawberry plantlets cultured in various systems under $CO_2$ enrichment and non-$CO_2$ enrichment with sugar containing medium.

| Culture vessel system | Light source | Plant height (cm) | Number of leaves | Fresh weight (mg) | | |
|---|---|---|---|---|---|---|
| | | | | Shoot | Root | Total |
| CP•RW | PGF[Y] | 4.0±0.1[Z] | 6.1±0.2 | 167.1±8.6 | 5.7±0.6 | 172.9±9.0 |
| CP•RW | LED | 4.2±0.2 | 7.2±0.3 | 212.0±12.4 | 7.6±1.0 | 219.5±13.0 |
| MP•RW | PGF | 4.6±0.2 | 5.9±0.1 | 185.8±7.2 | 6.9±0.6 | 192.7±7.6 |
| MP•RW | LED | 4.6±0.2 | 6.6±0.2 | 200.4±9.2 | 6.4±0.8 | 206.8±9.7 |
| BO•Agar | PGF | 5.4±0.2 | 5.5±0.2 | 191.6±9.0 | 18.3±2.3 | 210.1±10.3 |
| BO•Agar | LED | 5.2±0.2 | 5.8±0.2 | 211.5±13.5 | 13.8±1.5 | 225.3±10.6 |
| Control BO•Agar (+Suc) | PGF | 5.8±0.2 | 5.9±0.1 | 208.8±11.5 | 22.8±1.9 | 231.7±12.9 |

Z Average mean ± standard error
Y Plant growth fluorescent lamp (PGF)
CP: Culture Pack
MP: Miracle Pack
BO: bottle
RW: rockwool
Suc: sucrose

Subsequent growth of plantlets in the MP-rockwool system that were cultured under different light sources for 4 weeks and transferred to soil for 3 months is shown in Table 2. The number of leaves and roots of plants which were cultured *in vitro* under LEDs were higher than those of plantlets cultured under PGF. There was no difference in root length among treatments. Shoot and root fresh weight of plants cultured under LEDs was higher than that of plantlets cultured under PGF. The results obtained here demonstrate the effectiveness of a total radiation system by using LEDs for micropropagation of strawberry and subsequent acclimatization to *ex vitro* conditions.

*Table 2* Subsequent *ex vitro* growth of plantlets in the MP-rockwool system that were cultured *in vitro* under plant growth fluorescent lamps (PGF) or light emitting-diodes (LEDs) and transferred to soil for 3 months. Average means ± standard errors are shown.

| Culture vessel system | Light source | Number of leaves | Number of roots | Root length (cm) | Fresh weight (mg) | | |
|---|---|---|---|---|---|---|---|
| | | | | | Shoot | Root | Total |
| MP | PGF | 7.6±0.7 | 6.6±0.4 | 18.0±3.4 | 13977.0±1169.1 | 1300.6±175.3 | 15277.6±1289.3 |
| MP | LED | 9.6±0.7 | 9.4±0.7 | 19.2±1.3 | 16373.2±1427.5 | 1709.2±347.7 | 18082.4±1623.6 |

**References**

Barta, D. J., T. W. Tibbitts, R. J. Bula and T. W. Morrow 1992. Evaluation of light-emitting diodes characteristics for a space-based plant irradiation source. Advances in space Research. 12:141-149
Brown, C. S. and A. C Schuerger. 1993. Growth of pepper, lettuce and cucumber under light

emitting-diodes. Plant Physiol (Abstr.). 102:88.

Brown, C. S., A. C. Schuerger and J. C. Sagar. 1995. Growth and photomorphogenesis of pepper plants under red light-emitting diodes with supplemental blue or far-red light. Journal of American Society for Horticultural Science. 120:808-813.

Bula, R. J., T. W. Morrow, T. W. Tibbitts, D. J. Barta, R. W. Ignatius and T. S. Martin. 1991. Light-emitting diodes as a radiation source for plants. HortScience 26:203-205.

Hoenecke, M. E., R. J. Bula and T. W. Tibbitts. 1992. Importance of "blue" photon levels for lettuce seedlings grow under red-light-emitting diodes. HortScience 27:427-430.

Miyashita, Y., Y. Kitaya, T. Kozai and T. Kimura. 1995. Effects of red and far-red light on the growth and morphology of potato plantlets in vitro: Using light emitting diodes as a loght source for micropropagtion. Acta Hort. 393:710-715.

Murashige, T. and F. Skoog. 1962. A revised medium for rapid growth and bioassays with tabacco tissue cultures. Physiol Plant. 15:473-497.

Okamoto, K. and T. Yanagi. 1994. Development of light source for plant growth using blue and red super-bright LEDs. Shikoku-Section Joint Convention Record of the Institute of Electrical and Related Engineers. pp. 109.

Tanaka, M. 1991. Disposable film culture vessels. In: Biotechnology in agriculture and forestry, Vol. 17, High-Tech and Micropropagation I (Bajai, Y. P. S., Ed.). Springer-Verlag, Berlin. pp. 212-228.

Tanaka, M., M. Goi, and T. Higashiura. 1988. A novel disposable culture vessel made of fluorocarbon polymer films for micropropagation. Acta Hort. 226:663-670.

Tanaka, M., S. Nagae, T. Takamura, N.Kusanagi, M. Ujike and M. Goi. 1996. Efficiency and application of film culture systems in the in vitro production of planlets in some horticultural plants. J. Soc. High Tech. Agr. 8:280-285.

Tanaka, M., T. Takamura, H. Watanabe, M. Endo, T. Yanagi and K. Okamoto. 1998. In vitro growth of *Cymbidium* plantlets cultured under super red and blue light-emitting diodes (LEDs). Journal of Horticulture Science and Technology. 73:39-44.

Tanaka, M., T. N. Duong, T. Takamura, H. Watanabe and K. Okamoto. 1998. In vitro growth of strawberry plantlets cultured under superbright red and blue light emitting diodes (LEDs). Abstract of the XXVth International Horticultural Congress. pp. 407.

Tripathy, B. C. and C. S. Brown. 1995. Root-shoot interaction in the greening of wheat seedlings grown under red light. Plant Physiol. 107:407-411.

Schuerger, A. C. and C. S. Brown. 1994. Spectral quality may be used to alter plant disease development in CELSS. Advances in Spaces Research. 14:395-398.

Yanagi, T. and K. Okamoto. 1994. Super-bright light emitting diodes as an artificial light source for plant growth. Abstract of Third International Symposium on Artificial Lighting in Horticulture. pp. 19.

# APPLICATION OF RED LASER DIODE AS A LIGHT SOURCE FOR PLANT PRODUCTION

Aya Yamazaki, Hiroshi Tsuchiya, Hirofumi Miyajima, Takayoshi Honma and Hirofumi Kan
Hamamatsu Photonics K.K., 5000 Hirakuchi, Hamakita, Shizuoka 434-8601, Japan.
E-mail: yamazaki@crl.hpk.co.jp

**Abstract.** As a new type of light source for plant production, high-power and high electrical-to-optical power conversion efficiency AlGaInP laser-diode lamps with the continuous wave output power of 500 mW that have peak emission of 680 nm have been developed. To confirm the possibility of growing plants under this new light source, the effects of laser-diode light on growth of lettuce (*Lactuca sativa* L.) plants were studied. In experiment 1, lettuce plants were grown under 350 $\mu mol \cdot m^{-2} \cdot s^{-1}$ photosynthetic photon flux (PPF) of the laser-diode light with a 12-hour photoperiod. In experiment 2, lettuce plants were grown under laser-diode light supplemented with blue light with a total PPF of 350 $\mu mol \cdot m^{-2} \cdot s^{-1}$. The lettuce plants were able to grow even under the 680 nm laser-diode light. However, the leaves of the lettuce plants grown under the laser-diode light were long and thin, and their dry weight was low compared to lettuce plants simultaneously grown under high-pressure sodium lamps. By supplementation with the blue light, the shape of the leaves was much improved and the dry weights were much increased. These results indicate that red laser-diode lamps combined with blue LED light have a possibility for efficient plant production, including transplant production.

**Key index words.** blue light, electrical-to-optical power conversion efficiency, *Lactuca sativa* L., laser-diode light, 680 nm.

## 1. Introduction

Development of an effective, high-power, low-cost, artificial light source for use in plant-growing facilities would provide significant benefits for plant production.

Takatsuji et al. (1994) had proposed use of a laser-diode lamp (LD), which is a light source mainly used for applications of DVD, laser-beam printers, bar-code readers or optical disk systems, as a new type of light source for plant production. Their advantages over other light sources for use in plant production are its easy set-up for high power and pulse irradiation, low thermal radiation, small weight and volume, and selectivity for proper wavelength. In spite of these advantages, the LD hasn't been used in plant production because of its insufficiency of light power and wavelength for growing plants.

Recently, we have developed AlGaInP LDs with high-power and high electrical-to-optical power conversion efficiency for plant production. The peak wavelengths of those LDs are around 680 nm, which is in the red region of the photosynthetic action spectrum. The output power is 500 mW and the electrical-to-optical power conversion efficiency is as high as 40 percent, which is at least twice as efficient as that of LEDs.

Because the light emission of LD is monochromatic due to a characteristic of laser light, and differs from the light sources presently used in plant-growing, we started with growing lettuce (*Lactuca sativa* L.) plants to confirm the possibility of growing plants



*C. Kubota and C. Chun (eds.), Transplant Production in the 21st Century*, 119–124.

under this new light source.

## 2. Materials and Methods

### 2.1 Plant materials

Lettuce (*Lactuca sativa* L. cv. Okayama-saradana) seedlings at 5-6 leaf stage were used in this study. Seeds were individually sown on polyurethane cubes and grown hydroponically in a controlled environment. Air temperature, relative humidity and photosynthetic photon flux (PPF) were maintained at 20℃, 65 to 70% and 150 μmol $m^{-2}$ $s^{-1}$, respectively. The photoperiod was 12 hours per day and plants were grown for 16 days.

### 2.2 LDs, and the LD panel

LDs based on an aluminum-gallium-indium-phosphor substrate (AlGaInP) were used in this study. The continuous wave output power was 500 mW at a 700 mA drive electric current, and the wavelength of the peak emission was 680 nm.

As shown in Fig. 1, the emission pattern of LD is long and thin. Considering this emission pattern, 30 LDs were placed on a panel (300×240 mm) in the order shown in Fig. 2. Each LD was mounted on an aluminum heat sink (40×40×10 mm) to reduce the effect of the heat from the LD itself.

This LD panel was set in a growth chamber about 60 cm above the plant canopy level at the beginning of the experiment.

*Fig. 1* Two-dimensional intensity distribution of the laser light on the surface parallel to the emitting face of the LD.

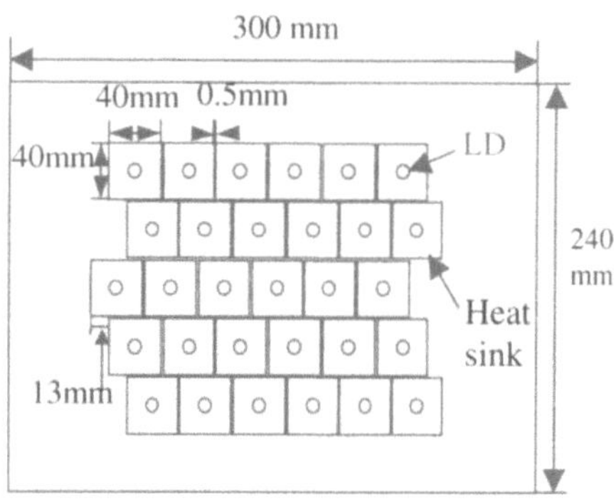


*Fig. 2* Schematic diagram of the LD panel.

### 2.3 Experiment 1

Six seedlings with 5-6 leaves each were transplanted into the growth chamber with the LD panel, and were cultured hydroponically. The pH and EC of the nutrient solution were kept at 6.0±0.5 and 20 mS $m^{-1}$, respectively. Air temperature, relative humidity and photoperiod were maintained at 20℃, 65 to 70% and 12 hours, respectively. The

PPF was maintained at 350 μmol·$m^{-2}$·$s^{-1}$ at the top of the plant canopy.

Two plants were harvested at 3 and 7 days after planting, and one was harvested at each time of 14 and 21 days after planting to measure the number of unfolded leaves. At the final harvest time (21 days after planting), the leaf length, leaf width/leaf length (W/L) ratio of the longest leaf, leaf area and dry weight were measured. Twelve samples were used for 2 replications.

### 2.4 Experiment 2

The general setup of this experiment was basically the same as that in experiment 1. The same plant materials, growing conditions (except the light condition) and ways of measurement were used.

The lettuce plants were grown under LD light supplemented with blue LED light. The peak emission of a blue LED was at 465 nm, with a typical spectral bandwidth of 25 nm, at half power. The blue LED array contained 400 blue LEDs (100×400×70 mm) and two of those arrays were placed at the side of the LD panel to supply approximately 35 μmol·$m^{-2}$·$s^{-1}$. The total PPF of the LD and blue light was maintained at 350 μmol·$m^{-2}$·$s^{-1}$ at the top of the plant canopy.

### 2.5 Treatment for control plants

For comparison purposes, lettuce plants were simultaneously grown under a high-pressure sodium lamp at each experiment, which is a light source presently used for plant growing facilities. Nine high-pressure sodium lamps were set into another growth chamber. To avoid the increase in temperature around the plants, an acrylic container with water was set under the lamps as a heat-absorbing filter. The PPF was maintained at 350 μmol·$m^{-2}$·$s^{-1}$ at the top of the plant canopy. The same plant materials, growing conditions (except the light condition) and ways of measurements used in experiment 1 and 2 were used.

## 3. Results and Discussion

### 3.1 Experiment 1

Lettuce plants grown under red LDs alone had significant difference in shape of leaves compared to plants grown under high-pressure sodium lamps (Fig. 3). W/L ratio at final harvest shows that the lettuce plants grown under LDs had thin and long leaves, while the plants grown under high-pressure sodium lamps had wide and round-shaped leaves (Table 1). Moreover, the leaf areas and dry weights were smaller than those of plants grown under the high-pressure sodium lamps (Table 1). However there were no significant differences in the numbers of unfolded leaves of lettuce grown under the two lamps (Fig. 4).

### 3.2 Experiment 2

By supplementing red LDs with blue light, the leaf shape of lettuce plants grown under the LDs was much improved and had no differences from the lettuce plants grown under the high-pressure sodium lamp (Fig. 5). Characteristics of plants such as the W/L ratio, leaf area and dry weight were produced closer to those of plants grown under high pressure sodium lamps (Table 2). Just as in experiment 1, there were no significant differences in the numbers of unfolded leaves of lettuce grown under the two lamps (Fig. 4).

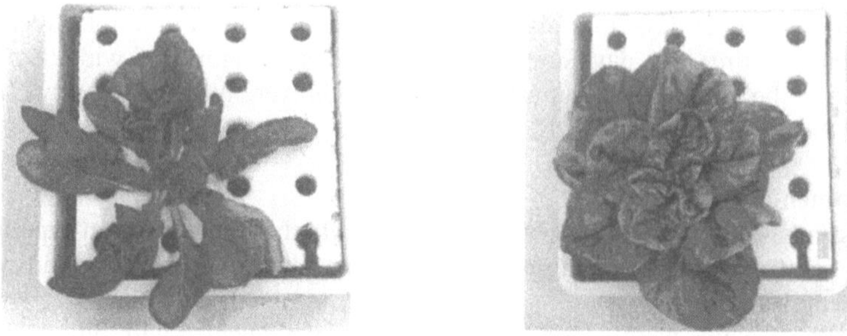

(a) LDs (b)High-pressure sodium lamps

*Fig. 3* Lettuce plants grown under the LDs (a) and the high-pressure sodium lamps (b) at 21 days after planting (Experiment 1).

*Table 1* Leaf length*, leaf width/leaf length (W/L) ratio*, leaf area and dry weights of lettuce plants grown under LDs and high-pressure sodium lamps at 21 days after planting (Experiment 1). Means ± S.E. are shown.

| | Leaf length (cm) | W/L ratio | Leaf area ($cm^2$) | Dry weight (g) | |
|---|---|---|---|---|---|
| | | | | Shoot | Root |
| LDs alone | 17.5 ± 1.5 | 0.65 ± 0.03 | 1366 ± 64 | 1.93 ± 0.06 | 0.26 ± 0.01 |
| High-pressure sodium lamps | 14.0 ± 0.0 | 1.02 ± 0.02 | 2204 ± 15 | 4.13 ± 0.17 | 0.48 ± 0.09 |

*Leaf length and W/L ratio are those of the longest leaf per plant.

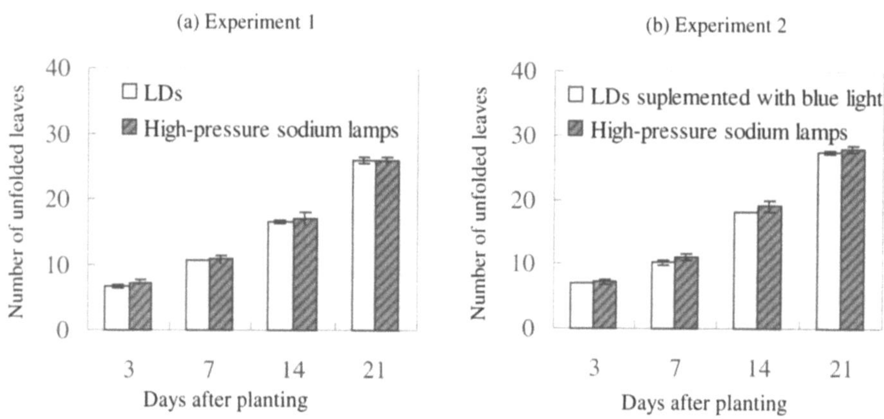


*Fig. 4* Changes with time in the number of unfolded leaves per plant.
Vertical bars represent ±S.D. of the means.

(a) LDs (b)High-pressure sodium lamps

*Fig. 5* Lettuce plants grown under the LDs supplemented with blue light (a) and high-pressure sodium lamps (b) at 21 days after planting (Experiment 2).

*Table 2* Leaf length*, leaf width/leaf length (W/L) ratio*, leaf area and dry weights of lettuce plants grown under LDs supplemented with blue light and high-pressure sodium lamps at 21 days after planting (Experiment 2). Means ± S.E. are shown.

| | Leaf length (cm) | W/L ratio | Leaf area ($cm^2$) | Dry weight (g) | |
|---|---|---|---|---|---|
| | | | | Shoot | Root |
| LDs supplemented with blue light | 15.0 ± 1.0 | 0.97 ± 0.03 | 1916 ± 288 | 3.13 ± 0.50 | 0.40 ± 0.02 |
| High-pressure sodium lamps | 14.8 ± 0.8 | 0.97 ± 0.07 | 2252 ± 92 | 3.77 ± 0.22 | 0.51 ± 0.00 |

*Leaf length and W/L ratio are those of the longest leaf per plant.

### 3.3 Discussion

Results of the present study proved that the lettuce plants were able to grow under the red LDs alone, although it is a monochromatic light. However, lettuce plants grown under red LDs require supplementation with blue light to obtain better-shaped and higher dry weight.

Lettuce plants grown under red light alone with a different light source, LED, had abnormally long and thin leaves (Yanagi et al., 1996). Thus, the abnormal shape seen under LDs may be a characteristic of plants grown under general red light or red monochromatic light. The results of the present study that the shape of leaves in lettuce plants improved by supplementing red LDs with blue light indicate that although red light corresponds to the region of the photosynthetic action spectrum, plants need light to control photomorphogenic response, as it is referred to in other papers (Hoenecke et al., 1992; Brown et al., 1995; Yanagi et al., 1996).

Lower dry weight in lettuce plants grown under LDs alone may be related to the lower photosynthetic rate caused by lower stomatal conductance. Stomata had been shown to be controlled more by blue light than red light, and in wheat plants grown under red and blue LED, stomatal conductance increased as the level of blue light increased (Goins et al., 1997). In the present study, lettuce plants grown under red LDs supplemented with 10% blue light produced dry weights close to those of plants grown under high-pressure sodium lamps, which indicates that blue light may have an important factor in control of the photosynthetic rate.

In the number of unfolded leaves, there were no significant differences between lettuce plants grown under LDs alone, LDs supplemented by blue light or high-pressure sodium lamps, during 21 days after planting. This indicates that red LD light alone may have the same effect on leaf development as high-pressure sodium lamps.

In this study, the possibility of using LD as a light source for plant production was confirmed. However further studies are needed to determine the effect of LD light on photomorphogenesis and photosynthesis.

With its high electrical-to-optical power conversion efficiency, low thermal radiation, small weight and volume, and selectivity for proper wavelength, LD would be a useful light source for plant production in a closed system. The electrical-to-optical power conversion efficiency of red LD is about 40% now, and can be expected to increase more than 60%. Thus, by combining with light of other wavelengths, more effective and lower-cost production will be possible, in the near future.

**References**

Brown, C.S., A.C. Schuerger and J.C. Sager. 1995. Growth and photomorphogenesis of pepper plants under red light-emitting diodes with supplemental blue or far-red lighting. J. Amer. Soc. Hort. Sci. 120:808-813.

Goins, G.D., N.C. Yorio, M.M. Sanwo and C.S. Brown. 1997. Photomorphogenesis, photosynthesis, and seed yield of wheat plants grown under red light emitting diodes (LEDs) with and without supplemental blue lighting. Journal of Experimental Botany 48:1407-1413.

Hoenecke, M.E., R.J. Bula and T.W. Tibbitts. 1992. Importance of 'blue' photon levels for lettuce seedlings grown under red-light-emitting diodes, HortScience 27:427-430.

Takatsuji, M. and M. Yamanaka. 1994. Possibility of laser plant factory. The Japan society of applied physics 63:1127-1130.

Yanagi, T., K.Okamoto and S. Takita. 1996. Effects of blue, red and blue/red lights of two different PPF levels on growth and morphogenesis of lettuce plants. Acta Horticulturae 440:117-122.

# EFFECTIVE VEGETABLE TRANSPLANT PRODUCTION PROGRAMS FOR CLOSED-TYPE SYSTEMS UNDER DIFFERENT LIGHTING REGIMES

Toru Maruo, Masahiro Tsuji, Hitomi Kida, Yutaka Shinohara and Tadashi Ito
Faculty of Horticulture, Chiba University, Matsudo, Chiba, 271-8510, Japan. E-mail: maruo@midori.h.chiba-u.ac.jp

**Abstract.** Since the high cost of electricity can limit the viability of artificial environment growing facilities, recent research has focused on the development of energy efficient lighting systems to reduce costs. In facilities designed for transplant production, however, reducing electricity usage *per se* is not enough to improve productivity. If plant growth can be increased and harvest stage reached more rapidly, both the fixed property cost per plant and the total production cost will decrease even if energy cost per plant remains constant. The present experiments test whether transplant production efficiency can be increased by shortening dark period length while maintaining a constant light period. Light treatments were initiated after cotyledon emergence in three crops (lettuce, cucumber and tomato). Light period was set at 10 hours in all regimes, while dark periods were set at 0, 2, 6, 8, or 14 hours. Lettuce plants grew faster with decreasing dark period length, requiring only 10 days to reach 1.2g in the 10:0 light/dark treatment, compared with 10, 13, and 17 days in 10:2, 10:8, 10:14 treatments, respectively. However, it should be noted that the 10:8 and 10:14 treatments received the approximately the same number of hours of light prior to harvesting. In contrast, shortening dark period length did not improve growth rates significantly in the tomato and cucumber plants, presumably because chlorosis occurred as a result of the shorter dark period. The facility efficiency can, therefore, be improved by shortening dark period and thereby promoting faster growth in some plants.

**Key index words.** cucumber, dark period, lettuce, plant factory, plug transplants, production cost, tomato.

## 1. Introduction

Although it has been proposed that artificial environment in plant growing facilities has been utilized for transplant production, high production cost remains a serious obstacle. To increase the utility of such facilities, recent research has focused on the development of energy efficient lighting systems to reduce costs (Hashimoto et al., 1987 and Ishii et al., 1995a). A closed-type plant factory system utilizing both different light and dark-periods and energy efficient lighting has also been investigated, but no remarkable progress has been made. This may be due to the fact that light- and dark-period ratios were kept constant and that only the light to dark cycle was changed in most of the previous research (Hayashi et al., 1993 and Ishii et al, 1995b).

Thanks to recent improvements in lighting and air conditioning systems, electricity costs now account for only a third of total production costs. This suggests that reducing electricity usage *per se* is not enough to improve productivity. If plant growth can be increased and harvest stage reached more rapidly, both the fixed property cost per plant and the total production cost will decrease even if energy cost per plant remains constant.



*C. Kubota and C. Chun (eds.), Transplant Production in the 21st Century,* 125–130.

In the present study, we explored the possibility of improving transplant production efficiency by shortening dark period length, while maintaining the same light period.

## 2. Materials and Methods

Lettuce "Green Leaf 2" (*Lactuca sativa* L., Mikado Seed Co. Ltd., Chiba, Japan), cucumber "Seiten" (*Cucumis sativus*, Nihon Engei Seisan Kenkyujo, Matsudo, Japan), and tomato "House Momotaro" (*Lycopersicon esculentum* Mill., Takii Seed Co. Ltd., Kyoto, Japan) were used in the experiment. Light period treatments (10 hours in all treatments) were initiated after the emergence of cotyledons in all crops. Dark period treatments were set at 0, 2, 8, and 14 hours, respectively.

Seeds were sown in cell trays, which were filled with Yanmer soil media (Yanmer Agricultural Equipment Co. Ltd., Osaka, Japan) for vegetable plug production. One tray was assigned to each treatment. Light was supplied via 660W HPS lamps (Japan Storage Battery Co. Ltd., Kyoto, Japan), with light intensity at the top of the canopy adjusted (300 $\mu$ mol·m$^{-2}$·s$^{-1}$ for lettuce and cucumber and 330 $\mu$ mol·m$^{-2}$·s$^{-1}$for tomato). Temperature and humidity were set at 25/16 °C and 60/90 % in light/dark, respectively, for lettuce. For the tomato and cucumber treatments, temperature and humidity were set at 25/16 °C and 75/90 %, respectively. Atmospheric carbon dioxide concentration was set at 500 ppm. One quarter (1/4) unit of Enshishoho hydroponic solution (Hori, 1966) was applied once daily during the experiment.

## 3. Results and Discussion

In lettuce, fresh shoot weight increased as the dark period decreased. Lettuce plants grew faster as the dark period decreased, requiring only 10 days to reach 1.2g in the 10:0 and 10:2 treatment light/dark treatment, compared to with 13 and 17 days in 10:8 and 10:14 treatment, respectively. However, in the 10:0 treatment, total hours of light period before the harvest was longer than that that in the other treatments (Fig. 2), indicating that the energy efficiency for plant growth in 10:0 treatment was lower than other treatments. Plants photosynthesize during light period and transport photo-assimilate during light and dark period. However, during the dark, plants will consume photo-assimilate by respiration after transportation is completed. Thus, dark period following the completion of photo-assimilate transportation (i.e., excess dark period) in not necessary for plant growth. Therefore the facility efficiency can be improved by reducing this excess dark period, which allows plants grow faster. In addition, the treatments with shorter dark periods produce seedlings of a higher quality (higher chlorophyll content, shorter leaf length, lower T/R and higher % dry matter). In contrast, shortening dark period length did not improve growth rates significantly in the tomato and cucumber plants (Fig. 3, 4). It was speculated that the shorter dark period caused leaf curling and chlorosis (Photo 1) as reported by Murage et al. (1995).

Our results indicate that low-cost transplant production may be achievable through the use of shorter dark period regimes, which lowers cost per transplant. We found that shortening dark period decreased production efficiency per unit of lighting period. However, the effect of changing light/dark schedule on economical efficiency depends largely on the proportion of electrical and fixed costs within the total expenditure. This is especially true unless transplants suffer from physiological disorders, such as chlorosis, which lower commercial value. As an example, cost breakdown of lettuce production in a typical growing facility is shown in Figure 5. In this example, dark period was set at 2 hours and offers an economic advantage since fixed cost is more than 60 percent of the total (Fig. 5).

Several different types of lamp such as HPS lamp, MH lamp, Fluorescent lamps, and LED (Yanagi et al, 1994; Tennessen et al, 1994) have been used in the growing facilities. Fixed costs vary considerably depending on the lamp type of and A/C systems installed. For instance, fixed costs will be relatively high in the system utilizing LED. Also, electricity costs vary widely between countries, within countries, and at different times of the day and/or year. Moreover, recently developed "mini-generation" and "co-generation" systems may be suitable for growing facilities and may help to reduce electricity costs dramatically. The appropriate dark period length should be determined for each crop, since the effect of shortening dark period length becomes more important when electricity cost is low and fixed cost is high.

Our results indicate that the method of shortening dark period length to increase the growth rate of transplants is applicable only to lettuce among the plants tested at this point. Future research should attempt to evaluate the effectiveness of the method on different plants. It may be possible to avoid the aforementioned physiological disorders through the use of shorter light periods.

**References**

Hashimoto, Y. et al. 1987. Characteristics in the photosynthesis for *Lactuca sativa* as affected by the pulsed light illumination, Environ. Control in Biol. 25:127-129.

Hayashi, M., T. Kozai, M. Tateno, K. Fujiwara and Y. Kitaya. 1993. Effects of the lighting cycle on the growth and morphology of potato plantlets in vitro under photomixotrophic culture conditions, Environ. Control in Biol. 31:169-175.

Hori, H. 1966. Gravel culture of vegetable and ornamental crops, Agri. And Hort. pp.210.

Ishii, M., T. Ito, T. Maruo, K. Suzuki and K. Matsuo. 1995. Growth and physiology of lettuce plants grown under artificial light of high intensity in short-day regime, Environ. Control in Biol. 33, 97-101.

Ishii, M., T. Ito, T. Maruo, K. Suzuki and K. Matsuo. 1995. Plant growth and physical characters of lettuce plants grown under artificial light of different irradiating cycles, Environ. Control in Biol. 33:143-149.

Murage, E.M., N. Watashiro and M. Masuda. 1995. Effect of carbon dioxide supply on carbon metabolism and leaf chlorosis in young eggplants under continuous illumination, Suppl. J. Japan. Soc. Hort. Sci. 64:326-327.

Tennessen, D.J., E.L. Singsaas and T.D. Sharkey. 1994. Light-emitting diodes as a light source for photosynthetic research, Photosynthetic Research 39:85-92.

Yanagi, T. and K. Okamoto. 1994. Super-bright light emitting diodes as an artificial light source for plant growth, The Third Inter. Sympo. Artificial Lighting in Horticulture pp. 19.

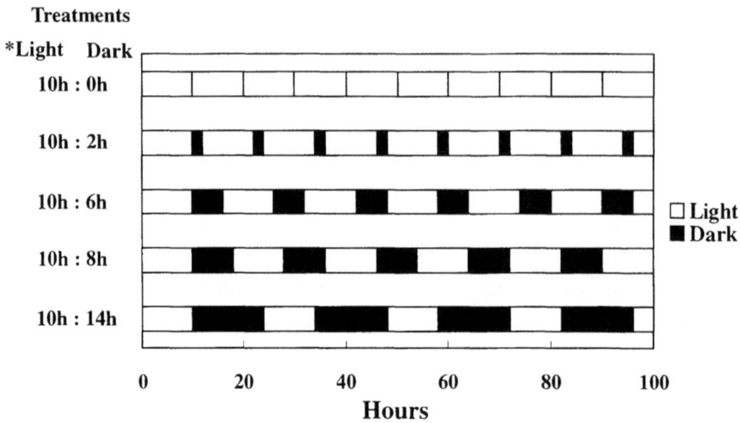


*Fig. 1* Lighting regimes in each treatment
*Light period was 10 hours in all treatments and dark period varied as 0 (continuous lightning), 2, 6, 8 or 14 hours.

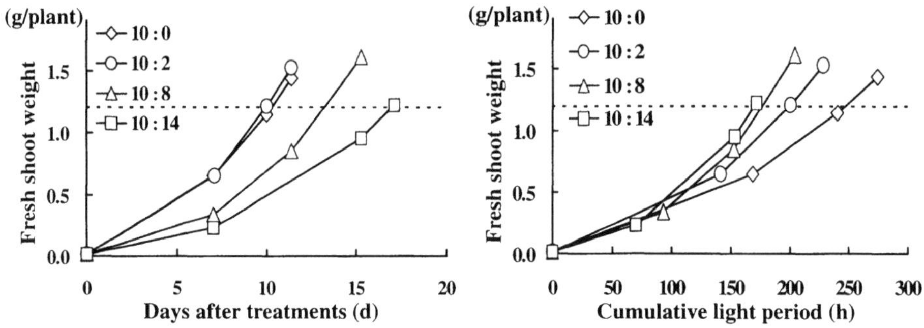


*Fig. 2* Effect of shortening dark period on growth of lettuce seedlings.

*Table 1* Effect of shortening dark period on growth of lettuce seedlings[z].

| Treatment Light: Dark (h) | Number of leaves | Relative content of chlorophyll [y] | Maximum leaf Length (cm) | Maximum leaf Width (cm) | Fresh Weight Top (g) | Fresh Weight Root (mg) | T/R | % of shoot dry matter (%) |
|---|---|---|---|---|---|---|---|---|
| 10 : 0 | 3.5 | 12.8 | 7.2 | 4.2 | 1.2 | 386 | 3.1 | 9.8 |
| 10 : 2 | 3.5 | 13.3 | 7.7 | 4.5 | 1.2 | 301 | 4.0 | 7.6 |
| 10 : 8 | 3.4 | 10.4 | 9 | 4.5 | 1.2 | 277 | 4.3 | 5.6 |
| 10 : 14 | 3.8 | 7.9 | 9.9 | 4.5 | 1.2 | 236 | 5.1 | 5.5 |

[z]: These data were estimated values at the point when top weights were 1.2g.
[y]: Color of the maximum leaf was measured by Minolta chlorophyll meter, SPAD-502.

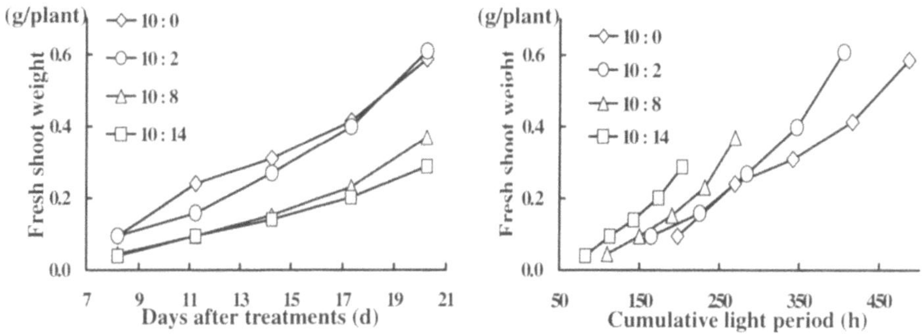


*Fig. 3* Effect of shortening dark period on growth of tomato seedlings.

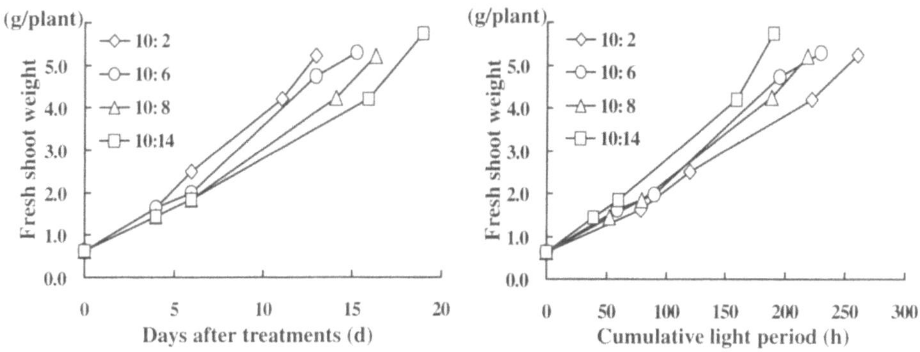


*Fig. 4* Effect of shortening dark period on growth of cucumber seedlings.

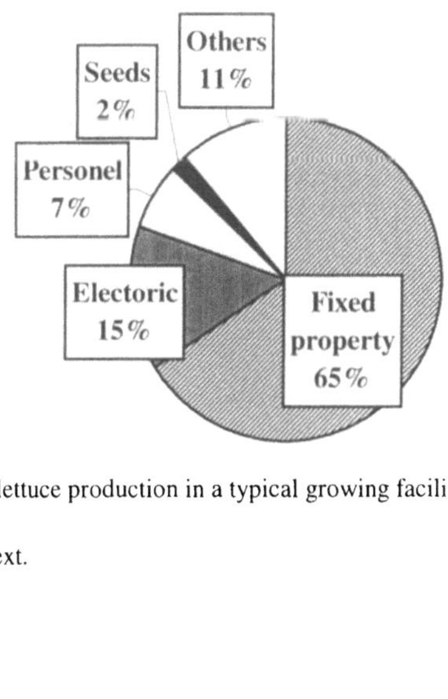


*Fig. 5* Cost breakdown of lettuce production in a typical growing facility.

* as shown in the text.

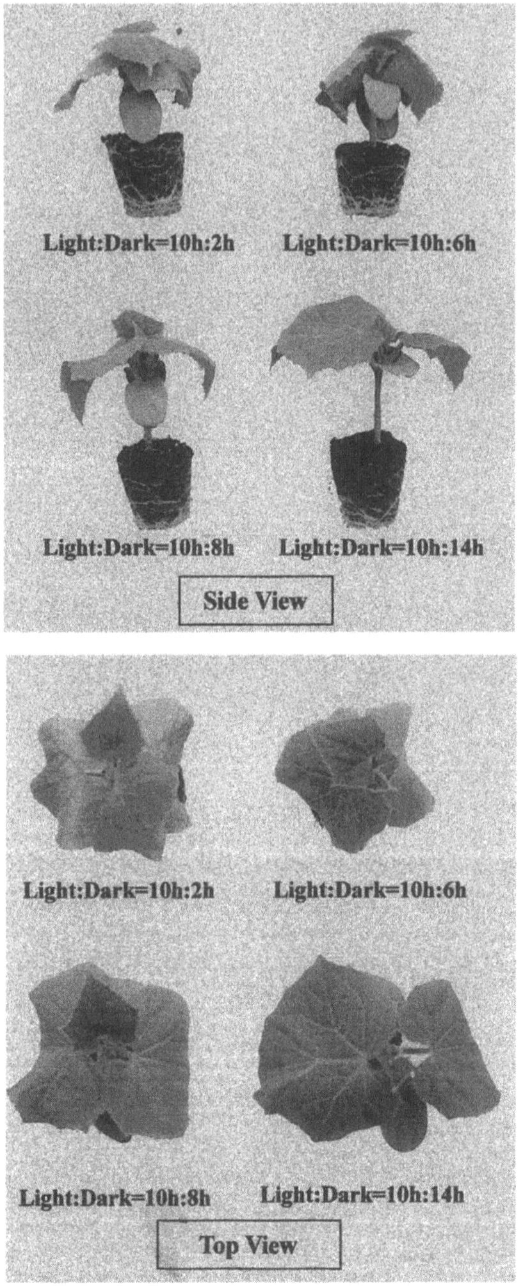


*Photo 1* Effect of shortening dark period on growth of cucumber seedling (190 hours of cumulative light period after treatments ).

# PHOTOAUTOTROPHIC MICROPROPAGATION IN A NATURAL LIGHT ENVIRONMENT

Jeff Adelberg, McNair Bostick, David Bishop and Robert Pollock
Clemson University, Clemson SC, 29634, USA. E-mail: Jadlbrg@clemson.edu

**Abstract.** Light intensity for optimal photoautophic growth is usually in excess of 100 μmol $m^{-2}$ $s^{-1}$ PAR, a level far exceeded in outdoor environments. Difficulties in utilizing natural light included heat load and non-uniform availability. Water-flow through acrylic panels allowed us to manage heat load while fluctuations in light intensity were monitored. Plants in these systems were provided moderate temperatures, saturated atmospheres and mineral nutrients under aseptic conditions. Angle of incidence, reflection and diffusion of direct light effect light transmittance and heat load distribution in model transplant canopies. Temperature control for shoot and root zones during plant growth was maintained ±2℃ of set point. $CO_2$ was rate limiting to plant growth without direct supplementation. $CO_2$-enriched conditions increased growth of *Hosta*, a shade-loving ornamental plant. NPR changed during the day based on variation in light intensity, illustrating the complexity of controlling $CO_2$ concentration in a dynamic environment.

**Key index words.** Acclimatization, *Hosta*, microhydroponic, net photosynthetic rate.

## 1. Introduction

Photoautotrophic culture during stages III and IV of the micropropagation process produces well acclimatized materials for a wide a variety of species (Deng and Donnely 1993; Parfitt and Almehdi 1994; Fujiwara et al 1988; Kozai and Watanabe 1988). Minimal environmental requirements for photoautotrophic growth include: moderate temperatures (15-30℃), light in excess of 40 μmol $m^{-2}$ $s^{-1}$ PAR, high relative humidity, $CO_2$ levels at or above compensation point, complete inorganic nutrient solutions, aerated root matrix, and readily available water. Potentially, pest populations could thrive in these conditions, and therefore inadvertent introduction of biological contaminants should be minimized by careful design of an environmental system. Reliability, efficient space utilization, and energy consumption must be balanced with high plant quality to offset facility cost. Closed production systems in electrically lit, insulated, air-conditioned, buildings represent the current consensus view of state-of-the-art photoautotrophic growth chambers.

In an open field, light varies by predictable seasonal changes in day-length, predictable hourly changes in solar angle, and is diffuse through somewhat unpredictable cloud-cover. Greenhouse glazings reflect light from their outer surface, and transmit or diffuse light, depending on angle of solar incidence. Likewise, transmittance into a sun-lit growth chamber is also affected by solar angle. In winter, direct light is effectively shaded or reflected from a horizontal surface such as a clear growth chamber lid within a greenhouse. Consequently, nearly all sunlight reaching plants under short-day conditions in a sunlit chamber is indirect, or diffuse light.

Clemson, SC, USA is located at 34.5° latitude, parallel to Osaka, Japan, where the shortest clear day provides nine hours of sunlight with a maximum altitude angle of 32° above the southern horizon. Acclimatron™ plant growth chambers were developed at



*C. Kubota and C. Chun (eds.), Transplant Production in the 21st Century,* 131–136.

Clemson University to acclimatize tissue cultures to natural light. There location in a roof-top greenhouse minimizes shading from trees and buildings.

Spectral filters, light quality and photomorphogenesis have been a long-standing focal point of horticultural research at Clemson University (Decoteau et al. 1993 ). Shoot canopies of green plants in nature evolved to make use of the solar spectrum, with chloroplasts functioning as the primary solar collector. Photosynthesis in the canopy is influenced by leaf angle and internal shading, where leaf angles and petiole length are regulated by phytochrome (Decoteau and Friend 1991). Blue light receptors play a role in stomatal function. The end-products of micropropagation are whole plants that use sunlight as their sole energy source. Acclimatization includes development of photosynthetic canopies that are functional under full solar flux.

This review 1) describes the physical environment within the Acclimatron™ system in natural light conditions, 2) demonstrates dry matter accumulation during acclimatization of a shade-loving (heat avoiding) perennial plant (*Hosta*) under $CO_2$-enriched and non-enriched conditions , and 3) analyzes interactive light and $CO_2$ effects on net photosynthetic rate (NPR).

## 2. Materials and Methods

Temperature control in transparent, channelized panels with flowing water under sunlight conditions was previously described (Pollock 1992). Thermocouples were placed on black-painted cork solar collector plates and placed at varying heights within rectangular polypropylene plant growth vessels. Temperatures were recorded.

Transition of plant material from photomixotrophic to photoautotrophic conditions, including removal of sugar from the rooted plantlet, was conducted in Acclimatron™ vessels and microhydroponic ebb and flow systems (Young and Adelberg, 1996; 1997; Adelberg et al. 1998). Design and implementation of $CO_2$-enrichment for an Acclimatron™ system and details on tissue culture of *Hosta* for growth experiments were previously described (Bostick 1999). Three $CO_2$ chambers with set points of 10,000, 2500 and 350 ppm $CO_2$ were filled with nine Acclimatron™ trays, each containing 90 *Hosta tokudama* var. 'Newberry Gold' plantlets. At regular intervals, two trays from each $CO_2$ chamber were destructively harvested to measure root and shoot dry weight. A second trial was conducted, with $CO_2$ chamber set-points of 9000, 5000 and 1000 ppm $CO_2$ using the *Hosta* , 'Blue Vision'. On both day 16 and day 30, NPR was estimated by measuring $CO_2$ concentration between the water filled panels (C-out) and within the Acclimatron vessel (C-in). Five, 1 ml samples per tray were taken over twenty minutes in morning, for each of the three $CO_2$ chambers. NPR was calculated by the steady-state method described by Fujiwara and Kozai (1995).

## 3. Results

Temperatures measured on flat black-painted cork were primarily influenced by (1) the quantity of sunlight transmitted into the vessel and (2) the temperature of water panels which formed the top and bottom of the Acclimatron™ growth chamber. Heat transfer analysis indicated that convection and thermal radiation in the growth chamber minimized heat gain for vessels 0.3-20.0 cm in height. Vessels containing plants had leaf zone temperature measurements 15-19℃ lower than collector temperatures in the simulated vessels of similar heights to the black-cork collectors.

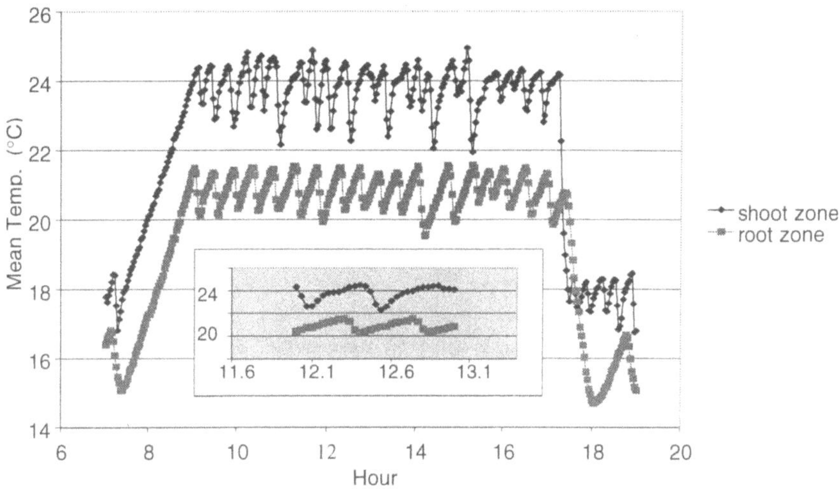


*Fig. 1* Means of temperatures at three points in the tempered water flow stream in the headspace (shoot zone) and hydroponic nutrient bay (root zone) of an Acclimatron™ system, on a sunny day in November between 7 a.m. and 7 p.m.

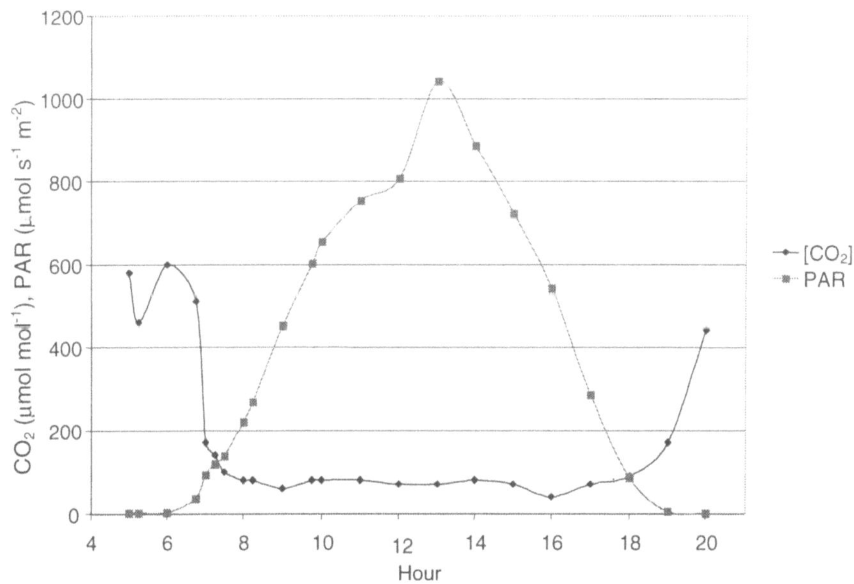


*Fig. 2* Diurnal variation in PAR and $CO_2$ concentration in the headspace of an Acclimatron™ vessel on a sunny day in July between 4 a.m. and 8 p.m.

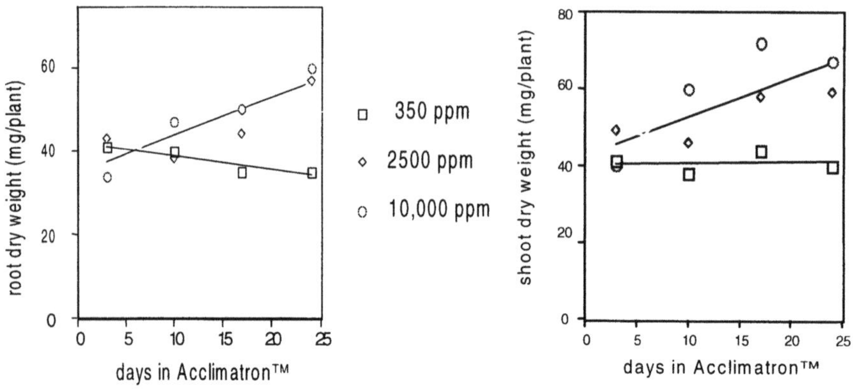


*Fig. 3* Growth of Hosta 'Newberry Gold' in Acclimatron™ system in December.

With dynamic control of water temperature, leaf and root zone temperatures were maintained ±2℃ of set-point (Fig. 1).

Inadequate light intensity is a risk during short days. In December, with a low horizon and cloud-filled skies, transmittance through the water-filled panel was approximately 90%. On clear days with some direct light in the greenhouse, transmittance through the panels was approximately 80%. During five weeks of November and December, an average of 7.4 mol $m^{-2}$ $d^{-1}$ PAR accumulated under the water-filled panel. This may be compared with 5.8 mol $m^{-2}$ $d^{-1}$ calculated for a hypothetical indoor growth chamber with 16-h day and 100 μmol $m^{-2}$ $s^{-1}$ PAR.

An early prototype vessel, with one plant per 27 $cm^3$ and 1.6 air exchanges per hour (N) was placed between the water-filled panels at high PAR. Approximately one hour after sunrise, $CO_2$ concentration in the vessel was under 100 ppm, and remained low until light levels approached zero (Fig. 2). A need for $CO_2$-enrichment during the photoperiod was clearly indicated.

A second vessel was developed for use in the remainder of the experiments. This vessel was measured to have approximately 8 air exchanges per hour and was planted a density of one plant per 11.25 $cm^3$. The $CO_2$-enriched Acclimatron™ was designed to maintain three distinct $CO_2$ environments. In a 24-day experiment with *Hosta* 'Newberry Gold', set-points of 350, 2500 and 10,000 ppm in the headspace within the water-panels and external to the vessels resulted in the following $CO_2$ levels inside the vessels: 350 (268-334 ppm), 2500 (1498-1721 ppm), and 10,000 (6597-7009 ppm), stated with 95% confidence. Both shoot and root biomass increased in the $CO_2$-enriched environments, but was not significantly different in the 2500 or 10,000 ppm chambers (Fig. 3). Shoots did not grow in the non-enriched environment, and there was likely partitioning of biomass from root to shoot resulting in a net loss of root biomass.

Information on biomass changes in sub-samples gathered over weeks leaves unanswered temporal questions as to relationships between light level, $CO_2$ and periods of most rapid growth. NPR measurements were made on a second clone of *Hosta*, 'Blue

Vision', in the Acclimatron™ chambers with set points of 1000, 5000 and 9000 ppm $CO_2$. An algebraic formula used to determine NPR was derived for

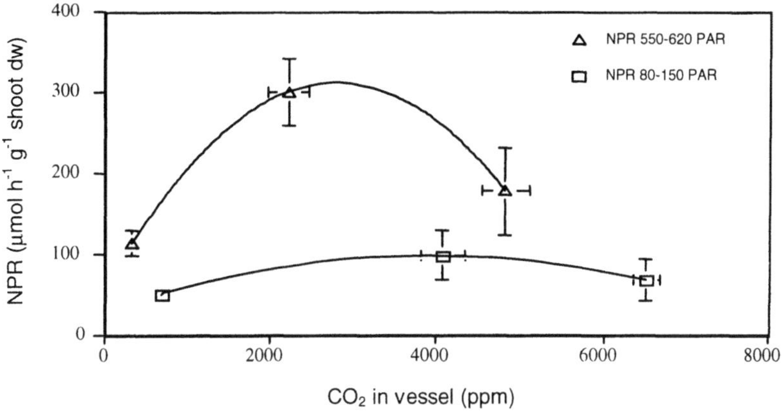


*Fig. 4* Net photosynthetic rate of Hosta 'Blue Vision' in Acclimatron™ vessels in morning (80-150 µmol $m^{-2}$ $s^{-1}$ PAR) and noontime (550-620 µmol $m^{-2}$ $s^{-1}$ PAR). Five samples were taken over 20-minute intervals, during both periods on days 16 and 30 of culture, and NPR was pooled for both days. Means with standard errors were presented for both variables.

steady-state conditions (Fujiwara et al. 1995) although quantitative limits to marginally steady-state conditions have not been established. In sunlit environments, fluctuations in light intensity and temperature made ideal steady-state seem unlikely. The Acclimatron™ permitted environmental control to approach steady-state, where changes were small over defined periods of time (Fig. 4). For example, standard errors for $CO_2$ levels during a 20-minute sample period were approximately 5% of the ranges tested. Shoot-zone temperature did not change during sunlight hours (24±2℃) and light levels during selected 15-minute intervals created discreet ranges. NPR in morning (80-150 µmol $m^{-2}$ $s^{-1}$ PAR under panel) was not strongly influenced by $CO_2$ level. In more intense midday light (550-620 µmol $m^{-2}$ $s^{-1}$ PAR under panel), NPR was greatly enhanced for all $CO_2$ levels (Fig. 4). Optimal NPR was interactively related to $CO_2$ and light levels, and during high light periods (approximately 600 µmol $m^{-2}$ $s^{-1}$ PAR) $CO_2$ levels should be maintained near 3000 ppm. When considering light effects on NPR under $CO_2$-enrichment, and the availability of light in this trial during short days in the winter, our light-limited scenario was at least comparable to artificially lit environments. Future experiments include examining NPR in other seasons, under more intense light load.

## 4. Discussion

At 34.5° latitude, the Acclimatron™ permitted concomitant acclimatization and growth of plantlets. In winter, plantlets receive mainly indirect sunlight, but this was adequate for transition from mixotrophic to photoautotrophic metabolism. In summer, there is likely benefit from partial shade, but the low temperatures of the surrounding water jacket provides a sufficient heat sink to control excessive temperature accumulation in an otherwise sealed transparent vessel. A real potential exists to improve plant growth during acclimatization by manipulating the physical environment. NPR increased during exposure to high levels of natural sunlight, when delivered to plants where leaf temperature, humidity and water availability were not limiting factors. Further study is needed to determine optimum conditions for growth of any given plant. Our experiences with crops besides *Hosta*, in the Acclimatron™ suggest there are many applications for acclimatization and growth in sunlit chambers. Our challenge, as growers, scientists and engineers, is to match materials and design to the environmental needs of plants.

**Acknowledgments.** The author's would like to thank Dr. Nihal Rajapakse for his thoughtful reading of this manuscript.

**References**

Adelberg, J., R. Pollock, N. Rajapakse and R. Young. 1998. Micropropagation, decontamination, transcontinental shipping and hydroponic growth of *Cattleya* orchid while sealed in semi-permeable membrane vessels. Scientia Hortic. 73:23-35.

Bostick, M. 1999. Carbon dioxide enrichment of an Acclimatron™ plant growth system. M. S. Thesis, Clemson University, Clemson SC, USA.

Decoteau, D., H. Hatt, J. W. Kelly, M. McMahon, N. Rajapakse, R. Young and R. Pollock. 1993. Applications of photomorphogenesis research to horticultural systems. HortSci. 28:974.

Decoteau, D. and H. Friend. 1991. Phytochrome-regulated growth of young watermelon. J. Amer. Hort. Sci. 116:512-515.

Deng, R., and D. J. Donnelly. 1993. In vitro hardening of red raspberry by $CO_2$ enrichment and reduced medium sucrose concentration. *Hortscience* 28(10):1048-1051.

Fujiwara, K. and T. Kozai. 1995. Physical microenvironment and its effects. In: J. Aitken-Christie, T. Kozai and M.A.L. Smith (eds.). Automation and Environmental Control in Plant Tissue Culture. Kluwer Academic Publishers, Dordecht,The Netherlands. pp. 319-369.

Fujiwara, K., T. Kozai and I. Watanabe. 1988. Development of a photoautotrophic tissue culture system for shoots and/or plantlets at rooting and acclimatization stages. *Acta Hort.* 23:153-158.

Kozai, T. and Y. Iwanami. 1988. Effects of $CO_2$ enrichment and sucrose concentration under high photon fluxes on plantlet growth of carnation (*Dianthus caryophullus* L.) in tissue culture during the preparation stage. *J. Jpn. Soc. Hort. Sci.* 57:279-288.

Kozai, T., Y. Koyama and I. Watanabe. 1988. Multiplication of potato plantlets in vitro with sugar free medium under high photosynthetic photon flux. Acta Hort. 230:121-127.

Parfitt, D.E. and A.A. Almehdi. 1994. Use of high $CO_2$ atmosphere and medium modifications for the successful micropropagation of pistachio. *Scientia Horticulturae* 56:321-329.

Pollock, R. 1992. Cooling characteristics of a greenhouse polycarbonate fluid roof panel. M. S. Thesis, Clemson University, Clemson SC, USA.

Young and Adelberg. 1996. Plant Propagation System and Method. US Patent 5,525,505.

Young and Adelberg. 1997. Plant Propagation System. US Patent 5,597,731.

# PRODUCTION OF VALUE-ADDED TRANSPLANTS IN CLOSED SYSTEMS WITH ARTIFICIAL LIGHTING

Hyeon-Hye Kim [1] and Toyoki Kozai [2]
[1] Department of Horticulture, Michigan State University, East Lansing, MI 48824-1325, USA. E-mail: khh@msu.edu
[2] Laboratory of Environmental Control Engineering, Faculty of Horticulture, Chiba University, Matsudo, Chiba 271-8510, Japan.

**Abstract.** The use of artificial light in commercial plant production is most economically favorable when plants are small, so a large number of plants can be simultaneously irradiated by a minimum of lighting units. Growing closely spaced plants from seeds or cuttings to transplants for field or greenhouse use is a major application of this approach. Transplants increase their economic value when their vegetative or reproductive responses are appropriately regulated by environmental manipulation. Closed transplant production systems with efficient artificial lighting are more cost-effective than using natural lighting alone. These systems can minimize energy and material consumption while, at the same time, optimize photosynthetic and photomorphogenic characteristics. In the future, more transplants will be produced under artificial light in closed systems. This paper reviews the impact of artificial lighting on plant growth and the production of quality transplants in closed systems with artificial lighting.

**Key intex words.** controlled environment, light quality, lisianthus, R : FR, spinach.

## 1. Introduction

There is potential in the commercial plant production of various transplant species using artificial lighting (Kozai, 1998, 1999; Kozai et al., 1990; Kozai et al., 1999). The use of artificial light in commercial plant production is most economically favorable when plants are small, so a large number of plants can be simultaneously irradiated by a minimum of lighting units (Vince-Prue and Canham, 1983). Growing closely spaced plants from seeds or cuttings to transplants for field or greenhouse use is a major application of this approach. There are many advantages of the transplant production in the closed systems with artificial lighting over open systems with natural lighting with respect to the environmental pollution due to the use of agro-chemicals, water consumption, labor and other costs, and quality of transplants (Kozai et al., 1999). This paper reviews recent studies related to the transplant production in closed systems with artificial lighting.

## 2. Closed systems and controlled environments

Closed systems are, in a broad sense, ecological systems covered or partially covered with films or other structural materials such as glass sheets and plastic nets. These materials prevent free exchange of energy between the inside and outside of the enclosed ecosystems (Kozai et al., 1997). In a narrow sense, closed systems are ecological systems covered with thermally insulated walls, which do not transmit light and are installed with artificial lights. Examples of closed plant production systems include tissue culture vessels, plastic tunnels, greenhouses, plant factories, and other



*C. Kubota and C. Chun (eds.), Transplant Production in the $21^{st}$ Century,* 137–144.

similar closed systems. By limiting the free exchange of energy, the environment in the closed ecosystems can be optimized to maximize the economic value of the plants.

Plant growth is greatly influenced by the environment. Several physical factors, namely temperature, light, carbon dioxide ($CO_2$), humidity, and media moisture, have great impact on plant growth and development. The ability to control the growth environment is the main advantage to growing plants in closed systems. For example, the temperature can be raised or lowered in a greenhouse to allow year-round production. Temperature control would allow for the use of natural sunlight, since during the warmer times of the year the interior temperature of air could be *ca.* 10 to 20°C above the exterior temperature (Styer and Koranski, 1997). The use of sunlight has economic advantages, but its variability in intensity, seasonal duration, and quality along with the cost of cooling must raise concerns about its use.

Closed transplant production systems with efficient artificial lighting are more cost-effective than using natural lighting alone. These systems can minimize energy and material consumption while, at the same time, can optimize photosynthetic and photomorphogenic characteristics. For example, particularly in the northern latitudes, a sufficiently extended growing season for many crops can only be obtained by producing them in glasshouses, which are usually heated for a substantial part of the year. Unfortunately, limitations in the natural light environment during winter, when heating costs are highest, result in growth which is slow and often unpredictable (Vince-Prue and Canham, 1983). The use of artificial lighting, either solely or as a supplement to natural light, can have economic advantages, but the lighting sources must be chosen to optimize plant growth characteristics.

## 3. Artificial lighting and plant growth

The use of artificial lighting for plant growth must be a compromise between the physiological processes of the plant and the economic concerns of the grower. Light can vary in quantity, quality, and duration and each of these characteristics influences plant growth. The total quantity of light that a plant receives during illumination has a direct impact on photosynthesis and plant growth and yield. This impact can be measured in terms of plant size, number of flowers, and other plant attributes. In many cases, plant morphology (height and shape) is primarily influenced by light quality. This refers to the intensity and distribution of the light over the energy spectrum, such as the amount of blue, red, and far-red light being emitted from the lamp. The duration of illumination has a great influence on the flowering morphology of photoperiodic plants, that is, short-day and long-day plants. Photoperiodic insensitive plants (day-neutral plants) are primarily influenced by light quantity (Styer and Koranski, 1997). Plants are highly responsive to lighting conditions, so plant growth and development can be enhanced by artificial lighting with appropriate characteristics.

Spectral differences among lamps are more important when lamps are used in growth rooms than it is when used for supplementing natural light (Moe, 1997). High-pressure sodium (HPS) lamps have come into increasing use as a supplement to solar energy. Their low emission in the blue region can be compensated by the natural radiation. When natural light is excluded, the morphogenic effect of lamps with restricted or unbalanced spectral distribution becomes more important and broad-spectrum lamps such as fluorescent lamps are more often used (Vince-Prue and Canham, 1983). Fluorescent lighting has been found most useful for seedling germination and initial growth or where low energy-requiring plants are grown in double tiers. However, fluorescent lamps are seldom used as supplemental lighting in greenhouses, since the units and their reflectors severely reduce solar radiation by casting a considerable amount of shade (Hanan, 1998; Tantau, 1997). Incandescent (INC) lamps are also commonly used to achieve longer photoperiods in greenhouses. The use of cool white fluorescent (CWF), HPS, or metal halide (MH) lamps in place of INC lighting offers benefits in energy savings as well. For example, to provide 1 μmol $m^{-2}$ $s^{-1}$, the electrical input that must be provided for a growing space is 2.4 $Wm^{-2}$ for CWF lamps, 1.5 $Wm^{-2}$ for HPS lamps, or 2.4 $Wm^{-2}$ for MH lamps, compared to 7.0 $Wm^{-2}$ for INC lamps (Cathey and Campbell, 1980). These lamps have different spectral characteristics, but it is important to examine their effects on plant growth.

When comparing the effects on poinsettia growth among fluorescent, MH and HPS lamps, the number of lateral shoots was highest with fluorescent and lowest with MH lamps (Moe, 1997). Also, many long-day plants do not initiate flowers or do so only slowly under fluorescent lamps because of their low far-red content (Thomas and Vince-Prue, 1997; Vince-Prue, 1976). These lamps can be used for long photoperiods without causing premature flower formation, which is commercially undesirable for some species.

The physiological impact of the different types of light and their individual lighting spectra are also critical factors. In general, environments with low red to far-red (R : FR) ratios, e.g., leaf-canopy shade, tend to promote stem elongation, while those with high R : FR ratios tend to suppress it (Morgan, 1981). Far-red light has several undesirable effects on plant morphology including promotion of stem elongation and suppression of lateral branching (Downs et al., 1958; Moe and Heins, 1990; Vince-Prue and Canham, 1983). INC lamps, which have a low R : FR ratio, frequently lead to stem elongation while fluorescent lamps, which have a high R : FR ratio, produce short and compact plants (Rajapakse et al., 1999; Whitman et al., 1998).

However, Tibbitts et al. (1983) reported greater hypocotyl elongation of lettuce (*Lactuca sativa* L.), spinach (*Spinacia oleracea* L.), and mustard (*Sinapis alba* L.) plants grown under HPS, which spectrum has a relatively high R : FR ratio and small amount of blue light compared to MH or MH plus tungsten-halogen lamps. The results could be correlated with the amount of blue light in the spectrum. Studies by Britz (1990) and Wheeler et al. (1991) showed that a reduction in stem length of soybean (*Glycine max* (L.) Merrill.) could be achieved by providing blue light. Grimstad (1991) compared the relative efficiency of six different types of fluorescent lamps for growth and development of young lettuce plants grown in growth rooms and greenhouses during winter. The differences between light sources with respect to dry weight and leaf production were more pronounced in the growth rooms than in the greenhouses, but the ranking of lamps regarding growth was nearly identical. The highest dry weight was associated with lamps that had a high emission of blue light, and red and far-red light as well. Plants grown using these lamps also had a higher chlorophyll content in their leaves. Studies using high-intensity red light-emitting diodes (LEDs) also noted the requirement of supplemental blue light to obtain normal shoot development of lettuce (Bula et al., 1991). Lighting has a profound impact on both plant growth and cost effective plant production, so appropriate systems must be chosen.

## 4. Production of value-added transplants

Transplants increase their economic value when their vegetative or reproductive responses are appropriately regulated by environmental control. Some transplants increase their economic value when their dormancy or rosette is broken by low temperature, daylength, water stress, etc. Others increase their economic value when the bolting and/or flower bud initiation are promoted or inhibited by environmental control. Other transplants increase their economic value when their stem and/or petiole elongation are prohibited by giving negative DIF (temperature is higher during the dark period than during the photoperiod), ultraviolet radiation, high air current speed, low humidity, etc. These types of environmental controls are easily achievable in a closed system with artificial lighting (Kozai, 1998).

Spinach (*S. oleracea* L.) production in hydroponic greenhouses is becoming popular in Japan. In these systems, seeds are first germinated in a growth cabinet, in which all environmental conditions are controlled. Once they have grown to an adequate size, the seedlings are transplanted into a greenhouse production area. Thus, transplants are produced under two different environmental regimes (Kim et al., 2000). Spinach is grown for its rosette leaves, which are harvested when they are fully expanded. Long day conditions are considered to result in stem elongation (bolting) and flowering (Hartmann et al., 1988). Short photoperiod conditions using artificial light during the transplant production process, which retarded floral development, also retarded bolting (Kim et al., 2000). This retardation occurred even though the growth conditions were changed to the natural long days and high temperatures of summer, under which spinach easily bolts and becomes unmarketable. By manipulating the day length and temperature during the transplant production process bolting could be prevented (Chun et al., 2000). Using short photoperiods during early development opens the possibility to produce high quality transplants of different species such as *Lactuca*, *Brassica*, and *Allium*.

Lisianthus (*Eustoma grandiflorum* (Raf.) Shinn.) has become the fastest growing segment of the new flower category worldwide (Harbaugh et al., 1997). To produce quality crops for cut-flowers or potted plants, many studies have been conducted. When grown at a day temperature above 30°C and with a 20°C night temperature minimum, seedlings remain in rosettes without bolting (Ohkawa et al., 1994; Ohkawa and Sasaki, 1999). Seedlings must be grown under conditions with an average temperature of about 25°C and with a minimum temperature of 20°C or lower from the time of sowing until two true leaf pairs expand to prevent the rosette induction (Ohkawa et al., 1991). However, once young seedlings form a rosette, they require low temperatures to induce bolting (Pergola, 1992; Takeda, 1988). To overcome this problem, various methods have been used, such as artificial cooling of seedlings at night, cultivating seedlings in the cooler highlands or, as is currently preferred, the artificial cooling of rosetted seedlings. These techniques guarantee success in breaking the rosette state, i.e., bolting, and in uniformity of flowering and have resulted in rapid commercial acceptance (Ohkawa and Sasaki, 1999).

Height control of transplants is also important in order to optimize efficient handling and rapid establishment in the field and/or greenhouses after transplanting. Many techniques are available, but chemical height control has been the standard practice in commercial operations (Murakami et al., 1995; Oyaert et al., 1999; Rajapakse et al., 1999). Due to perceived risks to humans and the environment, the use of some chemical growth regulators has recently been restricted (Kambalapally and Rajapakse, 1998; McMahaon et al., 1991; Rajapakse and Kelly, 1992). Several alternatives to the chemical control of growth have been developed. For example, studies on alternative methods such as day and night temperature management (Erwin and Heins, 1995; Heins and Erwin, 1990; Moe et al., 1992; Myster and Moe, 1995), gene manipulation (Jordan et al., 1995; Kusaba et al., 1998; Smith, 1992; Tennessen et al., 1997), and mechanical conditioning (Carlson, 1990; Latimer, 1990, Latimer, 1991, Latimer and Beverly, 1993) have been successfully demonstrated. Light quality manipulation as a non-chemical alternative for plant height control has also been developed.

Mortenson and Stromme (1987) demonstrated that under growth cabinet conditions an aqueous solution of copper sulphate ($CuSO_4$), which absorbs more far-red than red light, from a background with tungsten lamps reduced plant height. Under field conditions, chrysanthemums (*Dendranthema ×grandiflorum*) grown in $CuSO_4$ filtered sunlight had shorter internodes and increased chlorophyll. These results were similar to those induced by chemical growth regulators (McMahon et al., 1991; McMahon and Kelly, 1999; Mortensen and Stromme, 1987; Rajapakse and Kelly, 1991, 1995). $CuSO_4$ solutions are impractical as a filter for use with sunlight and artificial light, so solid spectral filters that absorb far-red radiation (FR filter) have been developed (Khattak et al., 1999). Murakami et al. (1997) showed that in transplant production of tomato (*Lycopersicon esculentum* Mill. cv. Saturn) and cucumber (*Cucumis sativus* L. cv. Hokushin), a FR filter could be used in preventing succulent growth and improving the reproductive state in the protected production. Runkel and Heins (2000) investigated how FR filtered sunlight influenced plug growth and subsequent flowering of pansy (*Viola ×wittrockiana* cv. Crystal Bowl Yellow), petunia (*Petunia ×hybrida* cv. Carpet Pink), impatiens (*Impatiens wallerana* cv. Accent Rose), snapdragon (*Antirrhinum majus* cv. Liberty Scarlet), and tomato (*L. esculentum* Mill. cv. Beefmaster). Compared to plants continuously grown under neutral-density filtered sunlight, stem length under the FR filter was significantly reduced in impatiens (by 11%), pansy (by 18%), petunia (by 34%), snapdragon (by 5%), and tomato (by 24%). Flowering from plugs under the FR filter was delayed by 2 to 3 days for snapdragon, petunia, and pansy. This filter did not effect the number of flowers or plant height at flowering. These results abtained in natural lighting could also be applied to artificial lighting systems, and illustrate the commercial potential of closed systems with artificial lighting. By using artificial lighting systems and manipulating the light quality, value-added transplants could be produced relatively easily.

## 5. Conclusions

The reaction of the plants (growth rate, quality, etc.) is most important for the economical optimization of artificial lighting systems (Tantau, 1997). Nevertheless the consideration of technical and energetic aspects of artificial lighting can improve the profitability. Closed systems with artificial lighting have a great deal of commercial potential. The environment can be optimized for different plant species and the lighting can be strategically manipulated to enhance plant marketability. As new and more cost efficient systems and lamps are developed, the flexibility to further optimize growth conditions and control plant development give closed systems with artificial lighting a great deal of potential for the future.

**Acknowledgements.** We would like to thank Dr. Joey H. Norikane, USDA-ARS, Michigan State University (MSU), for critically reviewing this paper and Mr. Erik S. Runkle, Dept. of Horticulture, MSU, USA, for providing useful information.

**References**

Britz, S.J. 1990. Photoregulation of root : shoot ratio in soybean seedlings. Photochem. Photobiol. 52:152-159.

Bula, R.J., R.C. Morrow, T.W. Tibbitts, R. W. Ignatius, T. S. Martin and D. J. Barta. 1991. Light-emitting diodes as a radiation source for plants. HortScience 26:203-205.

Carlson, W.H. 1990. Height control in vegetable transplants. Greenhouse Grower 8(2):16-17.

Cathey, H. and L. Campbell. 1980. Light and lighting systems for horticultural plants. *In:* J. Janick (ed.). Horticultural reviews, AVI Publishing, Westport, Conn. pp. 491-537.

Chun, C., K. Kozai, C. Kubota and K. Okabe. 2000. Manipulation of bolting and flowering in spinach (*Spinacia oleracea* L.) transplant production system using artificial light. Acta Hort. 515:201-206.

Downs, R., H. Borthwick and A. Piringer. 1958. Comparison of incandescent and fluorescent lamps for lengthening photoperiods. Proc. Amer. Soc. Hort. Sci. 71:568-578.

Erwin, J.E. and R.D. Heins. 1995. Thermomorphogenic response in stem and leaf development. HortScience 30:940-949.

Grimstad, S.O. 1991. The efficiency of fluorescent lamps in young lettuce plant production. Norw. J. Agric. Sci. 5:261-267.

Hanan, J.J. 1998. Greenhouses: advanced technology for protected horticulture. CRC Press.

Harbaugh, B.K., R.J. Mcgovern and J.P. Price. 1997. Potted lisianthus: secrets of success. Greenhouse Grower 15(12):28-37.

Hartmann, H.T., A.M. Kofranek, V.E. Rubatzky and W.J. Flocker. 1988. Plant science. Prentice-Hall, Inc., New Jersey. pp. 674.

Heins, R. and J. Erwin. 1990. Understanding and applying DIF. Greenhouse Grower 8(2): 73-78.

Jordan, E.T., P.M. Hatfield, D. Hondred, M. Talon, J.A. D. Zeevart and R.D. Vierstra. 1995. Phytochrome A overexpression in transgenic tobacco. Plant Physiol. 107:797-805.

Kambalapally, V.R. and N.C. Rajapakse. 1998. Spectral filters affect growth, flowering, and postharvest quality of Easter lilies. HortScience 33(6):1028-1029.

Khattak, A.M., S. Pearson and C.B. Johnson. 1999. The effect of spectral filters and nitrogen dose on the growth of chrysanthemum (*Chrysanthemum morifolium* Ramat. cv. Snowdon). J. Hort. Sci. & Biotech. 74(2):206-212.

Kim, H.H., C. Chun, T. Kozai and J. Fuse. 2000. The potential use of photoperiod during transplant production under artificial lighting conditions on floral development and bolting, using spinach as a model. HortScience 35:43-45.

Kozai, T. 1998. Transplant production under artificial light in closed systems, p. 296-308. *In:* H. Y. Lu, J. M. Sung, and C. H. Kao (eds.), Asian Crop Science 1998. Taichung, Taiwan.

Kozai, T. 1999. Development and application of closed-type transplant production system for solving the global issues on environmental conservation, food, resource and energy. Yokendo Co., Tokyo. Japan (in Japanese). pp. 191.

Kozai, T., C. Kubota and Y. Kitaya. 1997. Greenhouse technology for saving the earth in the 21$^{st}$ century. *In*: E. Goto, K. Kurata, M. Hayashi and S. Sase (eds.). Plant Production in Closed Ecosystems. Kluwer Academic Publishers, Dordrecht, The Netherlands. pp. 139-152.

Kozai, T., K. Ohyama, F. Afreen, S. Zobayed, C. Kubota, T. Hoshi and C. Chun. 1999. Transplant production in closed systems with artificial lighting for solving global issues on environmental conservation, food, resource and energy. Proceedings of ACESYS III Conference, New Brunswick, NJ, USA. pp. 31-45.

Kozai, T., S. Sase., G. Giacomelli and K.C. Ting., W. Roberts. 1990. The future of the transplant production system. Agriculture and Horticulture 65(1):97-103 (in Japanese).

Kusaba, S., M. Fukumoto, C. Honda, I. Yamaguchi, T. Sakamoto and Y. Kano. 1998. Decreased $GA_1$ content caused by the overexpression of OSH1 is accompanied by suppression of GA 20-oxidase gene expression. Plant Physiol. 117:1179-1184.

Latimer, J.G. 1990. Give plants the brush for height control. Greenhouse Grower 8(4):54-55.

Latimer, J.G. 1991. Mechanical conditioning for control of growth and quality of vegetable transplants. HortScience 26:1456-1461.

Latimer, J.G. and R.B. Beverly. 1993. Mechanical conditioning of greenhouse-grown transplants. HortTech. 3:412-414.

McMahon, M.J. and J.W. Kelly. 1999. $CuSO_4$ filters influence flowering of chrysanthemum cv. Spears. Scientia Hort. 79:207-215.

McMahon, M.J. and J.W. Kelly, and D.R. Decoteau. 1991. Growth of *Dendranthema* x*grandiflorum* (Ramat.) Kitamura under various spectral filters. J. Amer. Soc. Hort. Sci. 116(6):950-954.

Moe, R. 1997. Physiological aspects of supplementary lighting in horticulture. Acta Hort. 418:17-24.

Moe, R., T. Fjeld and L. M. Mortensen. 1992. Stem elongation and keeping quality in poinsettia (*Euphorbia pulcherrima* Willd.) as affected by temperature and supplementary lighting. Scientia Hort. 50:127-136.

Moe, R. and R. Heins. 1990. Control of plant morphogenesis and flowering by light quality and temperature. Acta Hort. 272:81-89.

Morgan, D. C. 1981. Shadelight quality effects on plant growth. *In:* H. Smith (ed.). Plants and the daylight spectrum. Academic Press, London. pp. 205-221.

Mortensen, L. M. and E. Stromme. 1987. Effects of light quality on some greenhouse crops. Scientia Hort. 33:27-36.

Murakami, K., H. Cui, M. Kiyota and I. Aiga. 1995. The design of special covering materials for greenhouses to control plant elongation by changing spectral distribution of daylight. Acta Hort. 399:135-142.

Murakami, K., H. Cui, M. Kiyota, T. Yamane, and I. Aiga. 1997. Control of plant growth by covering materials for greenhouse which alter the spectral distribution of transmitted light. Acta Hort. 435: 123-130.

Myster, J. and Moe, R. 1995. Effect of diurnal temperature alternations on plant morphology in some greenhouse crops – a mini review. Scientia Hort. 62:205-215.

Ohkawa, K., A. Kano, K. Kanematsu and M. Korenaga. 1991. Effects of air temperature and time on rosette formation in seedlings of *Eustoma grandiflorum* (Raf.) Shinn. Scientia Hort. 48:171-176.

Ohkawa, K. and E. Sasaki. 1999. *Eustoma* (Lisianthus) – its past, present, and future. Acta Hort. 482:423-426.

Ohkawa, K., T. Yoshizumi, M. Korenaga and K. Kanematsu. 1994. Reversal of heat-induced rosetting in *Eustoma grandiflorum* with low temperatures. HortScience 29(3):165-166.

Oyaert, E., E. Volckaert, and P.C. Debergh. 1999. Growth of chrysanthemum under colored plastic films with different light qualities and quantities. Scientia Hort. 79:195-205.

Pergola, G. 1992. The need for vernalization in *Eustoma russellianum*. Scientia Hort. 51:123-127.

Rajapakse, N.C. and J.W. Kelly. 1991. Influence of $CuSO_4$ spectral filters, daminozide, and exogenous gibberellic acid on growth of *Dendranthema* ×*grandiflorum* (Ramat.) Kitamura 'Bright Golden Anne". J. Plant Growth Regul. 10:207-214.

Rajapakse, N.C. and J.W. Kelly. 1992. Regulation of chrysanthemum growth by spectral filters. J. Amer. Soc. Hort. Sci. 117(3):481-485.

Rajapakse, N.C. and J.W. Kelly. 1995. Spectral filters and growing season influence growth and carbohydrate status of chrysanthemum. J. Amer. Soc. Hort. Sci. 120(1): 78-83.

Rajapakse, N.C., R.E. Young, M.J. McMahon and R. Oi. 1999. Plant height control by photoselective filters: current status and future prospects. HortTech. 9(4):618-624.

Runkle, E.S. and R.D. Heins. 2000. Stem extension and subsequent flowering of plugs grown under a far-red deficient film. HortScience (in press).

Smith, H. 1992. The ecological functions of the phytochrome family. Clues to a transgenic program of crop improvement. Photochem. Photobiol. 56:815-822.

Styer, R.C. and D.S. Koranski. 1997. Plug & transplant production: a grower's guide. Ball Publishing, Batavia.

Takeda, T. 1988. Rosette formation of *Eustoma grandiflorum*. Jpn. Soc. Hort. Sci. Autumn Mtg. pp. 574-575 (in Japanese).

Tantau, H. 1997. Technical and energetic aspects of artificial lighting. Acta Hort. 418:177-188.

Tennessen, D.J., P.S. Berlind, D.J. Weston and S.J. McIntosh. 1997. Phytochrome A as a biological growth retardant in transgenic chrysanthemum (*Dendranthema grandiflorum* cv. Nob Hill). Proc. Plant Growth Regul. Soc. Amer. 24:188-194.

Thomas, V. and D. Vince-Prue. 1997. Photoperiodism in plants. Academic Press, London.

Tibbitts, T.W., D.C., Morgan and I.J. Warrington. 1983. Growth of lettuce, spinach, mustard, and wheat plants under four combinations of high-pressure sodium, metal halide, and tungsten halogen lamps at equal PPFD. J. Amer. Soc. Hort. Sci. 108:622-630.

Vince-Prue, D. 1976. Phytochrome and photoperiodism. *In:* H. Smith (ed.). Light and plant development. Butterworth, London. pp. 347-369.

Vince-Prue, D. and A.E. Canham. 1983. Horticultural significance of photomorphogenesis, *In:* W. Shropshire., Jr., and H. Mohr (eds.). Photomorphogenesis. Springer-Verlag, Heidelberg. pp. 518-544.

Wheeler, R.M., C.L. Mackowiak and J.C. Sager. 1991. Soybean stem growth under high-pressure sodium with supplemental blue lighting. Agron. J. 83:903-906.

Whitman, C.M., R.D. Heins, A.C. Cameron and W.H. Carlson. 1998. Lamp type and irradiance level for daylength extensions influence flowering of *Campanula carpatica* 'Blue Clips', *Coreopsis grandiflora* 'Early Sunrise', and *Coreopsis verticillata* 'Moonbeam'. J. Amer. Soc. Hort. Sci. 123(5):802-807.

# HIGH QUALITY PLUG-TRANSPLANTS PRODUCED IN A CLOSED SYSTEM ENABLES POT-TRANSPLANT PRODUCTION OF PANSY IN THE SUMMER

Yoshitaka Omura[1], Changhoo Chun[1], Toyoki Kozai[1], Kei Arai[2] and Katuyoshi Okabe[3]
[1] Faculty of Horticulture, Chiba University, Matsudo, Chiba 271-8510, Japan. E-mail: omura@green.h.chiba-u.ac.jp
[2] Saitama Prefectural Nursery Center, Kita Saitama-gun, Saitama 365-0004, Japan.
[3] Taiyo Kogyo Co. Ltd., Taito-ku, Tokyo 111-0053, Japan.

**Abstract.** Pansy (*Viola* x *wittrockiana* Gams., cv. Fl Iona Yellow) seeds were sown on 288-plug trays on 13 August 1999. After the cotyledons were fully unfolded (14 days after sowing), the young plants were moved into a closed system (growth chamber) or an open system (greenhouse) and grown for 14 days. The plug-transplants produced in the closed system showed greater shoot fresh and dry masses and shorter stems than did those produced in the open system. Root dry masses of the plug-transplants produced in the closed system were almost 2.5-fold greater than those produced in the open system. The plug-transplants from each group were transplanted into plastic pots and grown for 42 days in either of two growth chambers at a constant air temperature of 23℃ or 30℃. For both chambers, the PPF was 250 $\mu$mol $m^{-2}$ $s^{-1}$ and the photo-/dark period was 16/8 h $d^{-1}$. At harvest, the survival percentage was 55% for the pot-transplants produced in the open system and the growth chamber at 30℃, and the surviving pot-transplants were of too poor quality to be sold. On the other hand, the survival percentage was 100 % for the pot-transplants grown in the growth chamber at 30℃ using the plug-transplants produced in the closed system, and all of the pot-transplants were of high quality. At 23℃, the shoot fresh and dry masses were greater in the pot-transplants grown using the plug-transplants produced in the closed system.

**Key index words.** closed system, open system, Pansy, plug-transplant, pot-transplant, survival percentage.

## 1. Introduction

Pansy (*Viola* x *wittrockiana* Gams.) is an annual or a short-lived perennial plant, which is mostly used as a bedding plant. Its importance as a bedding plant is especially high in the autumn through the spring in Japan and other countries. Generally, there are two stages for pansy transplant production. Seeds are sown in multi-celled plug trays and grown for about 30 days. And the produced plug-transplants are moved to pots and grown for an additional 30 to 60 days. Usually, pansies are sold as pot-transplants to general consumers or landscaping or construction companies.

To use a pansy in the autumn, its transplant production should be started in summer. Since pansies are cold-season plants, however, it is difficult to produce quality transplants in the summer in open systems such as greenhouses. Ikeda (1994) reported that the optimum range of temperature for pansy transplant production is between 10 and 20℃. Adams et al. (1997) showed that shoot dry mass was greatest at a temperature of approximate 20℃. Person et al. (1997) reported that the size of the pansy flower decreased linearly with increasing temperature between 9 and 31℃. It is also known that the stems of pansy transplants over-elongate and the % dry matter decreases under high temperature conditions.



*C. Kubota and C. Chun (eds.), Transplant Production in the 21st Century,* 145–148.

On the other hand, the commercial possibilities of producing various transplant species under artificial lighting in a closed system have been discussed by Kozai et al. (1998). In such a system, environmental conditions can be easily manipulated during transplant production regardless of the weather outside, and the economic value of some transplants can be increased by manipulating their physiological and morphological characteristics (Kozai, 1999). Chun et al. (2000a, 2000b) demonstrated that bolting of spinach can be inhibited through environmental control during transplant production.

In the present study, to test the feasibility of using a closed system for pansy transplant production, transplants were produced in a closed system with artificial lighting and the growth was compared with the growth of pansies produced in an open system (greenhouse). Furthermore, to estimate the performance of the plug-transplants from each group in the summer (when conditions are unfavorable for pansies), they were transplanted to growth chambers and cultured under high air temperatures.

## 2. Materials and Methods

Pansy (cv. F1 Iona Yellow, Takii Seed Co. Ltd., Japan) seeds were sown on 288-cell plug trays containing soil mixed with vermiculite and peat moss, on 13 August 1999. The plug-trays were placed in a dark germinating room (20℃) for 3 days and then were moved into the open system. After the cotyledons were fully unfolded (14 days after sowing (DAS)), the plants were moved into a closed system (growth chamber) or an open system (greenhouse) and grown for 14 more days. At 28 DAS, 25 plug-transplants were sampled from each system, and fresh and dry masses, stem length, number of true leaves and leaf area were measured.

On the same day, plug-transplants from each group were transplanted into 6 cm plastic pots containing peat moss-based media (Napurayodo, Yanmer Agricultural Equipment Co. Ltd., Japan), and grown for 42 days in either of two growth chambers at a constant air temperature of 23℃ or 30℃. For both chambers, the PPF was 250 μmol $m^{-2}$ $s^{-1}$ and the photo-/dark period was 16/8 h $d^{-1}$. The plants were irrigated as necessary and fed once a week with a solution of soluble fertilizer (Hi-spirit, Sumitomo Chemical Co., Ltd., Japan). At harvest (70 DAS), fresh and dry masses, survival percentage and flowering percentage of each group were measured.

## 3. Results and Discussion

Environmental conditions in the closed and open systems during plug-transplant production (15 to 28 DAS) are shown in Table 1. The weather during this period, which affected only the open system, was characterized by clear days and relatively high air temperatures.

Fresh and dry masses of shoot and root parts, stem length, number of true leaves of plug-seedlings produced in the closed and open systems are shown in Table 2. The plug-transplants produced in the closed system showed greater shoot fresh and dry masses and shorter stem lengths than did those produced in the open system, while the number of true leaves was not different between the two groups of plug-transplants. The root dry mass of plug-transplants produced in the closed system was almost 2.5-fold greater than the root dry mass of plug-transplants produced in the open system.

Plug-transplants produced in both the open and closed systems were cultured in growth chambers at 23 or 30℃ for 42 days. The survival percentage and flowering percentage of these pot-transplants are shown in Table 3. Table 4 shows fresh and dry masses of the plug transplants. The survival percentage was 55% for the pot-transplants produced in the open system and the growth chamber at 30℃. The quality of the surviving pot-transplants was too poor for the transplants to be sold. Many leaves of these transplants were brownish and most of them were not expected to flower. On the other hand, the survival percent was 100 % for the pot-transplants grown in the growth chamber at 30℃ using the plug-transplants produced in the closed system, and the quality of these transplants was high. For the pot-transplants that were cultured at 23℃, the shoot fresh and dry masses were greater in those that originated from plug-transplants produced in the closed system than in those that originated from plug-transplants produced in the open system.

The electric energy consumed in producing one pansy plug-transplant was estimated at 0.1 kWh or 0.38 MJ, using the method proposed by Ohyama and Kozai (1998). The cost of this energy is about 1 Yen (or 1 US cent) based on current rates in Japan. For comparison, the average price of a pansy pot-transplant in Japanese markets is 60 Yen.

These results suggest that environmental control (mainly air temperature) during plug-transplant production in closed systems enhances the quality of plug- and pot-transplants, and enables pansy transplant production in the summer season, when it is difficult to do this in open systems. Furthermore, since the cost of electric energy for producing pansy transplants was small compared to the total production cost, the production of quality transplants of pansies in closed systems should be feasible.

*Table 1* Environmental conditions in the closed and open systems during plug-transplant production.

| | Closed system | Open system |
|---|---|---|
| Air temperature | Photoperiod: 25℃ | Photoperiod: 30-40℃ |
| | Dark period: 21℃ | Dark period: 25-30℃ |
| PPF* | 250 $\mu$mol $m^{-2}$ $s^{-1}$ | 50% shading |
| Relative humidity | 40-70% | Uncontrolled |
| Photoperiod | 16 h $d^{-1}$ | 13 h $d^{-1}$ |

*Photosynthetic photon flux on empty plug trays.

*Table 2* Fresh and dry masses of shoot, root and total, stem length, number of true leaves of pansy plug-transplants produced in closed and open systems for 14 days.

| System | Fresh mass (mg) | | | Dry mass (mg) | | | Stem length (mm) | Number of true leaves |
|---|---|---|---|---|---|---|---|---|
| | Shoot | Root | Total | Shoot | Root | Total | | |
| Closed | 111a* | 98a | 209a | 19a | 7a | 26a | 7.3b | 3.9a |
| Open | 95b | 38b | 133b | 11b | 2b | 13b | 8.2a | 3.7a |

*Means in each column followed by the same letters are not significantly difference at P<0.05 level by t-test.

*Table 3* Survival percentage and flowering percentage of pansy pot-transplants cultured in growth chambers at 23 or 30℃ for 42 days using the plug-transplants produced in the closed or open system.

| Temp. in the chamber (℃) | System* | Survival percentage | Flowering percentage |
|---|---|---|---|
| 23 | Closed | 100 | 60 |
| 30 | Closed | 100 | 45 |
| 23 | Open | 95 | 55 |
| 30 | Open | 55 | 5 |

* System used for plug-transplant production.

*Table 4* Fresh and dry masses of shoot, root, and total of pansy pot-transplants cultured in growth chambers at 23 or 30℃ for 42 days using the plug-transplants produced in the closed or open system.

| Temp. in the chamber (℃) | System * | Fresh mass (g) | | | Dry mass (g) | | |
|---|---|---|---|---|---|---|---|
| | | Shoot | Root | Total | Shoot | Root | Total |
| 23 | Closed | 5.5a** | 2.2a | 7.7a | 1.20a | 0.41a | 1.61a |
| 30 | Closed | 4.1b | 0.9b | 5.0b | 0.91b | 0.14b | 1.05b |
| 23 | Open | 2.7c | 0.7bc | 3.4c | 0.49c | 0.11b | 0.60c |
| 30 | Open | 1.9c | 0.4c | 2.3c | 0.27c | 0.09b | 0.36c |

* System used for plug-transplant production.
**Means in each column followed by the same letters are not significantly difference at $P<0.05$ level by LSD test.

**References**

Adams, S. R., S. Pearson and P. Hadley. 1997. An analysis of the effects of temperature and light integral on the vegetative growth of pansy cv. Universal Violet (*Viola* x *wittrockiana* Gams.) Ann. Bot. 79:219-225.

Chun, C., K. Kozai, C. Kubota and K. Okabe. 2000a. Manipulation of Bolting and Flowering in Spinach (*Spinacia oleracea* L.) Transplant Production System using Artificial Light. Acta Horticulturae 515:201-206.

Chun, C., A. Watanabe, H.-H. Kim, T. Kozai and J. Fuse. 2000b. Bolting and growth of spinach (*Spinacia oleracea* L.) can be altered by using artificial lighting to modify the photoperiod during transplant production HortSciene (in press) 35(4).

Ikeda, Y. 1994. Nougyogijyututaikei. 8. Nousangyosonbunkakyoukai, Tokyo. (in press)

Kozai, T., C. Kubota, J. Heo, C. Chun, K. Ohyama, G. Niu and H. Mikami. 1998. Toward efficient vegetative propagation and transplant production of sweetpotato (Ipomoea batatas (L.) Lam.). Proceedings of International Workshop on Sweetpotato Production System toward the 21st Century. 201-214.

Kozai, T. 1999. Development and application of closed-type transplant production system for solving the global problems. Yokendo Co., Tokyo. (in press)

Ohyama, K. and T. Kozai. 1998. Estimating electric energy consumption and its cost a transplant production factory with artificial lighting: A case study. J. Society High Technology in Agriculture. 10:96-107.

Pearson, S., A. Parker, S. R. Adams, P. Hadley and D. R. May. 1995. The effects of temperature on the flower size of pansy (*Viola* x *wittrockiana* Gams.) J. Hort. Sci. 70:183-190.

# YIELD AND GROWTH OF SWEETPOTATO USING PLUG TRANSPLANTS AS AFFECTED BY THEIR AGES AND PLANTING DEPTHS

A.F.M. Saiful Islam, Changhoo Chun, Michiko Takagaki, Kosuke Sakami and Toyoki Kozai
Laboratory of Environmental Control Engineering, Faculty of Horticulture, Chiba University, Matsudo, Chiba 271-8510, Japan. E-mail: saiful@green.h.chiba-u.ac.jp

**Abstract.** The yield and growth of sweetpotato using plug transplants with roots and 3-6 unfolded leaves were compared with those of sweetpotato using conventional cuttings without roots and with 7-8 unfolded leaves (Control). The plug transplants were produced under artificial light using single node cuttings each with one unfolded leaf. The plug transplants of 11- and 15-day old were planted with 1 and 3 nodes inside the soil ridges (called 1- and 3-node depth, respectively, hereafter) to find out the optimum age of transplants and optimum depth of planting for higher yield of sweetpotato. The yield of storage roots 115 days after planting in the field was 40 ton $ha^{-1}$ when using 15-day old transplants planted with 3-node depth and was 17 ton $ha^{-1}$ greater than that in the Control. The number of storage roots per plant was 4.7 when using 11-day old transplants planted with 3-node depth and was 1.6 greater than that in the Control. The percent harvest index on day 115 was 81% when using 15-day old transplants planted with 3-node depth. The overall performance of the plug transplants was significantly higher than that of the conventional cuttings. The plug transplants planted with 3-node depth showed higher overall performance than did the plug transplants planted with 1-node depth.

**Key index words.** Harvest index, *Ipomoea batatas*, storage root, storage root diameter, storage root length.

## 1. Introduction

Generally, the terminal and lateral vine cuttings of about 0.3 m length with 6-8 unfolded leaves are used as plant materials for sweetpotato cultivation. Recently, Kozai et al. (1999) developed a closed-type system with artificial lighting for sweetpotato transplant production. The single nodal cutting with single leaf used as plant material is inserted in the root supporting materials in the multi-cell plug trays. The plug trays are placed inside the closed system for proper growth and development of the cuttings. The cutting develops the roots first and then produces a shoot (about 0.15 m long) with 4-6 unfolded leaves at the age of 2 weeks. At this stage the plug transplants are ready for planting in the field.

The performances of sweetpotato plug transplants produced in the closed system, however, are not yet evaluated in the field in respect of the growth and development of transplants and of storage roots. In the present study, the plug transplants were planted at different ages and planting depths into the soil ridge to find out the optimum transplant age and planting depth for high yield and quality of storage roots. Also the plug transplants were compared with the conventional cuttings in respect of the growth and development of plant, and storage roots yield of sweetpotato under field conditions.



*C. Kubota and C. Chun (eds.), Transplant Production in the 21st Century,* 149–153.

## 2. Materials and Methods

The experiment was conducted in the research farm of Chiba University, Japan (35° North latitude). The soil of the experimental land was Kanto loam type light colored Andosols. The description of the treatments is shown in Table 1. The plug transplants produced from single node cuttings each with one unfolded leaves were planted on 28 May 1999 at their ages of 11 and 15 days and placed 1 and 3 nodes inside the soil ridges at planting (Table 1). In case of Control treatment, the terminal and lateral vines of sweetpotato plants were excised without roots, with 7-8 unfolded leaves and about 0.3 m vine length. The experiment was laid out in a completely randomized block design with four replications.

The sweetpotato plants were harvested 73 and 115 days after planting (9 August and 20 September, respectively). After harvesting of the crops, eight plants from each treatment were randomly selected for measuring the number, length and diameter of storage roots, and fresh and dry masses of storage roots. The yield of storage roots per hectare was calculated from the fresh mass of storage roots per plant. The percent harvest index (HI) was determined following the expression, $HI$ = (storage root dry mass / whole plant dry mass) × 100. All the data were analyzed statistically and differences in mean values were evaluated by Duncan's Multiple Range Test at a 5 % level of significance.

## 3. Results and Discussion

The fresh mass of storage roots per plant is equal to the yield divided by the planting density (= 37,000 plants ha$^{-1}$). Storage root yield (fresh mass) at 73 days after planting (DAP) was about 14 ton ha$^{-1}$ in the treatment of A11D3 and A15D3, in which transplants were planted with 3-node depth, and was about 6 ton ha$^{-1}$ higher than that in the Control (Table 2). It was about 4 ton ha$^{-1}$ higher than in the treatment of A11D1 and A15D1, in which transplants were planted with 1-node depth. Storage root yield at 115 DAP was about 38-40 ton ha$^{-1}$ in the treatment of A11D3 and A15D3, in which transplants were planted with 3-node depth, and was about 15-17 ton ha$^{-1}$ higher than that in the Control. It was about 8-12 ton ha$^{-1}$ higher than in the treatment of A11D1 and A15D1, in which transplants were planted with 1-node depth.

*Table 1* Description of the treatments.

| Treatment code | Age of transplants | Depth of planting (Number of nodes inserted into the soil ridges) |
|---|---|---|
| Control | Cuttings[Z] | 3 nodes |
| A11D1 | 11 days[Y] | 1 node |
| A11D3 | 11 days | 3 nodes |
| A15D1 | 15 days | 1 node |
| A15D3 | 15 days | 3 nodes |

[Z] Cuttings (terminal and lateral vines without roots each, with 7-8 unfolded true leaves and about 0.3 m long) were excised from the sweetpotato plants grown in glasshouse.

[Y] Days from planting the single node cuttings in the closed system to grow plants with roots and a shoot with 4-6 leaves which were used as plug transplants.

The number of storage roots per plant of A11D3 was 4.7 and was 1.6 greater than that in the Control (Table 3). The maximum length of storage roots was 266 mm in A11D1 and was 36 mm greater than that in the Control. The maximum diameter of storage roots was 69 in A15D1 and was 24 mm greater than that in the Control.

The lower yield of storage roots when using the plug transplants of both ages planted with 1-node depth was related to the lower number of storage roots per plant than the plug transplants of both ages planted with 3-node depth. Generally, the storage roots of sweetpotato initiate from the nodes of the vines, which are in direct contact with soil. In case of the plug transplants of both ages planted with 1-node depth had the chance to initiate the storage roots from only one node. Due to the lower number of storage roots per plant when using the plug transplants of both ages planted with 1-node depth, the photosynthetic assimilates are considered to have lower sink capacity to accumulate in the storage roots. As a result, the photosynthetic assimilates might have preferentially translocated to the stem tips, growing branches, newly growing leaves, and thus resulted in greater growth and development of above-ground parts and roots excluding the storage roots in this study (data not shown in this paper). These results are in agreement with the results obtained by Islam et al. (1997a, 1997b and 1998). Some other reports also indicated a negative correlation between shoot mass and storage root mass (Mortley et al., 1991; Goswami, 1994). Although storage root growth depends on the shoot growth to a certain extent, excess shoot growth consumes a greater amount of photosynthates and less amount of photosynthates is consumed for storage root growth (Ravi and Indira, 1999).

*Table 2* Storage root yield (fresh and dry mass) per hectare of sweetpotato grown in the field using cuttings (Control) and plug transplants at 73 and 115 DAP. For treatment codes, see Table 1.

| Treatment code | Storage root yield (ton $ha^{-1}$) | | | |
|---|---|---|---|---|
| | 73 DAP | | 115 DAP | |
| | Fresh | Dry | Fresh | Dry |
| Control | 8.3±1.9 b | 2.9±0.8 b | 22.4±1.3 c | 8.7±0.8 c |
| A11D1 | 10.7±1.6 b | 3.7±1.0 a | 29.4±3.0 b | 10.2±1.0 b |
| A11D3 | 14.3±1.5 a | 4.0±2.2 a | 37.6±3.0 a | 14.3±1.4 a |
| A15D1 | 9.8±1.6 b | 3.2±2.0 b | 27.3±2.1 b | 9.9±1.0 b |
| A15D3 | 13.8±2.2 a | 4.2±0.6 a | 39.7±2.7 a | 15.2±1.4 a |

In each column, the figures with different letter(s) indicate a significant difference at the 5 % level by Duncan's Multiple Range Test.

The greater storage roots yield from the plug transplants was probably due to the fast growth and development of the plants in the field immediately after planting with less transplanting shock. The plug transplants did not wilted because they were planted with their intact root system with substrate. The lower yield in the conventional cuttings (Control) was probably due to the delay in growth and development in the field, because the cuttings were wilted after planting and it needed almost 2 weeks for recovery and establishment of the root system.

The percent harvest index at 73 DAP was 58 % in A15D3 and was 5 % greater than that in the Control (Table 3). The percent harvest index at 115 DAP was 81 % in A15D3 and was 8 % greater than that in the Control. The percent harvest indices ranged from 5 to 84 % in sweetpotato due to genotypic variation (AVRDC, 1990), and from 31 to 82 % in sweetpotato adapted to a hydroponic system in a greenhouse using nutrient film technique (Mortley et al., 1991). Percent harvest indices ranged from 14 to 80 % in sweetpotato grown in wet land field using indigenous organic materials such as rice straw, wheat straw and rice husks (Islam et al., 1997b) and rice husk charcoal inside the soil ridges as soil aerating materials (Islam et al., 1998 and 2000). In the present study, percent harvest indices ranged from 67 to 81 % and showed positive correlation (r = 0.79) with storage roots yield.

*Table 3* Morphological characteristics of storage roots and percent harvest index of sweetpotato grown in the field using cuttings (Control) and plug transplants. For treatment codes, see Table 1.

| Treatment code | Characteristics of storage roots | | | Percent harvest index [X] (%) | |
|---|---|---|---|---|---|
| | Number [Y] per plant | Length [Z] (mm) | Diameter [Z] (mm) | 73 DAP | 115 DAP |
| Control | 3.1±0.6 b | 230±17 b | 45±3.7 b | 53±2.6 b | 73±1.7 c |
| A11D1 | 3.6±0.6 b | 266±13 a | 61±3.2 ab | 51±3.2.b | 72±2.1 c |
| A11D3 | 4.7±0.8 a | 265±23 a | 63±3.8 a | 57±2.9 a | 76±1.5 b |
| A15D1 | 3.4±0.5 b | 225±14 b | 69±5.9 a | 54±2.3 b | 67±3.5 d |
| A15D3 | 4.6±0.5 a | 264±34 a | 65±4.7 a | 58±2.8 a | 81±1.3 a |

[X] Percent harvest index = (storage root dry mass / whole plant dry mass) × 100.
[Y] The storage roots those have more than 8 mm diameter were counted.
[Z] Mean value of maximum lengths and maximum diameters of storage root obtained from the individual plants. In each column, the figures with different letter(s) indicate a significant difference at the 5 % level by Duncan's Multiple Range Test.

In conclusion, the plug transplants produced in the closed system with artificial light resulted in higher yield of storage roots than the conventional cuttings when they were planted in the field. The storage root yield of sweetpotato using plug transplants was influenced by the depth of planting rather than the age of transplants in the ranges tested. The use of 11-15 day old transplants planted with 3-node depth gave the highest yield in the present experiment. Thus, the plug transplants can be used as transplants for higher yield with less labor in the field.

**References**

AVRDC. 1990. 1989 Progress Report. Asian Vegetable Research and Development Center. Shanhua, Tainan, Taiwan.

Goswami, R.K. 1994. Performance of sweetpotato cultivars in winter under Assam conditions. J. Root Crops. 20:132-134.

Islam, A.F.M.S., Y. Kitaya, H. Hirai, M. Yanase, G. Mori and M. Kiyota. 1997a. Growth characteristics and yield of sweetpotato grown by a modified hydroponic cultivation method under a wet land condition. Environ. Control in Biol. 35(2):123-129.

Islam, A.F.M.S., Y. Kitaya, H. Hirai, M. Yanase, G. Mori and M. Kiyota. 1997b. Effects of placing rice straw, wheat straw and rice husks in soil ridges on growth, morphological characteristics and yield of sweetpotato in wet lowlands. J. Agric. Meteorol. 53(3):201-207.

Islam, A.F.M.S., Y. Kitaya, H. Hirai, M. Yanase, G. Mori and M. Kiyota. 1998. Sweetpotato cultivation with rice husk charcoal as a soil aerating material under wet lowland field conditions. Environ. Control in Biol. 36(1):13-20.

Islam, A.F.M.S., Y. Kitaya, H. Hirai, M. Yanase, G. Mori and M. Kiyota. 2000. Effect of volume of rice husk charcoal masses inside soil ridges on growth of sweetpotato in a wet lowland. J. Agric. Meteorol. 56(1):1-9.

Kozai, T., C. Chun, K. Ohyama, T. Hoshi, F. Afreen, S. Zobayed and C. Kubota. 1999. Transplant production in closed systems with artificial lighting for solving global issues on Environment Conservation, Food, Resource and Energy. In Proc. ACESYS III Conference, Rutgers University, USA. pp. 31-45.

Mortley, D.G., C.K. Bonsi, P.A. Loretan, C.E. Morris, W.A. Hill and C.R. Ogbuehi. 1991. Evaluation of sweetpotato genotypes for adaptability to hydroponic system. Crop Sci. 3:845-847.

Ravi, V. and P. Indira. 1999. Crop physiology of sweetpotato. Hort. Rev. 23:277-338.

# YIELD AND GROWTH OF SWEETPOTATO USING PLUG TRANSPLANTS AS AFFECTED BY CELL VOLUME OF PLUG TRAY AND TYPE OF CUTTING

Dongxian He, Yee Hin Lok, Changhoo Chun and Toyoki Kozai
Laboratory of Environmental Control Engineering, Faculty of Horticulture, Chiba University, Matsudo, Chiba 271-8510, Japan. E-mail: he@green.h.chiba-u.ac.jp

**Abstract.** Sweetpotato (*Ipomoea batatas* (L.) Lam., cv. Beniazuma) plug transplants were produced in a closed-type transplant production system using single-node cuttings and shoot-tip cuttings each with one unfolded leaf. The two types of cuttings were grown for 11 days in two different multi-celled plug trays with cell volume of 35 or 55 mL. The four types of plug transplants (two cutting types × two cell volumes of plug trays) and the conventional vine-cuttings as a Control were transplanted in an experimental field on 23 May 1999, and were cultivated for 155 days. The planting density was 50,000 plants $ha^{-1}$ with intervals of 0.33 m between plants and 0.60 m between rows. After transplanting, the conventional vine-cuttings wilted and it took about 15 days to recover, while the plug transplants grew vigorously without wilting. Storage root yield in fresh mass harvested 125 days after transplanting (DAT) was about 26-31 t $ha^{-1}$, and was no significant differences among the treatments. It harvested 155 DAT was about 42-48 t $ha^{-1}$ in the treatments of plug transplants, and 1.5-1.7 times greater than that in the Control (29 t $ha^{-1}$). The cell volume of plug tray and type of cutting did not affect the storage root yield. Their date support conclusion that the plug transplants can be used as transplant materials and transplanted directly into the field. However, some of the storage roots harvested in the treatments of plug transplants were coiled or irregular in shape.

**Key index words.** *Ipomoea batatas* (L.) Lam., storage root, vine-cutting.

## 1. Introduction

Sweetpotato (*Ipomoea batatas* (L.) Lam.) is an important crop in Asia and Africa, and has been cultivated for food, animal feed and industrial raw material. This root crop will become more important in the 21st century, and is expected to be used in immense quantity as raw materials for biodegradable plastics and for fuel of automobiles (Kozai et al., 1997a, b). It is well known that the use of virus-free transplants, which are produced mostly in the form of vine-cuttings with several unfolded leaves, is essential for obtaining high storage root yield and quality. However, automatic the transplant production and transplanting process prevented from using vine-cuttings that have special shape or character, thus a large labor and time are needed. Therefore, we proposed and attempted to produce sweetpotato plug transplants derived from virus-free stock plants in closed system with artificial lighting for transplant directly into the field. In the present study, we investigated the yield and growth of sweetpotato in the field using plug transplants produced in the closed system as affected by cell volume of plug tray and type of cutting, and compared to using conventional vine-cuttings.

## 2. Materials and Methods

Sweetpotato (cv. Beniazuma) plug transplants were produced in a closed-type transplant production system with white fluorescent lamps as light resource using single-node cuttings and shoot-tip cuttings each with one unfolded leaf (SN and ST).



*C. Kubota and C. Chun (eds.), Transplant Production in the 21st Century,* 154–159.

The environmental conditions during the plug transplant production period in the closed-type system are shown in *Table 1*. The two types of cuttings were grown for 11 days in two different multi-celled plug trays (Nisshin Noko Sangyo Co., Gunma, Japan) that have cell volume of 35 or 55 mL (C35 and C55), respectively, and containing the mixed soil (Yanmar Agri. Equip. Co., Tokyo, Japan) as root supporting materials.

The four types of the plug transplant had about 2-3 unfolded leaves and root ball, and the conventional vine-cutting had about 5-6 unfolded leaves and without root was stored for 4 days as a Control are shown in *Table 2*. The above transplants were transplanted in an experimental field at Chiba, Japan, in which the rows were mulched with transparent polyethylene film (Thick 0.02 mm × Width 0.95 m, Mitsui Toatsu Chemi. Co., Chiba, Japan), on 23 May 1999. The planting density was 50,000 plants $ha^{-1}$ with intervals of 0.33 m between plants and 0.60 m between rows. A completely randomized design with 24 plants per treatment and 4 replications was used for the present experiment.

The soil of the field was Kanto loam type Andosols, and has been successively cropped with sweetpotato for 22 years. The soil sampled at a depth of 0.2-0.3 m from the surface of the row on the transplanting day showed the following properties: three phase distribution 32:33:35, hardness 2.2 kg $cm^{-2}$, porosity 65%, acidity 7.7, and electric conductivity 9.7 ms $m^{-1}$. The air-dried soil contained total nitrogen 2 mg $g^{-1}$, total carbon 23 mg $g^{-1}$, available phosphorus 116 mg $g^{-1}$, and available potassium 402 mg $g^{-1}$. During the cultivation period, the fertilizers and pesticides were not supplied, the polyethylene film was taken off at the end of June, the weeds in the field were wiped out twice, and the climatic conditions are shown in *Table 3*.

Fresh and dry masses of aboveground and underground parts, leaf area index (LAI) and shoot-root ratio were measured 0, 15, 30, 60, 90, 125 and 155 days after transplanting (DAT), and storage roots were harvested 125 and 155 DAT.

## 3. Results and Discussion

After transplanting, the conventional vine-cuttings wilted and it took about 15 days to recover, while the plug transplants grew vigorously without wilting. The increases in fresh mass of aboveground and underground parts measured 30 DAT were about 1.51-3.40 and 0.28-0.32 t $ha^{-1}$ in the treatments of plug transplants, and 1.5-3.4 and 2.8-3.2 times greater than that in the Control, respectively (*Fig. 1*). LAI measured 30 DAT was about 0.3-0.8 in the treatments of plug transplants and 1.2-2.8 times greater than that in the Control (*Fig. 2(a)*). Differences of growth between the treatments of plug transplants and Control in early growth stage were responsible for the characterized by the transplants with and without roots.

In middle growth stage, the aboveground part had a rapid growth period from 30 to 90 DAT in all the treatments. The underground part had a rapid accumulation period from 60 to 155 DAT in the treatments of plug transplants, while that in the Control was from 90 to 125 DAT. The fresh mass of aboveground part measured 90 DAT was 55 or 69 t $ha^{-1}$ in the C35-ST or C55-ST treatment, and 1.5 or 1.8 times greater than that in the Control, respectively. And it in the C35-SN or C55-SN treatment was 31 or 34 t $ha^{-1}$, and similar to the Control, respectively. LAI measured 90 DAT was about 4.3 in the C35-ST and C55-ST treatments and 1.5 times greater than that in the Control, and it was about 2.6 in the C35-SN and C55-SN treatments and similar to the Control. The fresh mass of underground part measured 90 DAT was about 11-14 t $ha^{-1}$ in the treatments of plug transplants, and 3.0-3.9 times greater than that in the Control. The plug transplants grew greater over conventional vine-cuttings, and the plug transplants from shoot-tip cuttings grew more vigorously regardless of the cell volume of plug tray than the plug transplants from single-node cuttings in middle growth stage. In fact, LAI of 2.8-3.2 can intercept 95% PAR and represents maximum canopy photosynthesis rate and weekly crop growth rate relates to the thickening property of storage roots were resulted (Fujise and Tsuno, 1962). Therefore, LAI of 4.3 in the C35-ST and C55-ST treatments were represented excessive vine growth.

In late growth stage, a major portion of dry matter is partitioned to storage roots. The fresh mass of aboveground part and LAI measured both 125 and 155 DAT were about 35-46 t $ha^{-1}$ and 2.5-3.8, and were no significant differences among the treatments. The fresh mass of underground part measured 125 DAT was about 27-32 t $ha^{-1}$, and was no significant differences among the treatments. And it measured 155 DAT was about 43-49 t $ha^{-1}$ in the treatments of plug transplants, and 1.4-1.6 times greater than that in the Control. The shoot-root ratio measured both 125 and 155 DAT were about 1.1-1.4 and 0.7-1.2 in all the treatments, respectively (*Fig. 2(b)*). The storage root yields in fresh and dry masses harvested 125 DAT were about 26-31 and 9-11 t $ha^{-1}$, respectively, and were no significant differences among the treatments (*Fig. 3(a)*). Those harvested 155 DAT were about 42-48 and 14-18 t $ha^{-1}$ in the treatments of plug transplants, and 1.5-1.7 and 1.4-1.8 times greater than that in the Control, respectively (*Fig. 3(b)*). On the other hand, compared to the conventional vine-cuttings, the plug transplants produced a similar or higher storage root yield harvested from 125 to 155 DAT regardless of the cell volume of plug tray and type of cutting.

Generally, the underground part of sweetpotato comprises non-storage and storage roots, and most is storage roots in the growth period after formation. The storage root yield is determined by the duration and rate of storage root growth those fluctuate over a long bulking period due to changes in the agroclimatic conditions. The sweetpotato (cv. Beniazuma) is classified into short-duration or early-maturing cultivator and is exhibited a maximum bulking rate during 12-17 weeks period. In the present experiment, compared to the treatments of plug transplants, higher rate of storage root growth for shorter bulking period (90-125 DAT) in the Control occurred similar storage root yield harvested 125 DAT. And compared to the Control, relativity higher rate of storage root growth for long bulking period (60-155 DAT) in the treatments of plug transplants occurred greater storage root yield harvested 155 DAT. However, mechanism of favorable changes duration of storage root growth under the same agroclimatic conditions including soil physical character and fertility, plant spacing, soil moisture, and climatic conditions is unknown. The plug transplants from shoot-tip cuttings produced storage root yields similar to transplants from single-node cuttings. This is responsible for the excessive vine growth in the C35-ST and C55-ST treatments consumed a greater among of photosynthates, and does not favored storage root growth and competed with storage roots growth for assimilates. The aboveground parts growth and storage root yield after transplanting were not affected by the cell volume of plug tray, because the 11 day-old plug transplants were not restricted by the cell volume due to the root balls were formed soon.

However, we were attended to some of the storage roots harvested in the treatments of plug transplants were coiled or irregular in shape, this is attributed to primarily the root balls of the plug transplants occurred not good formation. This does not have a negative influence on the sweetpotato production when cultivated for animal feed and industrial raw materials except for fresh vegetable.

## 4. Conclusion

The plug transplants can be used as transplant materials and be transplanted directly into the field. The advantages of using the plug transplants over the conventional vine-cuttings are shown as following points. 1) It could be obtained a similar or higher storage root yield in the field. 2) It can grow vigorously with a high survival ratio after transplanting even under unfavorable environmental conditions. 3) The automatic transplant production and transplanting process will be easy to realize and thus save labor and time. However, the further research is needed to solve the problem of coiled or irregular storage roots shaped for fresh vegetable.

**Acknowledgments.** The authors would like to thank C. Matsumoto, Shinmatsudo Kindergarten, for supplying the field used in the present experiment and thank the colleagues in laboratory of environmental control engineering, Chiba University, for their help and advice.

**References**

Bhagsari, Ajmer S. and Doyle A. Ashley. 1990. Relationship of photosynthesis and harvest index to sweetpotato yield. J. Amer. Soc. Hort. Sci. 115(2):288-293.

Bouwkamp, J.C. and M.N.M. Hassam. 1988. Source-sink relationships in sweet potato. J. Amer. Soc. Hort. Sci. 113(4):627-629.

Fujise, K. and Y. Tsuno. 1962. Studies on the dry matter production of sweetpotato. J. Jpn. Crop Sci. 31(1):145-149.

Fujiwara, T., H. Yoshioka and F. Sato. 1999. Effects of morphological and physical properties of cabbage plug seedlings on working precision of automatic transplanter. Jpn. J. of Farm Work Res. 34(2):77-84.

Hoque, A.S., Tanaka and M. Ito. 1999. Effect of soil compaction on plant growth in an Andisol. Jpn. Trop. Agri. 43(3):129-135.

Karen, M., T.S. and W.C. Wanda. 1986. Field performance and clonal variability in sweet potato propagated in vitro. J. Amer. Soc. Hort. Sci. 111(5):689-694.

Kozai, T., C. Kubota and Y. Kitaya. 1997a. Sweetpotato technology for solving the global issues on food, energy, natural resources and environment in the 21st century. Environ. Control in Biol. 34(2):105-114.

Kozai, T., C. Kubota and J. Heo et al. 1997b. Towards efficient vegetative propagation and transplant production of sweetpotato (*Ipomoea batatas* (L.) Lam.) under artificial light in closed ecosystems. International workshop on sweetpotato production system toward the 21st century. 34(2):105-114.

Ma, D. and H. Li. 1997. Sweetpotato production, utilization and research in China. International workshop on sweetpotato production system toward the 21st century. 34(2):129-135.

Mok, Il-Gin., D. Zhang and E.C. Edward. 1997. Sweetpotato breeding strategy of CIP. International workshop on sweetpotato production system toward the 21st century. 34(2):9-27.

*Table 1* Descriptions of the environmental conditions during the transplant production period in the closed-type system with white fluorescent lamps as light resource.

| Item | Description |
|---|---|
| Temperature | 29±1℃ |
| Relative humidity | 75±5% |
| *PPF*[x] | 140, 200 and 320 μmol $m^{-2}$ $s^{-1}$ (0-4, 5-9 and 10-11 days after planting) |
| Photo-/dark-period | 16/8 h $d^{-1}$ |
| $CO_2$ concentration | 950±30 μmol $mol^{-1}$ |

[x] Photosynthetic photon flux on the surface of the empty plug tray.

*Table 2* Number of unfolded leaves, leaf area, fresh and dry masses of shoot and root of the plug transplants and the conventional vine-cutting for each treatment. As the symbols of treatment codes, letters of C35 and C55 in the left represent cell volumes of 35 and 55 mL; letters of SN and ST in the right represent single-node cutting and shoot-tip cutting.

| Treatment codes | Number of unfolded leaves | Leaf area ($cm^2$) | Fresh mass (g/plant) Shoot | Fresh mass (g/plant) Root | Dry mass (g/plant) Shoot | Dry mass (g/plant) Root |
|---|---|---|---|---|---|---|
| C35-SN | 1.8±0.4[x)]c[y)] | 39±8 d | 1.27±0.27 c | 0.79±0.25 b | 0.15±0.04 c | 0.06±0.01 a |
| C35-ST | 2.5±0.5 b | 42±12 c | 1.10±0.34 c | 0.30±0.14 c | 0.11±0.03 c | 0.02±0.01 c |
| C55-SN | 2.3±0.9 b | 55±16 b | 1.89±0.62 b | 0.97±0.43 a | 0.20±0.08 b | 0.07±0.03 a |
| C55-ST | 2.3±0.8 b | 37±11 d | 1.32±0.46 c | 0.52±0.36 bc | 0.15±0.06 c | 0.04±0.03 b |
| Control | 5.5±1.5 a | 113±14 a | 6.88±1.87 a | 0±0 d | 0.67±0.22 a | 0±0 d |

[x)] Means±standard deviations are shown.

[y)] Means followed by different letters within a column are significantly different at $p<0.01$ by LSD test.

*Table 3* The climatic conditions during the cultivation period.

| Item | 125 DAT[x)] | 155 DAT |
|---|---|---|
| Accumulative hours over 18℃ (h) | 3006 | 3427 |
| Mean daily air temperature (℃) | 25.1 | 24.5 |
| Mean hours of sunshine (h) | 6.2 | 5.7 |
| Accumulative rainfalls (mm) | 699 | 768 |
| Average daily global solar radiation ($MJ\ m^{-2}$) | 14.9 | 13.9 |

[x)] Days after transplanting.

Source from national astronomical observatory, Japan, 1999.

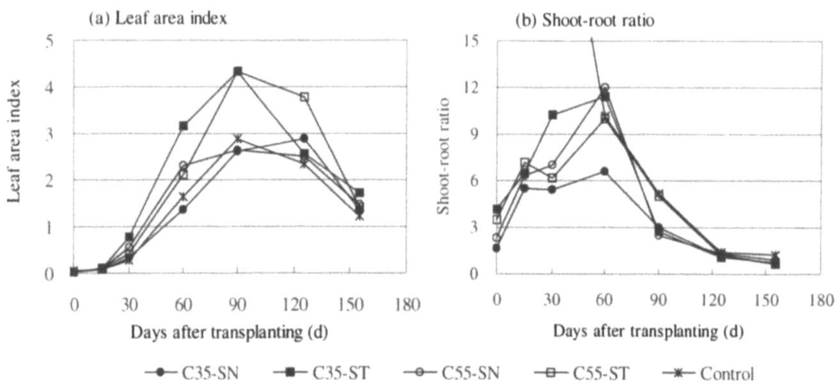


*Fig. 1* Time courses of the fresh mass of (a) aboveground and (b) underground parts during the cultivation period.

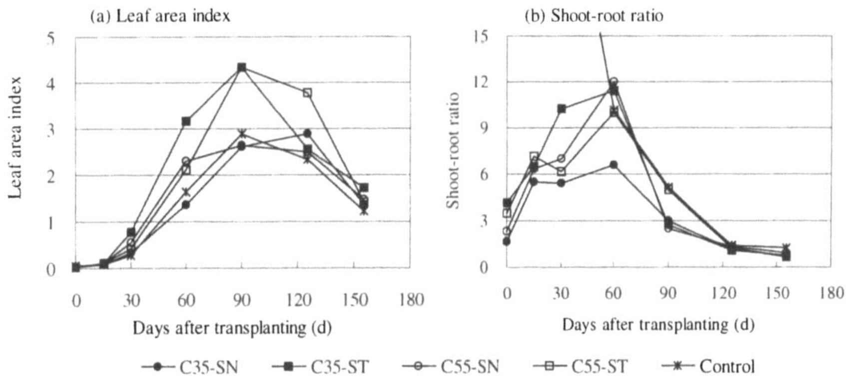


*Fig. 2* Time courses of (a) leaf area index and (b) shoot-root ratio during the cultivation period.

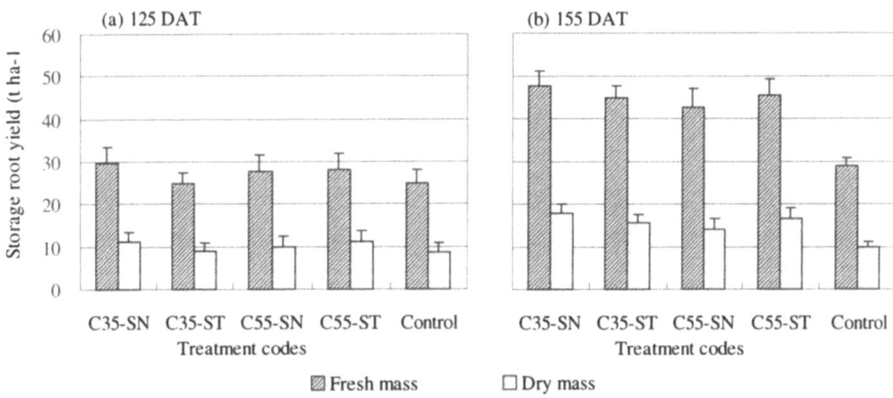


*Fig. 3* Storage root yields in fresh and dry masses of sweetpotato plants harvested (a) 125 and (b) 155 DAT. Vertical bars represent the standard error.

# PRODUCTION OF MEDICINAL PLANT SPECIES IN STERILE, CONTROLLED ENVIRONMENTS

Susan J. Murch, Sankaran KrishnaRaj and Praveen K. Saxena
Plant Cell Technology Laboratory, Department of Plant Agriculture, University of Guelph, Guelph, Ontario, Canada, N1G 2W1. E-mail: psaxena@uoguelph.ca

**Abstract.** The quality and efficacy of commercial medicinal plant preparations has been compromised by many factors including a lack of understanding of a) the physiology of medicinal plants b) the identity of medicinally active constituents, and c) the optimal conditions for growth and d) the adulteration of products with misidentified species, environmental pollutants, insects, bacteria and fungi. Micropropagation systems provide consistent sterile plant material under controlled conditions for high-quality plant-based medicines. We have recently developed *in vitro* culture protocols for the high-frequency multiplication of St. John's wort, (*Hypericum perforatum*), Huang-qin (*Scutellaria baicalensis*), *Echinacea* sp. and feverfew (*Tanacetum parthenium*) using the plant growth regulator thidiazuron [N-phenyl-N'-(1,2,3-thidiazol-yl)urea]. For St. John's wort, the optimal level of thidiazuron was 5 µmol $L^{-1}$ for a 6 or 9 day induction period. A bioreactor system was developed for the large-scale production of regenerated St. John's wort. The mammalian neurohormones serotonin and melatonin have been identified in *in vitro* grown St. John's wort plantlets. These data demonstrate the potential for the improvement of plant-based medicines via micropropagation technologies.

**Key index words.** *de novo* shoot formation, *Hypericum perforatum*, micropropagation, regeneration, organogenesis, St. John's wort, thidiazuron.

## 1. Introduction

One of the most popular medicinal plants in North America is St. John's wort (*Hypericum perforatum*). In 1998, 7.5 million Americans used St. John's wort for the treatment of neurological disorders and depression (Greenwald, 1998) based on its demonstrated efficacy in numerous clinical trials (Linde et al., 1996). In 1997, the National Institute of Health, Office of Alternative Medicine (USA) invested $4.3 million in a 3-year clinical trial to compare the effects of St. John's wort, a placebo and a standard anti-depressive drug in patients suffering from mild depression (NIH, 1997). The standard for St. John's wort preparations is "the whole fresh or dried plant or its components, including not less than 0.04% naphthodianthrones of the hypericin group calculated as hypericin" (St. John's wort Monograph, 1997). However, a recent clinical trial demonstrated that the therapeutic effect of St. John's wort were correlated to the concentration of a second compound, hyperforin (Laakmann et al., 1998). Similar to synthetic antidepressants the activity of hyperforin was found to be through the inhibition of uptake of serotonin, dopamine, noradrenaline, GABA and L-glutamate in animal cell cultures (Chatterjee et al., 1998). Further confounding the situation, studies into the unique physiology of St. John's wort have identified more than 25 additional compounds that may have medicinal activity (Miller, 1998; Nahrstedt and Butterweck, 1997; Evans and Morgenstern, 1997) including the recent report of the presence of relatively high levels of the mammalian neurohormone melatonin (Murch et al., 1997). In 1998, the Consumer Safety Symposium on Dietary Supplements and Herbs reported a 17-fold difference in the concentration of "active" constituents in commercial



*C. Kubota and C. Chun (eds.), Transplant Production in the 21st Century,* 160–165.

preparations of St. John's wort. Together these data are indicative of the dilemma faced by consumers, manufactures and regulators viz. a clearly demonstrated efficacy of plant-based medicines coupled with a high degree of variability in product quality. The *in vitro* propagation of St. John's wort can be utilized to obtain high-quality, consistent plant material since a) plants are grown under sterile, standardized conditions, b) individual superior plants can be identified and clonally multiplied, c) precise biochemical characterizations can be achieved and d) eventually protocols can be developed for the improvement of the crop through genetic manipulation. The main objectives of the current research program were to develop an *in vitro* culture system for the mass-propagation of St. John's wort and to further characterize the biosynthesis of mammalian neurohormones in sterile tissues.

## 2. Materials and methods

### 2.1 In vitro culture of St. John's wort

Seeds of St. John's wort (*Hypericum perforatum* L. cv Anthos) were surface sterilized and germinated on water agar (8 g $L^{-1}$; Laboratory Grade Agar, Fisher Scientific, Mississauga, ON) for 16 days in darkness in a growth cabinet at 24°C. Hypocotyl sections were excised from sterile etiolated seedlings and cultured on a basal medium containing MS salts (Murashige and Skoog, 1962), B5 vitamins (Gamborg et al., 1968), 30 g $L^{-1}$ sucrose and 3 g $L^{-1}$ gellan gum (Gelrite, Schweitzerhall Inc., South Plainfield, NJ) supplemented with varying levels (0, 5, 10, 15 and 20 μmol $L^{-1}$) of thidiazuron (TDZ: N-phenyl-N'-(1,2,3-thidiazol-yl)urea; Sigma Chemical Co., St. Louis, MO). The optimal time of exposure to TDZ was determined in explant cultures transferred onto MS medium after 3, 6, 9 and 12 d of incubation. The effect of TDZ on St. John's wort regeneration was quantified after 18 and 23 d of culture. Regenerating cultures of St. John's wort were subcultured into a 4 L bioreactor flask after 18 days of culture. The basal medium was pumped into the flask and drained for an exposure period of 30 min at 6 h intervals. The bioreactor flasks were aerated with a constant flow of sterile air throughout the culture period. All cultures were incubated in a growth cabinet with a 16 h photoperiod under cool white light at 40-60 μmol $m^{-2}$ $s^{-1}$ (Model F40/CW/RS/EW-II Philips Canada, Scarborough, ON).

### 2.2 Indoleamine Analysis

The endogenous concentrations of mammalian neurohormones melatonin and serotonin were quantified in field grown and *in vitro* cultured St. John's wort tissues as described previously (Murch et al., 1997). Briefly, frozen samples were mixed with 100 μL of Tris buffer (1 M Tris-HCl, pH 8.4) and homogenized with 300 μL of buffer (0.4 mol $L^{-1}$ perchloric acid, 0.05% sodium metabissulfate, 0.1% EDTA) (Poeggeler et al., 1994). After a 15 min incubation, samples were centrifuged at 12,000 x g for 15 min and 200 μL of the resulting supernatant was injected into the high performance liquid chromatography (HPLC) system. Tryptophan, indoleacetic acid, serotonin and melatonin were quantified on a Waters HPLC system (LCM1; Waters Chromatography Canada Inc., Mississauga, ON) with concurrent electrochemical (Waters 460 electrochemical detector (ECD); 2 namp, 0.85 V) and UV (Waters 484 variable UV detector; 278 nm) detection. The HPLC peak identification was confirmed by LC-MS-MS and the concentration of melatonin was confirmed by radioimmunoassay as described previously (Murch et al., 2000).

## 3. Results and Discussion

### 3.1 Micropropagation of St. John's wort

An optimized protocol was developed for thidiazuron-induced shoot organogenesis on etiolated hypocotyls explants of St. John's wort. Supplementation of the culture medium with 5 μmol $L^{-1}$ TDZ for 9 or 12 days resulted in the production of more than 40 regenerants per 1 cm tissue segment (Figure 1). Regenerants developed de novo from the subepidermal cells (Figure 2A) and these shoots formed roots and whole plants within 18 days of culture initiation (Figure 2B). The optimal duration of exposure of St. John's wort plantlets to liquid medium was assessed in a temporary immersion bioreactor system. Regenerating masses subcultured in bioreactors formed masses of sterile plant material within 2 months (Figure 2C).

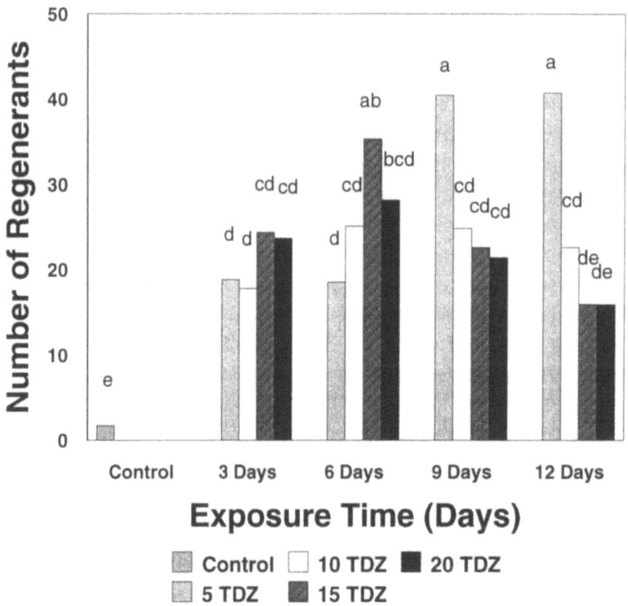


*Fig. 1* Effects of concentration and duration of exposure to thidiazuron on *de novo* organogenesis in St. John's wort. (Bars with different superscript letters are significantly different as assessed by the Student-Newman-Kuels means separation test at $P<0.05$).

*Fig. 2 In vitro* propagation of St. John's wort (*Hypericum perforatum* L. cv. Anthos). A. *De novo* shoot organogenesis following 9 days of exposure to thidiazuron and subculture on a basal medium for a further 9 days (bar = 0.166 cm). B. Regenerants developed into intact plantlets with fully formed shoots and roots within 30 days (bar = 1.44 cm). C. Prolific growth of regenerants subcultured into a bioreactor system with intermittent recirculating medium for a further period of 30 days (bar = 5 cm).

### 3.2 Indoleamine Analyses

Melatonin (N-acetyl-5-methoxytryptamine), a neural hormone secreted by the pineal gland in mammals, has recently gained prominence in the homeopathic market place as a cure for insomnia, jet lag and many diverse diseases including cancer. Two recent surveys of plant material have found melatonin in edible plant tissues (Hattori et al., 1995) and leaves of various plant species (Dubbels et al., 1995). In the present investigation we demonstrated the presence of melatonin in field-grown leaf and flower tissues and in vitro cultures of St. John's wort. There was no significant difference in the concentration of melatonin quantified in in vitro and in vivo grown stem or leaf samples (Table 1). However, etiolated hypocotyls explants excised from 18-day-old seedlings analysed without any exposure to light had significantly more melatonin than was observed in any other treatment group (Table 1). When the same hypocotyls were cultured on basal medium and incubated with a 16 h photoperiod, a significant decline was observed in the concentration of melatonin (Table 1).

*Table 1* Effect of tissue type and growth conditions on melatonin content in St. John's wort. Means with different letters are significantly different by Student-Newman-Kuels means separation test at P<0.05.

| Tissue | Culture Conditions | Melatonin Concentration µg melatonin / g DW |
|---|---|---|
| Leaf | Field grown | 1.8[b] |
| Flower | Field grown | 4.4[b] |
| Stem segments | *In vitro* culture | 1.9[b] |
| Etiolated hypocotyls | *In vitro* culture | 59.8[a] |
| Hypocotyl (day1) | *In vitro* culture | 12.7[ab] |
| Hypocotyl (day 3) | *In vitro* culture | 8.7[b] |
| Hypocotyl (day 5) | *In vitro* culture | 4.5[b] |

## 4. Conclusions

As the herbal market continues to expand, there will be an increasing demand for the production of consistent and standardized plant preparations. As well, it will become more important to fully characterize all of the biologically active compounds in plant-based medicines. Plant tissues contain a complex mixture of both potentially beneficial and harmful biochemicals. The findings reported here point to the need for a complete biochemical characterization of medicinal herbs and subsequent research is required to elucidate the effects of pharmacological doses of these chemicals. The in vitro technologies developed in the course of this study will provide consistent plant material for the biochemical characterization of St. John's wort in a sterile, controlled environment thereby facilitating the future development of novel phytopharmaceutical products. The adaptation of this micropropagation system will also allow for the improvement of St. John's wort through selection of genotypes with unusually high levels of pharmaceutically active compounds, and genetic engineering for novel and improved medicinal content. As well, the role of melatonin in the physiology and the regulation of growth and development of the plants remains to be elucidated. However, our findings provide early evidence for an interrelationship between melatonin and light exposure, thereby indicating new directions for research initiatives in plant morphogenesis.

**References**


Chatterjee, S.S., S.K. Bhattacharya, M. Wonnemann, A. Singer and WE Muller. 1998. Hyperforin as a possible antidepressant component of *Hypericum* extracts. Life Sci. 6:499-510.

Dubbels, R., R.J. Reiter, E. Klenke, A. Goebel, E. Schnakenberg, C. Ehlers, H.W. Schiwara and W. Schloot. 1995. Melatonin in edible plants identified by radioimmunoassay and by high performance liquid chromatography-mass spectrometry. J. Pineal Res. 18:28-31.

Evans, M.F. and K. Morgenstern. 1997. St. John's wort: an herbal remedy for depression? Can. Fam. Physician. 43:1735-1736.

Gamborg, O.L., R.A. Miller and K. Ojima. 1968. Nutrient requirement of suspension cultures of soybean root cells. Exp. Cell Res. 50:150-158.

Greenwald, J. 1998. Herbal healing. Time, November 23, 1998. 48-58.

Hattori, A., H. Migitaka, M. Iigo, M. Itoh, K. Yamamoto, R. Ohtani-Kaneko, M. Hara, T. Suzuki and R.J. Reiter. 1995. Identification of melatonin in plants and its effects on plasma melatonin levels and binding to melatonin receptors in vertebrates. Biochem. Mol. Biol. Int. 35:627-634.

Laakmann, G., C. Schule, T. Baghai and M. Kieser. 1998. St. John's wort in mild to moderate depression: The relevance of hyperforin for the clinical efficacy. Pharmacopsychiat. 31:54 59.

Linde, K., G. Ramirez, C.D. Mulrow, A. Pauls, W. Weidenhammer and D. Melchart. 1996. St John's wort for depression--an overview and meta-analysis of randomized clinical trials. British Medical J. 313:253-258.

Miller, A.L. 1998. St. John's wort (*Hypericum perforatum*): Clinical effects on depression and other conditions. Altern. Med. Rev. 3:18-26.

Murashige, T. and F. Skoog. 1962. A revised medium for rapid growth and bioassays with tobacco tissue cultures. Physiol. Plant. 15:473-497.

Murch, S.J., C.B. Simmons and P.K. Saxena. 1997. Melatonin in feverfew and other medicinal plants. Lancet. 350:1598-1599.

Murch, S.J., S. KrishnaRaj and P.K. Saxena. 2000. Tryptophan is a precursor for melatonin and serotonin biosynthesis in *in vitro* regenerated St. John=s wort (*Hypericum perforatum* L. cv. Anthos) plants. Plant Cell Rep. (In Press)

Nahrstedt, A. and L. Butterweck. 1997. Biologically active and other chemical constituents of the herb of *Hypericum perforatum* L. Pharmacopsychiat. 30:129-134.

Poeggeler, B. and R. Hardeland. 1994. Detection and quantification of melatonin in a dinoflagellate, *Gonyaulax polyedra*: Solutions to the problem of methoxyindole destruction in non-vertibrate material. J. Pineal Res. 17:1-10.

# EFFECT OF AIR TEMPERATURE ON TIPBURN INCIDENCE OF BUTTERHEAD AND LEAF LETTUCE IN A PLANT FACTORY

Ki Young Choi[1], Kee Yoeup Paek[1] and Yong Beom Lee[2]
[1]Research Center for the Development of Advanced Horticultural Technology, Chungbuk National University, Cheong-ju, Chungbuk 361-763, Korea. E-mail: kychoi0706@hanmail.net
[2]Dept. of Environmental Horticulture, The University of Seoul, Seoul 130-743, Korea.

**Abstract.** In a plant factory with completely controlled environments, incidence of tipburn in the middle growth stage was observed only in butterhead lettuce at 30/25℃ (day/night) when head development was in progress. At 20/15℃ in later stage, tipburn was observed 100% in butterhead lettuce and 50% in leaf lettuce. Growth of both lettuces was affected by air temperature, which increased photosynthesis and ion leakage. Highest photosynthesis and growth were observed at 30/25℃ until the middle stage. In later stage, on the other hand, photosynthesis was the highest at 20/15℃ in both lettuces, while growth of leaf lettuce was the fastest at 30/25℃, and butterhead lettuce did not show any difference between 30/25℃ and 20/15℃. Ion leakage in both lettuces was 2.2-2.6 times higher at 30/25℃ than that at 10/7℃. Butterhead lettuce was more sensitive to tipburn compared to leaf lettuce. Mass production of quality lettuce in the plant factory can be achieved through optimal temperature management since growth rate is closely related to tipburn incidence. In this experiment, optimum day temperature for butterhead and leaf lettuces was 22 to 26℃ in the initial and middle growth stages and 20 to 24℃ in the later growth stage. Optimum night temperature was 15 to 20℃.

**Key index words.** growth rate, growth stage, ion leakage, optimal temperature management, photosynthesis.

## 1. Introduction

There are increasing demand for mass production of quality crops in the plant factory with controlled environment to supply fresh vegetables year-round. Above all, it is considered to necessary to conversion of culture system in the plant factory for leafy vegetables, such as lettuce, which is much in demand for consumer together with functionality, stability and quality of food.

Tipburn, a physiological disorder of lettuce, is a serious limitation to the production of high-quality crops both in the field and the greenhouse (Barta and Tibbitts, 1991). Environmental factors affect tipburn incidence of the lettuce (Bangerth, 1979; Choi, 1999). Increasing temperature is associated with production of large leaves, increasing photosynthesis per unit leaf area with increasing water uptake and translocation of the nutrient, result in a rapid growth rate (Graves, 1983).

Wien (1998) reported temperature as the main factor determining the rate of growth of lettuce during the early seeding periods. Marsh (1987) reported that optimal day temperature of 'Ostina' lettuce identified 24℃ in greenhouse experiments, and Wurr et al. (1992) identified temperature ranges of 17 to 28℃ (day) and 3 to 12℃ (night) as suitable temperatures for outdoor production of lettuce. Cox et al. (1976) reported that the increased severity of tipburn in controlled environments has been attributed to the accelerated plant growth, with young leaves more sensitive because



*C. Kubota and C. Chun (eds.), Transplant Production in the 21st Century,* 166–171.

their rapid growth increases demand for calcium.

This experiment was conducted to examine the effect of temperature on tipburn incidence of lettuce between butterhead and leaf lettuce. The objective was also to determine the optimal temperature management of lettuce according to growth stage in controlled environments, which enables high quality production year-round.

## 2. Materials and Methods

Seed of 'Omega' butterhead and 'Grand Rapids' leaf lettuces (*Lactuca sativa* L.) were sowed in July 25, 1998 and their seedlings were grown in nutrient solution in the Venlo-type greenhouse with artificial light source. Both lettuces were selected a suitable leafy vegetables for the plant factory (Lee, 1997). After 23 days, seedlings were transferred to the plant factory with a completely controlled environment. Uniform seedlings with 6-7 true leaves were transplanted and grown under 10/7℃, 20/15℃, or 30/25℃ of air temperatures (day/night). In the plant factory, PPFD was maintained at 200-300 $\mu$mol $m^{-2}$ $s^{-1}$ with 16-hour photoperiod using metal halide lamps and high-pressure sodium lamps. Relative humidity was 60±5 % and $CO_2$ concentration was 700-800 mg $L^{-1}$. Carbon dioxide was supplied by $CO_2$ controller (DGT volmatic 3600, Saeki, Korea). Root zone pH and EC were controlled to 5.5-6.0, 1.3-1.5 dS $m^{-1}$, respectively. Fresh weight, tipburn incidence, relative growth rate, photosynthesis, transpiration rate and water use efficiency were observed at the 15th (initial growth stage), 25th (middle growth stage) and 35th days (later growth stage) after treatment. Photosynthesis and transpiration rates of the expanded leaves (12th-16th from the apex) were measured using a Li-6200 potable photosynthesis system (Li-cor Inc., USA) and a steady state porometer (Li-1600, Li-cor Inc, USA). Water use efficiency measurement was calculated using Malmstom and Field methods (1997). Ion leakage of the leaf was measured using an EC meter (CM-20E, TOA, Japan) 21 days after treatment. Leaf discs of 0.5g fresh weight (approximately 5 $mm^2$) was added to ∅ 25×25 test tube containing 10ml distilled water and shaked for 2 hours using a shaker (VS-8480SR, Vision, Korea). The mean values were from 3 sample and mean difference were analyzed using analysis of variance procedures (ANOVA) of the statistical analysis system (SAS) program (University of Alkansas).

## 3. Results and Discussion

Fresh weight of butterhead 'Omega' and leaf lettuce 'Grand Rapids' were lowest at 10/7℃ and highest at 30/25℃ during growing period (Table 1). Both lettuces grown under 10/7℃ showed shorten leaf lengths and leaf widths and less leaf numbers, which resulted in visibly greener and thicker leaves in texture than those grown under 20/15℃ or under 30/25℃, until middle growth stage. In the middle growth stage, incidence of tipburn was observed only in butterhead lettuce 'Omega' at higher temperatures, 30/25℃ when head development was in progress. Barta and Tibbitts (1991) suggested that tipburn incidence of butterhead lettuce resulted from inadequate supply of water and nutrients to inner tissues due to rapid growth rate during head development. Although relative growth rate of leaf lettuce 'Grand Rapids' at 30/25℃ was faster than that of butterhead 'Omega', tipburn was not observed in the leaf lettuce because of the little growth increment and morphological differences between them. In the later growth stage, fresh weight of butterhead lettuce under 20/15℃ was not significantly different from that under 30/25℃. On the other hands, fresh weight of leaf lettuce increased with increasing temperature. Differences in growth pattern between species were depended on each growth stage. The fastest growth rates in both lettuces were observed under 10/7℃, which corresponded to growth increment in the middle stage. Therefore, tipburn incidence was not affected. But growth increment of lettuces at 20/15℃ was 3.2-4.2 times higher than that in the middle growth stage. Incidence of tipburn was observed at 100% of the butterhead lettuce and 50% of the leaf lettuce. Although growth increment of butterhead and leaf lettuces at 30/25℃ showed 2.1-2.9 times higher than that in the middle growth stage, tipburn was not observed. According to Thompson et al. (1998), maximum dry weight was produced under 24/24℃ (air/root temperature), while final dry weight under 31/24℃ did not differ significantly from that under 24/24℃. Similar tendency was obtained in butterhead. If optimum root temperature was maintained, water balance of crop production could have been enhanced by the reduced root resistance to water flow even if air temperature was over optimum range (Challa et al., 1995).

The lowest photosynthesis and transpiration rate were observed when the lettuces were grown at 10/7℃ during the growing period, whereas the highest at 30/25℃ on 25 days and 20/15℃ on 35 days after treatment (Fig. 1). These results were the same as those obtained from Table 1. Low temperature treatment of 10/7℃ prevented mineral nutrient and water uptake through the root, resulting in decrease in photosynthesis and growth. The lowest transpiration rate and the highest water use efficiency were observed at 10/7℃ on 15 days after treatment. It indicated that the increase of the water use efficiency was induced by low temperature treatment and that of 10/7℃ may have contributed to the increase of the water retention ability of the leaves. Hirohumi and Ogata (1987) reported the increase of that was induced by the deposition of wax on the leaf surface and the reduction of cuticular evaporation by means of thick mesophyll per unit leaf area. The water use efficiency of the lettuces at 30/25℃ were increased because of the lowest transpiration rate in spite of the highest photosynthesis 25 days after treatment. Transpiration rate has been regarded as the main driving force for calcium transport to transpiring organs such as leaves (Clarkson, 1984). Therefore, a rapid growth and slow transpiration rate would be typical condition of developing leaves around the growing point, which resulted in tipburn incidence. In the later growth stage, the highest transpiration rate was observed at 20/15℃ in butterhead and at 10/7℃ in leaf lettuce. But photosynthesis and transpiration rates in both lettuces decreased at high temperature of 30/25℃. These results suggested that lettuce production could be achieved through optimum temperature management according to each growth stage.

Ion leakage of lettuces was 2.2-2.6 times higher at 30/25℃ than that at 10/7℃ and 20/15℃ (Table 2). Ion leakage of lettuce was higher in butterhead lettuce than leaf lettuce. Although the faster growth rate in both lettuces was affected by high temperature of 30/25℃, tipburn in the middle stage was observed only in butterhead lettuce. As a result, ion leakage between species was different. Therefore, butterhead lettuce was more sensitive to tipburn than that of leaf lettuce. Although Marschner (1995) reported membrane permeability increased in the calcium–deficient tissues, calcium concentration of nutrient solution in this experiment was suitable for lettuce growth. Therefore, increase of ion leakage considered that the membrane permeability varied with proceeding in cell expansion and tissue differentiation, slow calcium movement to the tissue and loose bound of cell wall due to the rapid growth rate caused by high temperature, which resulted in tipburn.

Considering the results of this experiment, factors determining tipburn incidence between species according to the growth stage were not clear. They may be the rapid growth rate or morphological or genetic differences. Production of quality lettuce in the plant factory can be achieved through optimal temperature management since growth rate is closely related to tipburn incidence. In this experiment, optimum day temperature for butterhead and leaf lettuce was 22 to 26℃ in the initial and middle growth stage and 20 to 24℃ in later growth stage. Optimum night temperature was 15 to 20℃.

**Acknowledgements.** This research was funded by the SGRP/HTDP (High –Technology Development Project for Agriculture and Forestry) in Korea.

**References**


Bangerth, F. 1979. Calcium-related physiological disorders of plants. Ann. Rev. Phytopathol. 17:97-122.

Barta, D.J. and T.W. Tibbitts. 1991. Calcium location in lettuce leaves with and without tipburn; Comparison of controlled-environment and field-grown plants. J. Amer. Soc. Hort. Sci. 116(5):870-875.

Challa, H., E. Heuvelink and U. van Meeteren. 1995. Crop growth and development. In: J.C. Bakker, G.P.A. Bot, H. Challa, and N.J.van de Braak (eds.). Green- house climate control. Wageningen Press. pp. 62-84.

Choi, K.Y. 1999. Environmental factors in a plant factory affecting tipburn incidence of the lettuce. Ph.D. Diss., The University of Seoul, Seoul.

Clarkson, D.T. 1984. Calcium transport between tissues and its distribution in the plant. Plant Cell Environ. 7:449-456.

Cox, E.F., J.M.T. McKee and A.S. Dearman. 1976. The effect of growth rate in tipburn of head lettuce. HortScience 6:19-20.

Graves, C.J. 1983. The nutrient film technique. Horticultural Reviews. 5:1-44.

Hirohumi, S. and S. Ogata. 1987. Relationship between water use efficiency and cuticular wax deposition in warm season forage crops grown under water deficit conditions. Soil Sci. Plant Nutrtion 33(3):439-448.

Lee, Y.B. 1997. Development of optimum nutrient management system in a plant factory. The 2nd report of Agricultural R&D Promotion Center. Korea.

Malmstrom, C.M. and C.B. Field. 1997. Virus-induced differences in the response of oat plant to elevated carbon dioxide in plant. Plant Cell and Environ. 20:178-188.

Marschner, H. 1995. Mineral nutrition of higher plants. Academic press. London. pp. 99-105.

Marsh, L.S. 1987. A model of greenhouse hydroponic lettuce production: Daily selection of optimum air temperatures and comparison of greenhouse covers. Ph.D. Diss., Cornell Univ., Ithaca, N.Y.

Thompson, H.C., R.W. Langhans, A.Both and L.D. Albrldght. 1998. shoot and root temperature effects on lettuce growth in a floating hydroponic system. J. Amer. Soc. Hort. Sci. 123(3):361 364.

Wien, H.C. 1998. Veg. Crop Physiol. CAB International.

Wurr, D.C.E., J.R. Fellows and A.J. Hambridge. 1992. Environmental factors influencing head density and diameter of crisp lettuce cv. Saladin. HortScience 67(3):395-401.

*Table 1* Effect of air temperature on fresh weight, incidence of tipburn and relative growth rate (RGR) of butterhead and leaf lettuce at 15, 25 and 35 days after treatment in the plant factory.

| Air temp. (°C) (day/night) | Fresh wt. ($g \cdot plant^{-1}$) 15 days | 25 days | 35 days | Tipburn (%) 15 days | 25 days | 35 days | RGR ($g\ g^{-1}\ day^{-1}$) I[z] | II[y] |
|---|---|---|---|---|---|---|---|---|
| | | | | Butterhead lettuce 'Omega' | | | | |
| 10/7 | 3.2 c[x] | 12.5 c | 57.8 b | 0 | 0 | 0 | 0.092 | 0.134 |
| 20/15 | 8.4 b | 36.8 b | 116.5 a | 0 | 0 | 100 | 0.103 | 0.117 |
| 30/25 | 14.7 a | 57.3 a: | 120.6 a | 0 | 67 | 0 | 0.120 | 0.076 |
| | | | | Leaf lettuce 'Grand Rapids' | | | | |
| 10/7 | 1.8 c | 5.1 c | 25.7 c | 0 | 0 | 0 | 0.069 | 0.175 |
| 20/15 | 3.4 b | 15.9 b | 66.7 b | 0 | 0 | 50 | 0.114 | 0.142 |
| 30/25 | 6.1 a | 37.1 a | 105.6 a | 0 | 0 | 0 | 0.147 | 0.119 |

[z]Middle growth stage; from August 31 to October 8, 1998.
[y]Laster growth stage; from October 8 to October 23, 1998.
[x]Mean separation within columns by Duncan's multiple range test, 5% level.

*Table 2* Effect of air temperature on ion leakage of butterhead and leaf lettuce at 3 weeks after treatment in the plant factory.

(unit: $dS\ m^{-1}$)

| Air temp.(°C) (day/night) | Butterhead lettuce 'Omega' | Leaf lettuce 'Grand Rapids' |
|---|---|---|
| 10/7 | 0.096 b[z] | 0.056 b |
| 20/15 | 0.085 b | 0.060 b |
| 30/25 | 0.219 a | 0.132 a |

[z]Mean separation within columns by Duncan's multiple range test, 5% level.

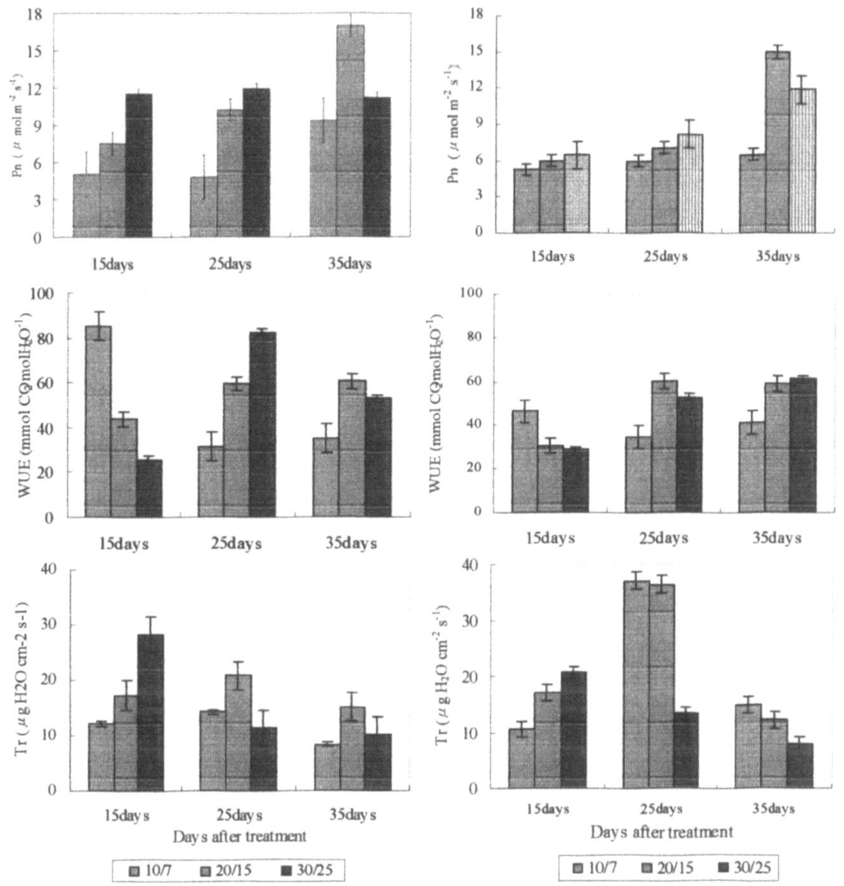


*Fig. 1* Effect of air temperature on photosynthesis (Pn), water use efficiency (WUE) and transpiration rate (Tr) of butterhead 'Omega' (left) and leaf lettuce 'Grand Rapids' (right) at 15, 25 and 35 days after treatment in the plant factory. Vertical bars represent standard error.

# EVALUATION OF LETTUCE CULTIVARS SUITABLE FOR CLOSED PLANT PRODUCTION SYSTEM

Masahisa Ishii[1, 2], Toru Maruo[2], Yutaka Shinohara[2] and Tadashi Ito[2]
[1]Research Fellow of the Japan Society for the Promotion of Science, Present address: National Research Institute of Agricultural Engineering, Kannondai 2-1-2, Tsukuba, Ibaraki 305-8609, Japan. E-mail: masaisii@nkk.affrc.go.jp
[2]Faculty of Horticulture, Chiba University, Matsudo, Chiba 271-8510, Japan.

**Abstract.** There are many cultivars in butterhead lettuce (Lactuca sativa L.) and if some of them can perform well in the relatively low light and high temperature condition, the commercial lettuce production become feasible because of the lower electric consumption in the plant factory. Eighteen cultivars were grown in the summer greenhouse under 50% shading condition, and the growth indices of seedling stage were examined in relation to the quality of final products (Expt.1). These cultivars were classified into two groups, namely unsuitable (lower qualities) and suitable (higher qualities). The ratio of leaf length to width and chlorophyll content in the seedling stage closely related to the quality of the final products. Twelve cultivars of both groups were selected from the previous experiment and grown in a closed production system to evaluate the productivity and the quality (Expt.2). The performance of the plants showed a similar tendency with those under the greenhouse condition. Thus, the ratio of leaf length to width and chlorophyll content in the seedling stage were found to be an effective index for the evaluation of the performance of lettuce cultivars.

**Key index words.** chlorophyll contents, high temperature, *Lactuca sativa* L., low light intensity, plant growth and quality, seedling stage.

## 1. Introduction

Intensive plant factories have been constructed for the production of vegetable plants under artificially controlled conditions in the since last 16 years in Japan. However, due to high initial and running costs commercially, only a few succeeded. From practical point of view, the charge for electricity have to be given consideration, because it covers over 90% of running costs of plant factories (Ito, 1986). Since high electricity cost is inevitable issue in plant factory, research has been focused on efficient method of lighting to lower the cost (Goto and Takakura, 1987; Ito, 1989; Ishii et al., 1995). Therefore, many researches were conducted to get the optimum environmental conditions to cultivate plants in plant factories using various types of cultivars for different cropping. The objective of this study was to find out the appropriate evaluation method for lettuce cultivars fitted for the closed plant production system using relatively low light and high temperature conditions.

## 2. Materials and methods

### 2.1 Experiment 1

Eighteen butterheaa lettuce cultivars were used for this experiment. Lettuce seeds were sown in plug trays with 200 cells (Yammer Co. Ltd., Japan) filled with substrate (Yasaiyodo; Yammer Co. Ltd., Japan) on June 12, 1997. Seedlings of 4-6 true leaf stages were transplanted in the 50% shaded greenhouses on July 3, 1997. The planting density was 20mm between the rows and 20mm between the plants. After 21 days of planting, 10 plants of each cultivar were sampled and the fresh weight, ratio of leaf length to



C. Kubota and C. Chun (eds.), *Transplant Production in the 21st Century*, 172–177.

width, chlorophyll content, of each stage and the quality index (Fig.1) and leaf area at harvest were measured. Chlorophyll contents were expressed as relative values read with Green meter (SPAD 502, Minolta Co. Ltd. Japan).

**2.2 Experiment 2**

Twelve cultivars were selected from the results obtained in Expt.1 and grown in a closed plant production system. Lettuce seeds were sown on the watered polyurethane cube. After one week when the first true leaf appeared, the plants were transferred to hydroponic systems with planting density of 45mm×25mm (staggered row planting). When the seedling became ninth to tenth true leaves stages, the plants were transplanted to the same hydroponic systems. The growing conditions of the closed system were 15 hrs-photoperiod by High Pressure Sodium Lamp (HPSL) with approximately 224 $\mu$ mol·$m^{-2}$·$s^{-1}$ Photosynthetic Photon Flux (PPF), 25/20 ℃ (light/dark) temperature and 70/95% (light/dark) relative humidity. Carbon dioxide concentration was maintained at 400 ppm. The density of all lettuce cultivars was 105mm between rows and 80mm between plants. After 8 days of planting, the plants were transplants again with a distance of 105mm×160mm (staggered row planting). The formula of the 1 nutrient solution was $NO_3$-N: 12, $PO_4$-P: 4, K: 8.3, Ca: 5, Mg: 2, $SO_4$: 2 me·$liter^{-1}$ (EC: 1.8 dS·$m^{-1}$). Five plants of each cultivar were sampled at 16 days after planting and quality indices at harvest were evaluated and other growth factors were also determined.

**3. Results and discussion**

There were significant differences in the growth characteristics of lettuce cultivars under low light intensity and high temperature condition (50% shaded greenhouse). Nine cultivars were selected as suitable and remaining nine as unsuitable by the quality index at the harvest stage (Fig.1, Photo.1). High negative correlation between the ratio of leaf length to width (seedling stage) and the quality index (harvest stage) were obtained in 50% shaded greenhouse ($y = -2.992x + 11.055$; $R = -0.755$), where x = the ratio of leaf length to width (seedling stage) and y = the quality index (harvest stage) (Fig.2). On the other hand, high positive correlation were found between the relative content of chlorophyll (seedling stage) and the quality index (harvest stage) in 50% shaded greenhouse, where x = relative content of chlorophyll (seedling stage) and y = quality index (harvest stage) (Fig.3). When lettuce cultivars were developed by the ratio of leaf length to width (seedling stage) and the relative content of chlorophyll (seedling stage), then the cultivars were classified into suitable and unsuitable group (Fig.4-Expt.1). Under greenhouse low light intensity and high temperature conditions, suitable cultivars (those quality indices were distributed within 3.0 to 5.0) were classified by two growth indices at seedling stage, namely the ratio of leaf length to width (<2.7) and the relative content of chlorophyll (>22).

Similar development also found in closed plant production system, where eight suitable cultivars had low ratio of leaf length to width (<1.8) and high relative content of chlorophyll (>29) compared with those in four unsuitable cultivars (Fig.5-Expt. 2). These results indicated that the appropriate evaluation method for lettuce cultivars might be realized by the classification with two seedling indices, such as the ratio of leaf length to width and the relative content of chlorophyll.

The quality evaluation in closed plant production system could also be possible by the above greenhouse lettuce seedling indices (Fig.6). In general, lettuce leaf length and

SLA (specific leaf area) begin to increase immediately after shading treatment with the decrease in solar radiation intensity (Noguchi et al., 1978). This means the seedling indices should be effective for the final products of lettuce.

Compared with the suitable cultivars, the unsuitable cultivars had large fresh weight, but the quality indices were found poor than the suitable cultivars (Fig.7-Expt.1, 2). This indicate that suitable cultivars have high quality index and with considerable fresh weight of products.

The products qualities (shapes, leaf color and so on) were very important factors for commercial lettuce production in Japan. Therefore, it was suggested that lettuce cultivars suitable for growing with low light and high temperature conditions could be evaluated using the ratio of leaf length to width and chlorophyll content in the young seedling stage. This result may contribute to the possibility of lettuce production in the closed production system with lower electric power cost.

**References**

Goto, E. and T. Takakura. 1987. Relationship between light intensity and light energy requirement in growth of leaf vegetables under artificial light (in Japanese). J. Agric. Meteorol. 43:229-232.

Ishii, M., T. Ito, T. Maruo, K. Suzuki and K. Matsuo. 1995. Growth and physiology of lettuce plants grown under artificial light of high intensity in short-day regime (in Japanese text with English summary). Environ. Control in Biol. 33:97-101.

Ito, T. 1986. Commercial approach to artificially controlled plant production facilities (in Japanese). Agric. and Hortic. 61:174-180.

Ito, T. 1989. More intensive production of lettuce under artificially controlled conditions. Acta Hortic. 260:381-389.

Noguchi, M., M. Kikkawa, K. Hoshino, S. Ikeda and K. Kobayashi. 1978. Analysis of the factors determining the vegetable crops yield II The effect of solar radiation on the growth and the dry matter production in lettuce (in Japanese text with English summary). Bull. Natl. Res. Inst. Veg. Jpn. A.4:55-76.

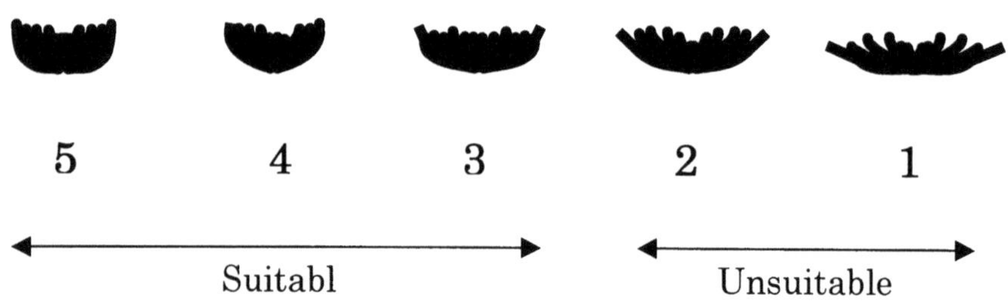


*Fig.1* Explanation of Quality Index.
Commercial value (5: very high, 4: high, 3: normal, 2: low and 1: very low).

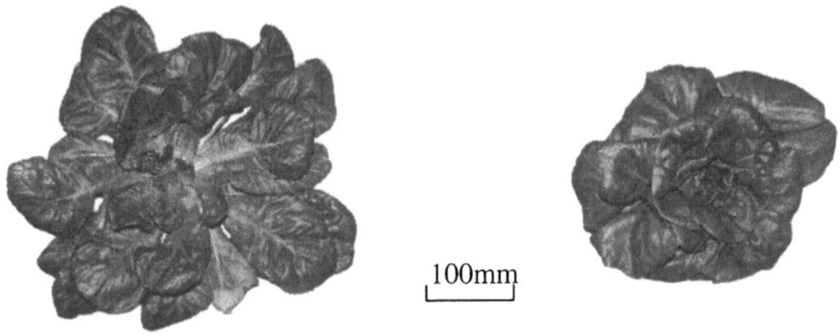


*Photo. 1* Lettuce plants at harvesting grown under closed plant production system (Expt. 2) . In the photographs of 'okayama saradana'(left sides; in unsuitable) and 'M shiki fuyu'(light sides; in suitable).

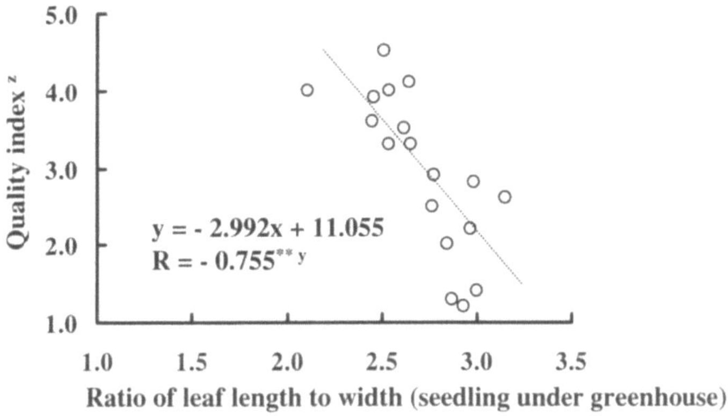


*Fig. 2* Relationship between ratio of leaf length to width (seeding) and quality index (harvest) of lettuce cultivars grown under greenhouse (Expt. 1).

z: Quality index, see Fig. 1.

y: ** Significant at 1% level.

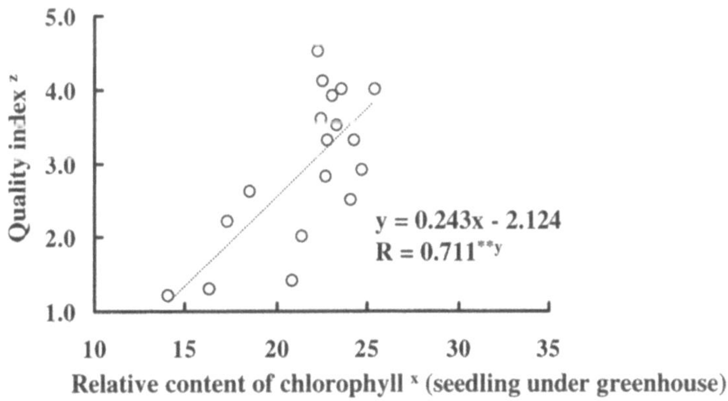


*Fig. 3* Relationship between relative content of chlorophyll (seeding) and quality index (harvest) of lettuce cultivars grown under greenhouse (Expt. 1).

z: Quality index, see Fig. 1.

y: **Significant at 1% level.

x: Values measured by Minolta chlorophyll meter, SPAD-502.

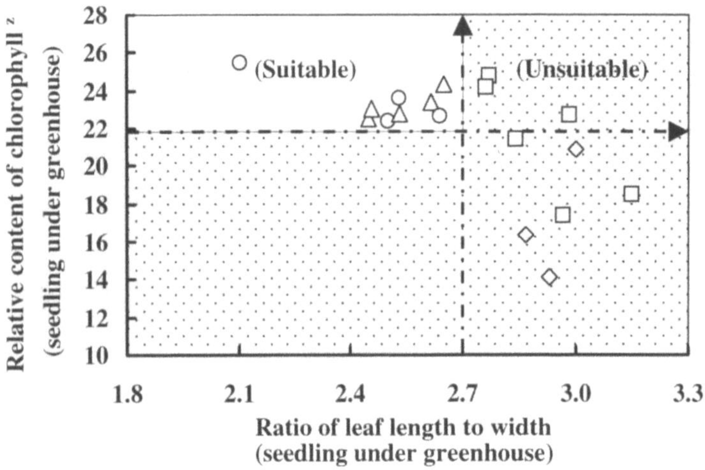


Grading of quality index y (greenhouse):
○ (4.0-5.0), △ (3.0-3.9), □ (2.0-2.9), ◇ (1.0-1.9)

*Fig. 4* Relationship between ratio of leaf length to width (seeding), relative content of chlorophyll (seedling) and quality index (harvest) of lettuce cultivars grown under 50% shaded greenhouse (Expt. 1).

z: Values measured by Minolta chlorophyll meter, SPAD-502.

y: Quality index, see Fig. 1.

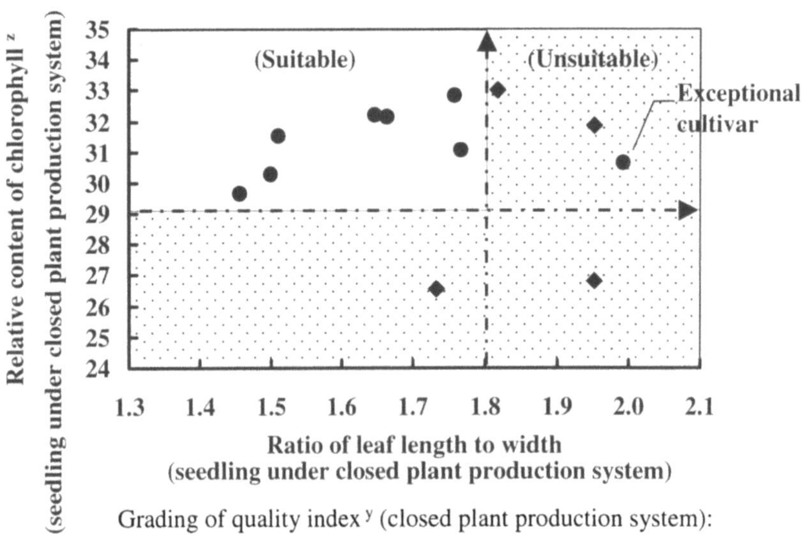


Grading of quality index y (closed plant production system):
● (4.0-5.0), ▲ (3.0-3.9), ■ (2.0-2.9), ◆ (1.0-1.9)

*Fig. 5* Relationship between ratio of leaf length to width (seeding), relative content of chlorophyll (seedling) and quality index (harvest) of lettuce cultivars grown under closed plant production system (Expt. 2).

z: Values measured by Minolta chlorophyll meter, SPAD-502.

y: Quality index, see Fig. 1.

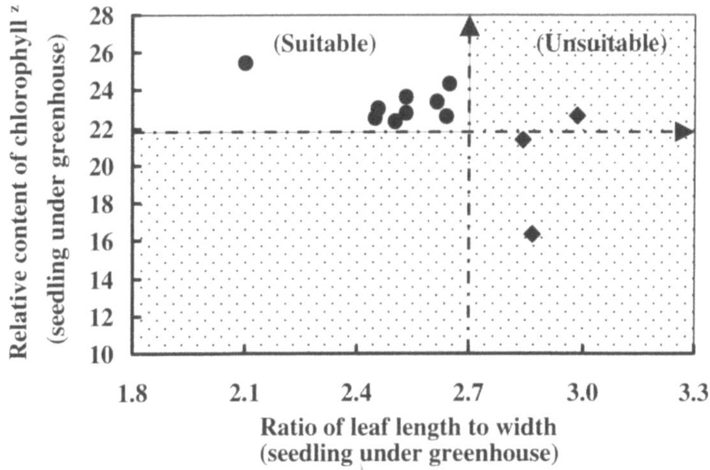


Grading of quality index y (closed plant production system): ● (4.0-5.0), ▲ (3.0-3.9), ■ (2.0-2.9), ◆ (1.0-1.9)

*Fig. 6* Relationship between ratio of leaf length to width (seeding under greenhouse), relative content of chlorophyll (seedling under greenhouse) and quality index (harvest under closed production system) of lettuce cultivars grown under 50% shaded greenhouse and closed plant production system (Expt. 1, 2).

z: Values measured by Minolta chlorophyll meter, SPAD-502.

y: Quality index, see Fig. 1.

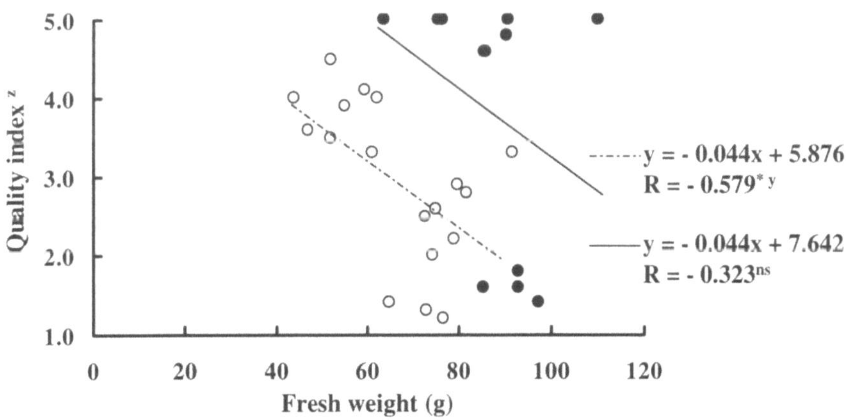


○: 50% shaded greenhouse ●: Closed plant production system

*Fig. 7* Relationship between fresh weight and quality index of lettuce cultivars grown under greenhouse and closed plant production system (Expt. 1, 2-harvest).

z: Quality index, see Fig. 1.

y: *and ns indicate significant difference at 5% and non-significant respectively.

# ROOT GROWTH SUBSEQUENT TO TRANSPLANTING IN PLUG-GROWN CABBAGE SEEDLINGS

Satoshi Yoshida
Biotron Institute, Kyushu University, Fukuoka 812-8581, Japan. E-mail: yochi@agr.kyushu-u.ac.jp

**Abstract.** For analysis of root growth in plug-grown seedlings just after transplanting, cabbage (*Brassica oleracea* L.) seedlings were transplanted in pots filled with sand and grown for two weeks under controlled environment in a phytotron glass room. Subsequent growth of new roots sticked out of a root clump was examined by image analysis of the root system dug out from sand. The delay of transplanting depressed elongation and development of vigorous "anchoring" roots after transplanting, although volume of new roots was supplemented with a rapid increase in fine roots by branching frequently. Thus, the seedlings overgrown in a plug tray could not become deeply-rooted plants after transplanting, and they might be intolerant of drastic changes in root environment. Coating the tray with a copper compound for preventing formation of a root clump improved subsequent root growth of overgrown seedlings to some extent.

**Key index words.** *Brassica oleracea* L., image analysis, overgrowth, plug tray, root system, transplant.

## 1. Introduction

In transplant production on a commercial basis, a plug tray is used regularly (Styer and Koramski, 1997). A seedling is grown in a tightly-restricted root zone under higher plant density in the tray. Therefore, the plug-grown seedling can be easily overgrown in the tray, and it results in depressed growth after transplanting (Sato et al., 1999). To maintain seedling quality, growers must rush to transplant the seedlings during a short period. Recently, the period suitable for transplanting can be prolonged by growth suppression in low temperature storage (Kozai et al., 1996).

In the restricted root zone in cells of plug tray, roots of the seedling intertwine in a clump. The root-clump formation is necessary for handling plug seedlings efficiently, particularly in the case of machine transplanting. However, the root-clump formation brings the loss of seedling quality. To promote plant growth after transplanting, a large amount of water has to be supplied until new roots are developed widely and deeply into the soil. Furthermore, Yoshioka et al. (1998) have reported that a cabbage plug seedling with a tight root clump cannot form a root system in a deep layer of the soil even in a long growing period after transplanting. Such shallowly-rooted plants could be intolerant of water stress or drastic changes in soil temperature in the field. From these facts, it is necessary to grow new roots out of the root clump as soon as possible and form a wide and deep root system in the soil for application of plug seedlings to various field conditions.

The present study deals with relationship between overgrowing in the tray and root growth just after transplanting. In this experiment, plug-grown cabbage seedlings were transplanted in sand, and subsequent growth of new roots elongated into sand was observed by digging up the root system.



*C. Kubota and C. Chun (eds.), Transplant Production in the 21$^{st}$ Century,* 178–182.

## 2. Materials and Methods

Cabbage (*Brassica oleracea* L. cv. Okina) seeds were sown in a plug tray with 128 cells (25 ml/cell; Yammer, Japan) and germinated at 20℃ and 70%RH in a phytotron glass room. Another tray coated with a copper compound (Spinout™, Griffin Co., Ltd., U. S. A.) was used for preventing the formation of a tight root clump. The plug-grown seedlings were used as transplants at two, four and six weeks after sowing. One seedling was transplanted into a pot (4 L) filled with sand, and put it in the phytotron glass room continuously. Subsequent root growth was examined at one and two weeks after transplanting. A root system of the seedling was dug out and spread on a flat plate. An image of the root system was photographed and inputted into a personal computer for image analysis, using a software "Cosmos 32" (Library Co. Ltd., Japan). The digital image analysis makes them possible to measure root length and to observe branches in roots sticked out of the plug, although a three-dimensional structure of the root system in the pot cannot be obtained.

## 3. Results and Discussion

A seedling grown for two weeks after sowing in the tray was transplanted before forming a root clump. On the other hand, a seedling grown for four weeks after sowing had a root clump at transplanting, which can be used as a transplant commonly in practical cabbage production. Furthermore, the seedling grown for six weeks after sowing had symptoms of overgrowth, i. e., a tight root clump, a succulent stem and chlorotic leaves. Just after transplanting, new roots were generated and elongated into sand, and many fine roots and several vigorous roots with many branches were sticked out of the root clump. The vigorous roots would be "anchoring roots" spreading widely and deeply into sand in order to support the whole plant body and absorb water and nutrients.

Figure 1 shows time course increase in total length of new roots sticked out of the clumped root. A rapid increase in total length of new roots was found in the seedling transplanted later. Therefore, volume of the new roots became larger with a delay of transplanting. Figure 2 shows the root image of the seedlings grown for two weeks after transplanting. In the seedling grown for two weeks after sowing, a seminal root and/or some of the 2nd-order branches were elongated vigorously after transplanting. They would be able to become "anchoring roots". In the seedling grown for four weeks after sowing, several "anchoring roots" were found, but the root system had a tendency to branch frequently. In the seedling grown for six weeks after sowing, most of the new roots were fine, branching of roots occurred frequently, and elongation of the respective roots appeared to decline. Figure 3 shows a distance from a root clump to a root tip in the largest "anchoring root" of the seedlings grown for two weeks after transplanting. It means the possible area where the "anchoring root" was able to reach. The distance in the seedling grown for six weeks in the tray and two weeks after transplanting (8-week-old) was almost same as that in the seedling grown for four weeks in the tray and two weeks after transplanting (6-week-old). Figure 4 shows the root image of the seedlings grown for four weeks after sowing in the untreated and coated trays and two weeks after transplanting. Using the tray coated with the copper compound for preventing the formation of a tight root clump, several vigorous "anchoring roots" were found, and volume of fine roots appeared to be reduced. Thus, subsequent root growth of the overgrown seedling was improved to some extent by using the coated tray.

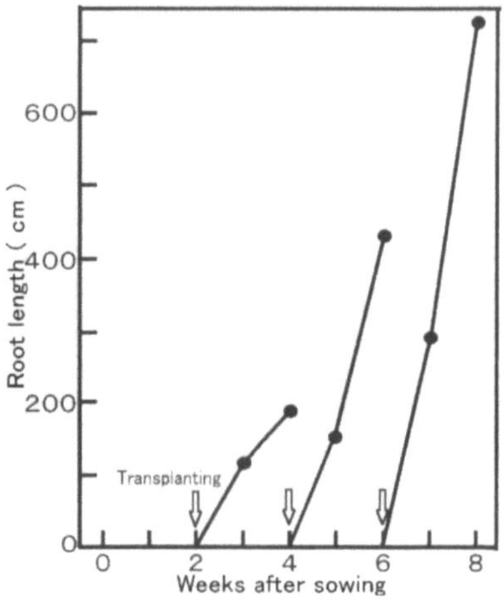


*Fig. 1* Time course increase in total length of new roots elongated into sand after transplanting in cabbage plug seedlings (a mean of measured values of six plants is shown): The seedlings were transplanted at two, four and six weeks after sowing and grown for two weeks after transplanting.

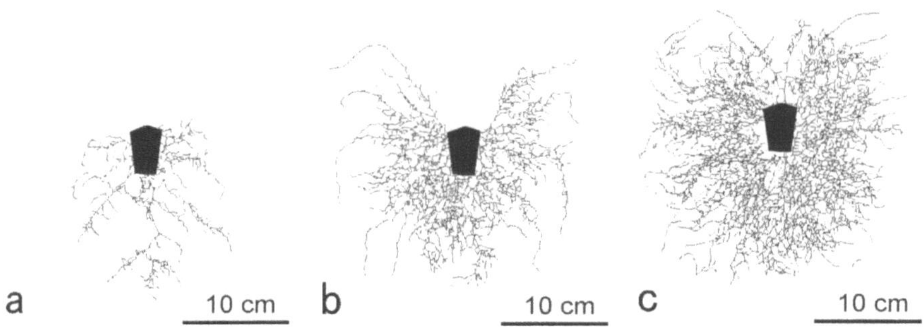


*Fig. 2* Typical thinned image of a root system in cabbage plug seedlings grown for two weeks after transplanting: The seedlings were transplanted at two (a), four (b) and six (c) weeks after sowing.

From the results, delay of transplanting in the plug-grown cabbage seedling resulted in losing vigor of "anchoring roots", although the root volume was supplemented with a rapid increase in branching roots. The overgrown seedling could not become a deeply-rooted plant. Kato and Lou (1987) and Leskovar and Cantliffe (1993) have reported that a decrease in pot size in transplant production reduces root volume in a deep soil layer, shoot growth and yield after transplanting. Thus, a root system developed only in a shallow soil layer might be intolerant of drastic changes in root environment, such as water stress in drought field. In addition, coating a plug tray

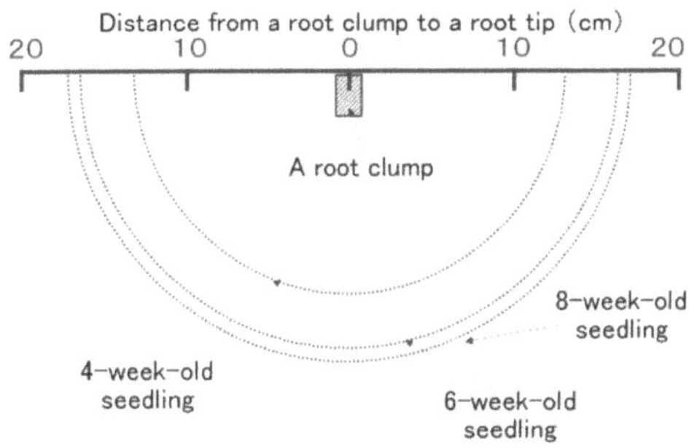


*Fig. 3* Distance from a root clump to a root tip in the largest "anchoring root" of cabbage plug seedlings grown for two weeks after transplanting (a mean of measured values of six plant is shown): The seedlings were transplanted at two, four and six weeks after sowing.

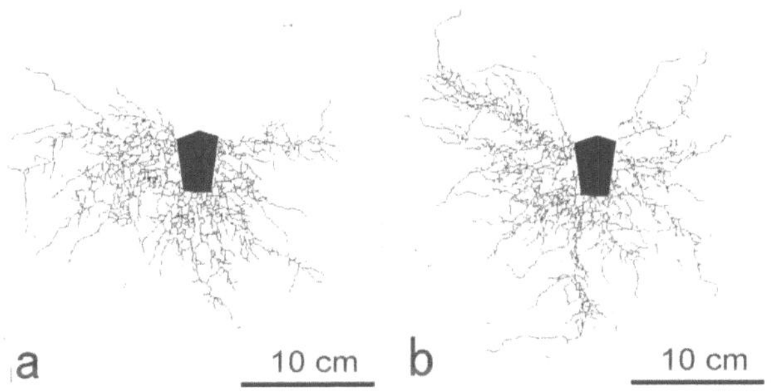


*Fig. 4* Typical thinned images of a root system in cabbage plug seedlings transplanted at four weeks after sowing on untreated (a) and coated (b) trays and grown for two weeks after transplanting.

was considered to be one of the applicable methods for activating root elongation after transplanting in order to obtain the deeply-rooted plant, and therefore such a new methodology may help to make a plug technique applicable to wide regions in the world for agricultural plant production.

**References**

Kato, T. and Lou, H. 1987. Studies on the characteristics of seedlings raised in pot under various conditions and their productivity in eggplant and sweet pepper. 3. Effects of pot size. Environ. Control in Biol. 25:19-23 (Japanese text with English summary).

Kozai, T., Kubota, C., Sakami, K., Fujiwara, K. and Kitaya, Y. 1996. Growth suppression and quality preservation of eggplant plug seedlings by low temperature storage under dim light. Environ. Control in Biol. 34:135-139 (Japanese text with English summary).

Leskovar, D. I. and Cantliffe, D. J. 1993. Comparison of plant establishment method, transplant, or direct seeding on growth and yield of bell pepper. J. Amer. Soc. Hort. Sci. 118:17-22.

Sato, F., Yoshioka, H. and Fujiwara T. 1999. Effects of storage temperature on carbohydrate content and seedling quality of cabbage plug seedlings. Environ. Control in Biol. 37:249-255.

Styer, R. C. and Koramski, D. S. 1997. Plug and transplant production: a grower's guide. Ball Publishing, Illinois, U. S. A.

Yoshioka, H., Kobe, T., Fujiwara, T. and Sato, F. 1998. Comparison between root systems of plug-grown cabbage seedlings and conventionally grown ones after transplanting. J. Japan. Soc. Hort. Sci. 67:459-461 (Japanese text with English summary).

# EFFECTIVE STORAGE CONDITIONS FOR SUBSEQUENT GROWTH ENHANCEMENT OF *FICUS CARICA* L. CUTTINGS

Michiko Takagaki[1], Yoshihiro Murata[1], Naoto Sakurai[1], Hideo Enomoto[1] and Yuji Udagawa[2]
[1] University Farm, Faculty of Horticulture, Chiba University, Matsudo, Chiba 271-8510, Japan. E-mail: mtgaki@midori.h.chiba-u.ac.jp
[2] Foundation Seed and Stock Farm of Chiba Prefecture, Midori-ku, Chiba 266-0007, Japan.

**Abstract.** For fig plants propagation in Japan, cultural environments require harvested branches as primary cuttings to be stored from February until late March or early April to avoid low temperature and moisture in the ground. Although it has been known that environmental conditions during the storage period of cuttings affect subsequent growth of seedlings after planting secondary cuttings to the field, the relationship between the storage conditions and the growth is still unclear. Thus in the present study, the effects of different storage conditions and planting date of hardwood cutting on the growth of fig (*Ficus carica* L. cv. Masui Dauphin) seedlings were investigated. Storage conditions were controlled in the environmentally controlled chamber. The results indicated that longer the storage period in the chamber increased shoot dry weight and number, length and dry weight of root. The cuttings moved from field condition to environmentally controlled chamber within two weeks after the harvest had the highest values in the number of root, root dry weight and shoot length in both planting dates.

**Key index words.** hardwood cutting, storage period, temperature, water contents.

## 1. Introduction

It has been reported that fig plant is one of the easiest plants for propagation by the hardwood cutting method (Kabumoto, 1984). In Japan, branches have to be cut from stock plants as primary cuttings at middle of February, before stock plants become biologically and physiologically active in spring. Then, the primary cuttings need to be stored under field condition until late March to early April to avoid low temperature and moisture in the ground. Conventionally, primary cuttings are made into proper length as secondary cuttings at just before planting to the field. Although the environmental conditions during storage period of primary cuttings seem to affect subsequent growth of seedlings after planting to the field, the relationship between the environmental conditions and subsequent growth of the seedlings is not well documented yet. Understanding the relationship between the storage environments of cutting and seedling growth will also help uniform transplant production of fig plants in large scale. Also, occasional low temperatures for root growth during late March to early April require the technique to delay and to adjust cutting date.

Takagaki et al. (1997) indicated that the storage of secondary cuttings under wet condition up to 20 days improved shoot and root growth in fig. Honda (1979) and Minamizawa (1984) also reported that mulberry cuttings stored under wet conditions for 20 days did not restrain their root growth. In this study, we explored the better storage conditions and to clarify the most important environmental factor(s) for root and shoot growth of *Ficus carica* L. cv. Masui Dauphin cuttings. The effects of delaying the planting of date the cutting on root and shoot growth were also investigated to explore



*C. Kubota and C. Chun (eds.), Transplant Production in the 21st Century,* 183–188.

the possibility of extending storage period.

## 2. Materials and Methods

Treatment codes are shown in Table 1. The branches were cut from stock plants as the primary cuttings on 12 Feb. 1997, and separated into two groups. First and second groups were stored for 6 and 8 weeks, respectively. First group of cutting was planted on March 27 as control (conventional storage period) , and second group was on April 10 as longer storage treatment. First and second groups were separated into four and five subgroups, respectively (Table 1). Each subgroup constituted with the combination of first and second storage. In first storage, primary cuttings were stored at field condition. Basal parts of the primary cuttings were staked into the ground 40 to 50 cm deep and covered with soil and rice straw (Fig.1). In second storage, secondary cuttings were placed in environmentally controlled storage chamber. Temperature and humidity in the chamber was set at 20 $^{O}$C and 95 %, respectively. In all treatments, the primary cuttings were made into proper length with two nodes as the secondary cuttings after first storage and were marked the numbers, measured diameters and fresh weights respectively. The secondary cuttings were put into sealed plastic box filled with sand of 30 % water contents as second storage (Fig.1).

At the end of each storage treatment (6 and 8 weeks), each of the secondary cutting was planted on plastic tray filled with sand with 5 to 7 cm depth after fresh weight was measured to investigate subsequent growth in the glasshouse without heating equipment. Dry matter content was measured on 10 cuttings of each treatment every two weeks for estimating water contents and dry weights. Length and dry weights of root, shoot and leaves of the cuttings were also measured at 42 days after planting.

## 3. Results and Discussion

Chronological changes in average of the daily maximum, minimum and average air temperatures in the field during storage treatments (i.e., two weeks) are shown in Table 2. Average temperature was 10 $^{O}$C or above after March 27. Figure 2 shows the chronological changes of water content after storage treatments started. Water contents continuously decreased as the period of storage got longer in T8/0 and T6/0 treatments. The change of water contents was relatively small in T8/0 and T0/6 treatments. In the treatments which the cuttings were moved from outside into the chamber, the water content decreased than they were in field condition but increased after moving into chamber. Within two weeks after moving into chamber, water contents recovered as same level as before the storage started.

The percentage of root emergence, root and shoot growth is shown in Table 3. T0/8, T2/6, T0/6 and T2/4 treatments showed higher rooting rate, longer root and higher value of root and shoot dry weight than other treatments. Also, treatments planted at April 10 showed longer root and heavier dry weight of root than treatments planted at March 27 (Table 3). Two factorial analyses, the length of chamber storage and the date of planting, showed significance of both factors and interaction of two factors. Usually, shoot starts to grow earlier than root in the fig plant cuttings (Kabumoto, 1984) because nutrients stored in stem is primarily used for shoot development. Day matter ratio of root and newly emerged shoot was calculated from the dry weight. The shoot held 70 to 80 % of total dry matter in the treatments in which the cuttings were planted at March 27, however only 10 to 30 % in treatments planted at April 10, indicated that the delayed planting promoted growth of root than shoot. The results may imply that delayed transplanting may be favorable for fig cultivation in long run, because of the fine root system is essential for vigorous shoot growth and nutrient supply. Also, low temperature and humidity before planting might restrain following root growth.

Results also showed that storage of cuttings in environmentally controlled chamber improved root growth. It has been reported that high water contents of cuttings promotes rooting of hardwood and leafy cuttings (Graves and Zang, 1996; Honda 1979; Minamizawa 1984). In present experiment, strong correlation was found between the water contents and the root and shoot growth (Table 4), meaning water contents of fig cuttings would be one of the indicator to affect root growth without relation to the planting date.

Correlation analysis between the cumulative daily average and minimum temperatures during the storage treatments and subsequent growth are shown in Table 5. The cumulative average and minimum temperatures were highly and positively correlated with root dry weight and poorly and negatively with newly developed shoot dry weight indicating that higher storage temperatures would promote root growth after planting in the temperature range of present study (Haissing, 1985).

It has been known that stored nutrients in hardwood cutting affect following root growth and development (Machida and Fujii, 1969; Pinheiro and De Oliveira, 1973). Conversion of starch into sugar, rather than photosynthetic activity, is relatively important to supply energy required for root growth at the time of planting of the cuttings. Also, total carbohydrate level, which can be represented by cuttings dry weight, may affect root growth. In the present experiment, however, the difference of total carbohydrate level, which can be calculated from water contents and fresh weight data, among the treatments was not significant. This may indicate storage treatments did not change the amount of carbohydrate lost by respiration but affected the carbohydrate metabolism, so the type and the amount of sugars converted from starch might have been different by the treatment. Although storage temperatures also affects respiration of the cuttings, respiration might not have been affected by the temperature rage of current study since no difference in dry weight of the cutting was observed. Higher temperatures during storage may increase respiration and cause loss of carbohydrate, however.

The results indicated that growth of fig transplants in hardwood cutting methods can be well controlled by using environmentally controlled growth chamber. The storage of the cutting in the chamber will be also applicable to achieve uniform growth of transplants. Further research will be necessary to clarify how water and temperature conditions affect nutritional components of the cutting, related with root initiation and growth.

**References**

Graves, W.R. and H. Zang 1996 Relative water content and rooting of subirrigated stem cutting in four environments without mist, Hort. Sci., 31:866-868.

Haissing, B.E. 1985 Metabolic processes in adventitious rooting in non-woody cuttings, In New root formation in plants and cuttings, (Michael, B.J. edited.), Kluwer academic publishers, Boston, 141-190.

Honda, T. 1979 Study on the cutting of mulberry, Tech. Bull. Sericulture Exp. Sta., 24:133-245 (Japanese text with English abstract).

Kabumoto, T. 1984 Production of the seedlings, in Complete works of fruit tree, (Noubunkyo edited), Noubunkyo, Tokyo, 341-342 (Japanese text).

Machida, H. and T. Fujii 1969 Studies on the propagation of the rooting in cuttings and the formation of adventitious roots, Bull. Tokyo Univ. of Education, 15:48-92 (Japanese text with English abstract).

Minamizawa, Y. 1984 Propagation of mulberry, In Study on cultivation of mulberry, (Minamizawa, Y.

edited.), Meiousha, Tokyo, 313-331 (Japanese text).
Pinheiro, R.V.R. and L.M. de Oliveira 1973 The influencee of fig cutting length on strilling, rooting and branch and leaf development, Rev. Ceres, 20:35-43.
Takagaki, M., Y. Udagawa and E. Takahashi 1997 Effect of pretreatment on rooting and shoot growth of *Ficus carica* L. cuttings, Tech. Bull. Fac. Hort. Chiba Univ., 51:227-230 (Japanese text with English abstract).

*Table 1* Treatment codes of the experiment.

| Treatment Code | Duration of 1st storage[z] | Duration of 2nd storage[y] | Date of plant the cuttings | Date of Measurement |
|---|---|---|---|---|
| T0/6 | - | 6 weeks | | |
| T2/4 | 2 weeks | 4 weeks | March 27 | May 7 |
| T4/2 | 4 weeks | 2 weeks | | |
| T6/0 | 6 weeks | - | | |
| T0/8 | - | 8 weeks | | |
| T2/6 | 2 weeks | 6 weeks | | |
| T4/4 | 4 weeks | 4 weeks | April 10 | May 21 |
| T6/2 | 6 weeks | 2 weeks | | |
| T8/0 | 8 weeks | - | | |

[z] Storage primary cuttings in field
[y] Storage secondary cuttings in environmental controlled chamber

*Table 2* Temperature and precipitation during the experiment.

| Date | Temperature (°C) | | | Precipitation |
|---|---|---|---|---|
| | Maximu[z] | Average | Minimum | (mm) |
| 2/12～ 26 | 10.7 | 4.6 | -1.5 | 5.0 |
| 2/27～3/12 | 15.1 | 9.3 | 3.6 | 0.0 |
| 3/13～ 26 | 13.2 | 8.2 | 2.9 | 32.0 |
| 3/27～4/ 9 | 17.3 | 12.7 | 8.3 | 97.0 |
| 4/10～ 23 | 18.8 | 13.4 | 7.9 | 18.5 |
| 4/24～5/ 7 | 23.1 | 17.2 | 11.0 | 14.0 |
| 5/ 8～ 21 | 22.7 | 18.0 | 13.5 | 31.0 |

[z] Averages of the daily maximum, average and minimum temperatures.

*Table 3* Factorial analysis of storage treatments on subsequent growth of fig cuttings.

| | Rooting (%) | Number of root | Root length (mm) | Shoot length (cm) | | Dry weights (g) | |
|---|---|---|---|---|---|---|---|
| | | | | Upper node | Lower node | Root | Shoot[z] |
| T0/6 | 100 | 20.9±0.8[y] | 346±21 | 3.1 ±0.06 | 4.2±0.09 | 0.16±0.01 | 0.67±0.01 |
| T2/4 | 100 | 32.8±1.6 | 611±32 | 4.1 ±0.14 | 4.4±0.05 | 0.27±0.02 | 0.66±0.01 |
| T4/2 | 90 | 13.3±0.8 | 160±8.2 | 1.7±0.10 | 3.6±0.05 | 0.12±0.01 | 0.43±0.01 |
| T6/0 | 65 | 8.0±0.6 | 82±7.3 | 1.5±0.06 | 1.8±0.07 | 0.06±0.01 | 0.24±0.01 |
| T0/8 | 100 | 46.5±1.0 | 1245±36 | 3.2±0.09 | 4.8±0.08 | 0.90±0.01 | 0.36±0.01 |
| T2/6 | 100 | 42.3±2.8 | 1381±111 | 3.8±0.14 | 5.7±0.19 | 1.10±0.05 | 0.47±0.05 |
| T4/4 | 80 | 10.7±1.0 | 233±20 | 2.7±0.08 | 3.5±0.10 | 0.55±0.02 | 0.17±0.01 |
| T6/2 | 85 | 7.8±1.0 | 86±12 | 2.5±0.20 | 2.7±0.11 | 0.45±0.02 | 0.04±0.00 |
| T8/0 | 75 | 12.6±1.1 | 251±27 | 3.1 ±0.19 | 3.0±0.07 | 0.45±0.02 | 0.11 ±0.01 |
| A[x] | - | **[v] | ** | ** | * | ** | ** |
| B[w] | - | ** | ** | ** | ** | ** | ** |
| A×B | - | ** | ** | ** | NS | ** | ** |

[z]; Dry wrights of leaves and shoots
[y]; mean±S.E.
[x]; A: Dates of planting fig cuttings (March 27 and April 10)
[w]; B: Storage under field condition and in environmental controlled chamber
[v]; **, *, NS: indicate significantly differences 1%, 3% levels and non significant by F-test.

*Table 4* Correlation coefficients between water contents at the planting time and subsequent growth of fig cuttings.

| | March 27[z] | April 10[y] |
|---|---|---|
| Total root length | 0.66 | 0.79 |
| Total shoot length | 0.87 | 0.74 |
| Dry weights of roots | 0.72 | 0.76 |
| Dry weights of shoots | 0.94 | 0.83 |

[z]; Data of T0/6,T2/4,T4/2,T6/0
[y]; Data of T0/8,T2/6,T4/4, T6/2,T8/0

*Table 5* Correlation coefficients between cumulative daily average and minimum temperatures during the treatments and subsequent growth of fig cuttings.

| | Average | Minimum |
|---|---|---|
| Total root length | 0.56 | 0.77 |
| Total shoot length | 0.48 | 0.76 |
| Root dry weights | 0.82 | 0.67 |
| Shoot dry weights | -0.28 | 0.40 |

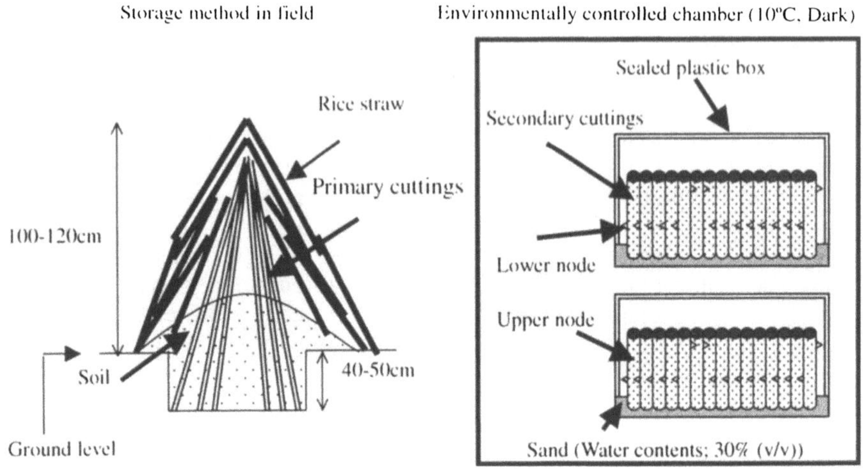


*Fig.1* Storage method of fig primary cuttings in field and secondary cuttings in environmentally controlled chamber.

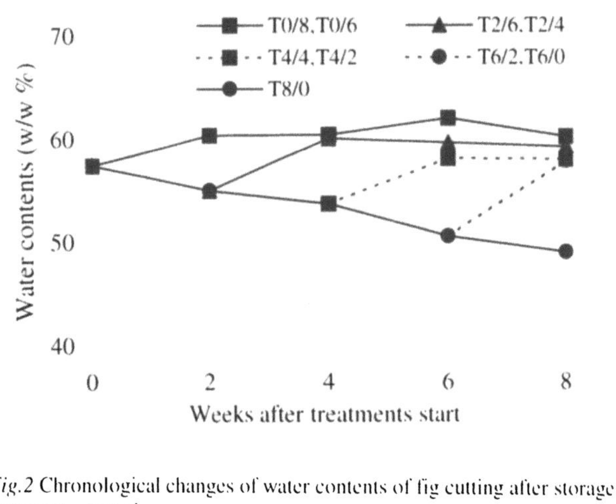


*Fig.2* Chronological changes of water contents of fig cutting after storage treatments started.

# 3. BIOTECHNOLOGY FOR TRANSPLANT PRODUCTION

# CHARACTERIZATION OF TRANSFORMED POPLAR FORMED BY THE INHIBITION OF PEROXIDASE

Noriyuki Morohoshi
Graduate School of Bio-Application and Systems Engineering Tokyo University of Agriculture & Technology, 2-24-16 Nakamachi, Koganeishi, Tokyo 184-8588, Japan.
E-mail: moronori@cc.tuat.ac.jp

**Abstract.** The objective of our research was to form a super tree using new biotechnological and genetic engineering techniques. The primary aim was to produce a tree with a lower lignin content by controlling lignin biosynthesis genes using antisense RNA. The isolation and sequencing of some peroxidase (PO) genes from hybrid aspen (*Populus kitakamiensis*), and also the isolation of the promoter region of these genes was achieved. From the results of Northern hybridization and expression of these genes, it was concluded that the PO gene involved in lignification was *prxA3a*. A new vector containing the original peroxidase gene (*prxA3a*) promoter and the antisense *prxA3a* gene involved in lignification was constructed. The transgenic poplars had lower total peroxidase activity compared with that of the control. After peroxidase isozyme analysis by the isoelectric focusing, it was clear that a peroxidase band (pI 3.8) disappeared in the transgenic plants. The lignin content in transgenic plants decreased 20-60% compared with control parts when the potassium permanganate oxidation method was used. This result was supported by ultraviolet microscopy results. The amount of glucose, determined by alditol acetate method, in transformants increased 5-10% compared with non-transformants. The production of transgenic poplars with a lowered lignin content was achieved by controlling the PO gene expression using the antisense RNA method. The use of biochemical analyses and ultraviolet microscopy confirmed this result.

**Key index words.** Acidic peroxidase, *Agrobacterium tumefaciens* LBA4404, Antisense RNA, Hybrid poplar (*Poplus kitakamiensis*), Lignification, Permanganate oxidation method, Ti-plasmid.

## 1. Introduction

The exhaustion of energy and resources, and aggravation of the environmental situation on the earth are serious problems and will continue in the near future. One way to solve these problems is to achieve an increased biomass yield and to develop useful techniques for biomass utilization, in order to take the place of energy and resources at present. As trees are a major biomass, it is very important to improve trees with genetic engineering techniques and to establish some useful conversion systems for biomass in near future.

The lignin biosynthetic pathway occupies a huge secondary metabolism system in trees. Inhibition of lignin biosynthesis to form a useful tree with a lower lignin content and being able to use polysaccharide materials was the focus of this research. *Populus kitakamiensis* was used as the plant material. A target gene for the lignin biosynthetic pathway was the peroxidase gene involved in lignification, which is inhibited by antisense RNA.

Three technical pre-requisites are essential to form an improved plant by genetic engineering. These are as follows: 1) to establish the developmental and regenerative techniques from callus initiation to a mature plant; 2) to establish a stable transformation



*C. Kubota and C. Chun (eds.), Transplant Production in the 21st Century,* 191–196.

system for foreign gene integration into the plant tissue; and 3) to isolate and analyze target genes and their promoters.

## 2. Materials and Methods

### 2.1 Plant materials, regeneration and growth conditions

Hybrid aspen (*Populus kitakamiensis*) plantlets were grown from the shoot meristem using methods developed by Ebinuma et al. (1997). A 18-hr fluorescent light period was used. The callus of the hybrid aspen grew on Murashige-Skoog (MS) medium containing 2,4-dichlorophenoxyacetic acid (2,4-D: 0.5 mg/l). Formation of adventitious buds from the callus was carried out in MS medium containing benzyladenine (BA: 0.1 mg/l) and zeatin (1.0 mg/l). Rooting of shoots was achieved in MS medium containing1-naphthaleneacetic acid (NAA: 1.5 mg/l).

### 2.2 Formation of transgenic plants

PO genes, *prxA1, prxA2a, prxA2b, prxA3a, prxA4a* and *HPOX14* were isolated and *prxA3a* was selected as the PO gene involved in lignin biosynthesis based on Northern hybridization and expression analysis (Osakabe et al., 1994). The vector used for the transduction was pBI121. Transformation of plant tissue was achieved using Ti-plasmid method (Kajita at al., 1998). The transformants (6 lines) were grown to calli, to shoots in vitro (about 10 cm), and to young plants ex vitro (about 4 m).

### 2.3 Chemical and biochemical analyses of transformants

Fresh plant materials of the shoots were weighed and frozen with liquid nitrogen, and then ground with mortar and pestle. Samples were pre-extracted with Tris buffer (100mM Tris-HCl, pH7.0). The content of protein was determined using a PROTEIN ASSAY KID (BIO-RAD) (Bradford 1978).

Peroxidase activity was determined using guaiacol (13 mM) as a substrate. Isoelectric focusing patterns were produced on gel plate colored with guaiacol and $H_2O_2$ (5 mM).

Extractives of the samples were obtained using the protein-free wood meals that were air-dried, weighed and extracted with an ethanol-benzene (1: 2) solution.

Lignin content was determined by the acetyl bromide method (Johnson et al., 1961; Iiyama and Wallis, 1988) and permanganate oxidation analysis, which is performed according to Morohoshi's method (1979) and a modified Miksche's method (Erickson et al., 1973) in order to give higher yield of degradation products. Thioacidolysis according to Lapierre et al. (1986) was used. Sugar analysis was done using the Alditol acetate method according to Blakeney (1983).

For the visible and ultra-violet microscopic observation of samples, 10μm thick transverse sections were cut using a sliding microtome following extraction with benzene-ethanol (2:1). They were then used for measuring the UV-absorption spectra of the secondary wall of fiber cells (Takabe et al., 1992) and visible light absorption spectra for both fiber and vessel secondary walls after the Maule and phloroglucinol staining (Srivastava, 1966). These absorption data were obtained by both quantitative and qualitative methods

## 3. Results and Discussion

### 3.1 Chemical properties of the transformed and untransformed poplars

Peroxidase is a final-step enzyme in the synthetic pathway from monolignols to a highpolymer lignin. This enzymatic step has no other bypass and could be a rate-limiting step for dehydropolymerization during lignification. Therefore, it was hypothesized that to inhibit the lignin biosynthesis effectively, the peroxidase gene must be controlled in order to control lignin content.

An antisense vector with the original promoter *prxA3a* (Fig. 1) was constructed and transformants were produced. These transformants appeared to grow as well as controls (no data shown). Peroxidase activity in the stem of the transformants decreased 75-90% compared with the control and in the leaf of transformants decreased 20-80%. This result suggested that the *prxA3a* gene was specifically expressed in the stem. The transformants did not contain a peroxidase isozyme, pI 3.8 that was expressed in the control stem from the results of isoelectric focusing.

In order to analyze the lignin chemical structure of transformants in detail, the transformants were subjected to potassium permanganate oxidation and thioacidolysis. The results showed that the lignin content of transformants decreases 20-60% in comparison with the control (Fig.2) and that the characteristic chemical structure of the transformants was the decrease of p-hydroxyl, guaiacyl and syringyl units evenly (data not shown). It was concluded that the inhibition of peroxidase in transformants had a major effect on decreasing lignin content in plant.

To determine the amount and quality of polysaccharides in the transformed poplar, the composition of monosaccharide degraded by alditol acetate method was analyzed. This result indicated an increasing production of glucose, which forms cellulose and a decreasing production of xylose, which forms hemicellulose. These results could mean that transformed poplar had increased cellulose content and decrease hemicellulose fractions in comparison with the wild. However, further research is needed to substantiate this.

### 3.2 Histochemical properties of the transformed and untransformed tissues

Transformants had a lowed lignin content than the control following chemical analysis (Fig.2). To ascertain this using histochemical methods, these transformants were subjected microscopy observations after maule and phloroglycinol color reactions and of ultra-violet microscopy.

Transformants appeared to have smaller vessel and fiber cells, and the secondary walls of the vessels were also thinner than the control (Fig.3). From the observation of UV-microspectrophotometry, the absorbance of transformants decreased in the secondary wall of vessel and fiber compared with the control (Fig. 4). It showed that transformants had a lower lignin content.

## 4. Conclusion

In order to produce a tree with lowed lignin content, peroxidase genes involved in lignin biosynthesis were isolated from the hybrid poplar. Five peroxidase genes were isolated and the DNA sequences and their promoters were analyzed. As the promoter of prxA3a was expressed in the stem specifically, it was selected as being involved in lignin biosynthesis. By using the antisense RNA method, some transformed poplar plants were formed. These transformants had a lower peroxidase activity (70-90% in stem) and lower lignin content (20-60%).

Furthermore, the transformants were subjected to visible microscopy after color reactions and to ultra-violet microscopy. It was ascertained that transformants had a lowed lignin content, smaller vessels and fiber cells, and especially the secondary wall of the vessel in the transformant appeared to be thinner. Further research is necessary to determine specific effects on the vessels.

These results will give us hope that we may succeed in forming useful trees by means of genetic engineering techniques.

**References**

Blakeney A.B., P.J. Harris, R.J. Henry et al. A simple and rapid preparation of alditol acetates for monosaccharide analysis. 1983. Carbohydr.Res, 113: 291-299

Bradford M. M., 1976. Anal. Biochem., 72, 249-258

Ebinuma H., K. Sugita, E. Matunaga et al. 1997. Selection of marker-free transgenicplant using the isopentenyl transferase gene. J Proc Natl Acad Sci USA , 94: 2117-2121.

Erickson M., S.Larsson and G.E.Miksche. 1973. Gaschromatografishe Analyse von Lignin oxidations produkten.VII. Ein verbessertes Verfahren zur Charakterisierung von Ligninen durch Methylierung und oxydativen Abbau. Acta Chemica Scandinavica, 27: 127-140

Iiyama K. and A.F.A.Wallis. 1988. An improved acetyl bromide procedure for determining lignin in woods and wood pulps. Wood Sci. Technol, 22: 271-280

Johnson D.B., W.E.Moore and L.C.Zank, 1961. The spectrophotometric determination of lignin small wood samples. Tappi. 44: 793-798

Kajita S., K. Osakabe, Y. Katayama et al. 1998. Agrobacterium-mediated transformation of poplar using a disarmed binary vector and the overexpression of a specific member of a family of poplar peroxidase gene in transgenic poplar cell. J Plant Science, 103: 231-239.

Lapierre C., B.Montiies and C.Rolando. 1986. Thioacidolysis of poplar lignin- identification of monomeric syringyl products and characterization of guaiacyl-syringyl lignin fractions. Holzforschurrg, 40: 113-118

Morohoshi N. and W. G. Glasser. 1979. The structure of lignins in pulps. Part 4–Comparative evaluation of five lignin depolymerization techniques-. Wood Sci. Technol., 13: 165-1781. N. Morohoshi and W. G. Glasser.1979. The structure of lignins in pulps. Part 5 -Gas and gel permeation chromatography of permanganate oxidation products-, Wood Sci. Technol., 13: 249-264

Osakabe K., H. Koyama, S.Kawai et al. 1994. Molecular cloning and the nucleotidesequences of two novel cDNAs that encode anionic peroxidases of Populus kitakamiensis. J. Plant Science , 103:167-175.

Osakabe K., H. Koyama, S.Kawai et al. 1995. Molecular cloning of two tandemly arranged peroxidase genes from Populus kitakamiensis and their differential regulation in the stem. J. Plant Molecular Biology 28:677-689.

Stivastava L. M., 1966. Histochemical studies on lignin, Tappi, 49:173-183.

Takabe K., S. Miyauchj, R. Tsunoda and K.Fukazawa, 1992. Sistribution of guaiacyl and syringyl lignin in Japanese beech ( Fagus crenata): variation within an annualring. 13:105-112.

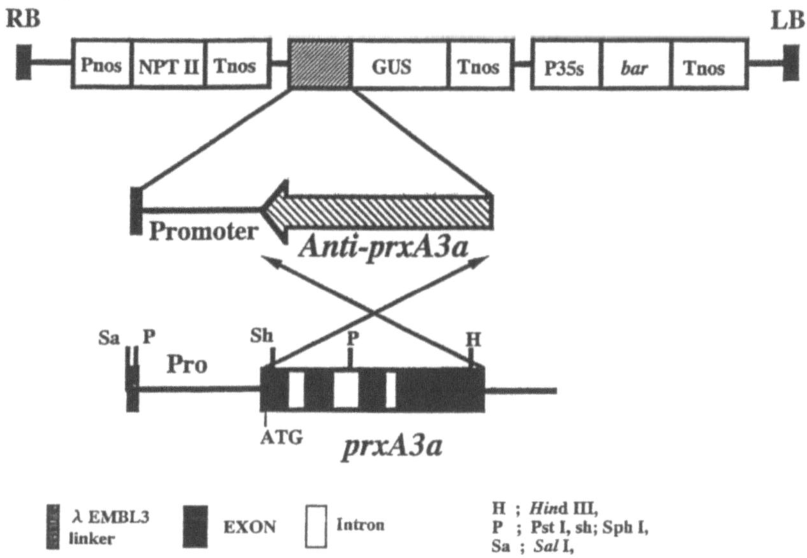


*Fig. 1* Structure of a plasmid containing anti-*prxA3a*.

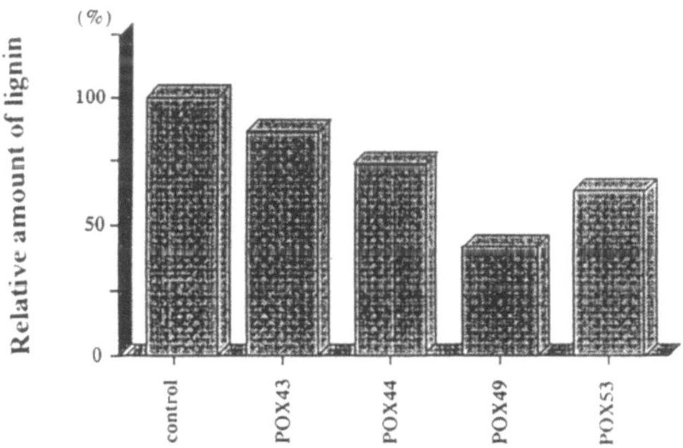


*Fig. 2* Lignin content of the transformed hybrid aspen estimated by thepermanganate oxidation method.
Transformed shoots in vitro are labelled POX43, POX44, POX49, POX53. Control is the untransformed shoot.

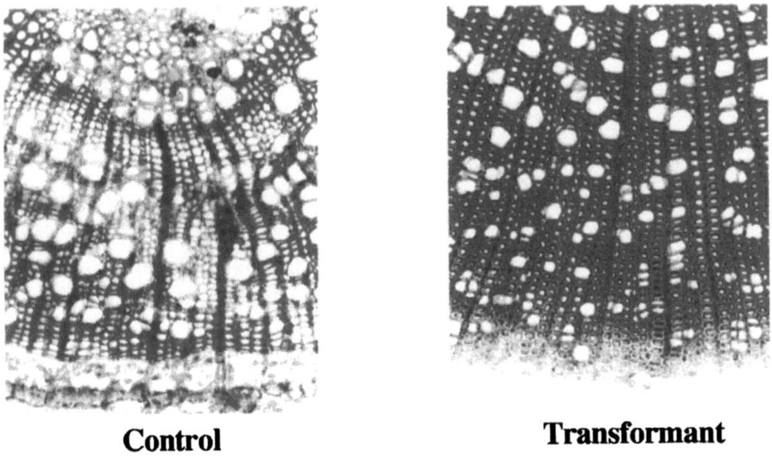


*Fig. 3* A cross-section of the control and the transformed poplar (young plant ex vitro, POX49) after the Maule treatment.

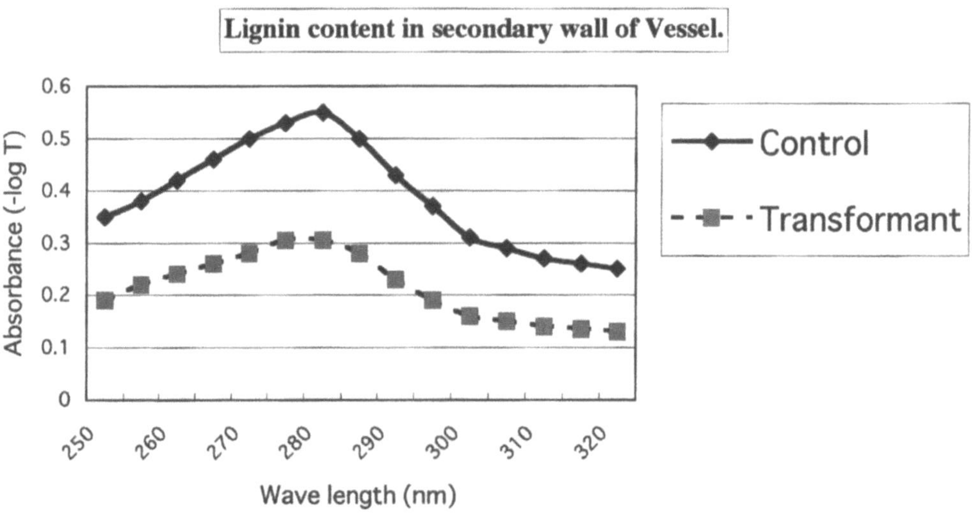


Fig. 4 Ultra-violet absorption curve of the secondary cell wall of the vessel.
The transformant is young plant ex vitro of POX49
Control is the untransformed young plant

# MICROPROPAGATION OF CANADIAN SPRUCES (*PICEA* SPP)

Trevor A. Thorpe and Indra S. Harry
Plant Physiology Research Group, Department of Biological Sciences, University of Calgary, 2500 University Drive NW, Calgary, Alberta, T2N 1N4, Canada. E-mail: tthorpe@ucalgary.ca

**Abstract.** The spruces are a major component of the Canadian forest, and are harvested for pulp and lumber. Reforestation is currently done with seedlings, but clonal material could play a role. Micropropagation of native spruces was first achieved via the multi-stage organogenic process in the early 1980s. Since the former method produces too few propagules per explant for clonal propagation, most of the recent effort has been with somatic embryogenesis. Somatic embryogenic tissue of the spruces can be induced on immature and mature zygotic embryos in the presence of auxin and cytokinin. This tissue develops into somatic embryos in the presence of abscisic acid, but some desiccation treatment via partial drying or the use of a non-permeating osmoticum such as polyethylene glycol is needed for optimum germination and conversion into vigorous somatic seedlings. Synthetic seed and large-scale propagation are being achieved. Somatic seedlings of interior spruce have performed similarly to zygotic seedlings in both the nursery and in reforestation sites. Thus it appears that reforestation with somatic seedlings of spruces may be feasible in the near future.

**Key index words.** *P. engelmannii*; Engelmann spruce, *Picea glauca*; white spruce, *P. glauca x engelmannii*, *P. mariana*; black spruce, *P. rubens*; red spruce, *P. sitchensis*; Sitka spruce, interior spruce, organogenesis, somatic embryogenesis, plantlet regeneration.

## 1. Introduction

The spruces (*Picea* spp) are a significant part of the Canadian forest. Five of the seven species native to North America are found in Canada: red spruce (*P. rubens*) in the east, Sitka (*P. sitchensis*) and Engelmann (*P. engelmannii*) spruces in the west, with white (*P. glauca*) and black (*P. mariana*) spruces being transcontinental (Farrar, 1995). Hybridization is common where the species overlap, with interior spruce (*P. glauca x engelmannii* complex) being the most important. Spruces are a source of wood pulp and first among Canadian species in volume of lumber produced. They are used in general construction, mill work, interior finishings, and for plywood, boxes and musical instruments. Genetic improvement work has been carried out in Canada, and over 400 million seedlings are planted yearly on reforestation sites.

It is recognized that there are benefits in the use of clonal planting stock in reforestation programs, as it is a reliable method for capturing both additive and non-additive genetic gain. The traditional methods of vegetative or clonal propagation of woody species are rooted cuttings and grafting. However, these methods do not produce enough propagules for clonal reforestation (Thorpe and Biondi, 1984). Micropropagation is a newer approach that is more efficient in plantlet regeneration, as it produces more propagules per explant per unit time. In spruces, as in conifers in general, it has been achieved by both organogenesis and somatic embryogenesis (Thorpe and Biondi, 1984; Thorpe et al., 1991; Dunstan et al., 1995).



*C. Kubota and C. Chun (eds.), Transplant Production in the 21st Century,* 197–204.

Organogenesis leads to the formation of unipolar adventitious shoot buds, which must be elongated into shoots, multiplied (often through many cycles) and then rooted to produce plantlets, which must be hardened before transfer to the greenhouse (Thorpe and Patel, 1984; Thorpe and Harry, 1991). This process is thus a multi-staged one and the requirements for each stage múst be empirically determined for each species. In contrast, somatic embryogenesis leads to the formation of a bipolar embryo, with rudimentary shoot and root axes. This process is also multi-staged and superficially resembles zygotic embryogenesis, as early and late stage filamentous embryos, early and late cotyledonary stages, etc., are observed. Such embryos must often be given maturation treatments to enhance germination and plantlet growth. In both forms of regeneration, seed embryos and seedling parts are the materials of choice.

## 2. Micropropagation by Organogenesis

Lab scale micropropagation protocols have been developed for Canadian spruces using this approach.

### 2.1 Culture Initiation

Following preculture (in some cases) and seed sterilization, mature embryos may be excised or seeds germinated (aseptically) to give seedling parts such as epicotyls, cotyledons and hypocotyls. A variety of explants have been used for culture initiation and the induction of adventitious buds in the spruces (Table 1). Various salt formulations have been used and a cytokinin, usually $N^6$-benzyladenine (BA) has been required. In addition, an energy source (usually sucrose), vitamins, and reduced nitrogen (normally amino acids) are needed.

*Table 1* Conditions required for shoot induction and development in Canadian spruces.

| Species | Explants | Salt Formulation Sucrose (%) Induction | Growth Regulators Induction (μM) | Salt Formulation + Sucrose (%) Shoot Development | References |
|---|---|---|---|---|---|
| P. engelmannii | Mature embryos & Seedling parts | AE[1], 3% | 10 BA | AE + 2% +0.1% AC | Patel & Thorpe, 1986 |
| | Mature embryos | AE, 3% | 10 BA, 10 NAA | AE + 3% + 25 BA | Lu & Thorpe, 1988 |
| P. glauca | Cotyledons | 1/2GMD, 2% | 2 BA | 3/4GMD + 2% | Toivonen & Kartha, 1988 |
| | Epicotyls | SH, 2% | 10 BA+1 2iP | SH + 2% + 0.2% AC | Rumary & Thorpe, 1984 |
| | Vegetative buds | GMD, LM, AE, 3% | 13.2 BA | GMD + 3% | Mohammed et al., 1986 |
| P. mariana | Epicotyls | SH, 2% | 10 BA + 1 2iP | SH + 2% + 0.2% AC | Rumary & Thorpe, 1984 |
| P. rubens | Mature embryos | 1/2MCM, 3% | 10μM BA | 1/2AE + 2% + 0.1% AC | Lu et al., 1991 |
| P. sitchensis | Mature embryos | 1/2AE, 1% | 2h pulse, 250 BA | 1/2AE + 1% | von Arnold & Woodward, 1988 |

[1]AE (von Arnold & Eriksson, 1981); GMD (Mohammed et al., 1986); LM ( Litvay et al., 1981); SH (Schenk & Hildebrandt, 1972); MCM (Bornmann, 1983).

### 2.2 Shoot Development and Multiplication

The second phase of the micropropagation process allows for the development of the nodular bud primordia into elongated shoots with primary needles and the multiplication of these shoots by axillary or adventitious budding (Thorpe and Patel, 1984). The ease and frequency of shoot multiplication determines the potential of the species to form large numbers of plantlets in a reasonable period of time and therefore the cost of the process.

Shoot development in the spruces generally requires transfer to a medium with altered nutritional, e.g., half-strength mineral salts, no or reduced cytokinin, and reduced sucrose (1-2%) levels (Table 1). Often activated charcoal (AC, 0.1-0.5%) must be included for optimum development (Thorpe et al., 1991). Several culture cycles may be needed to obtain rootable shoots, e.g., in *P. glauca* and *P. mariana* (Rumary and Thorpe, 1984).

Multiplication of shoots is achieved by separating individual shoots and culturing to allow axillary shoot development. This often requires the addition of a cytokinin, e.g., 1 μM kinetin (K) for *P. rubens* (Lu et al., 1991).

### 2.3 Rooting of Shoots and Hardening of Plantlets

Rooting is generally acknowledged to be the most difficult stage of micropropagating conifers (Mohammed and Vidaver, 1988), but rooting in spruces is generally easier than pines. Rooting has been carried out under in vitro conditions or ex vitro, in which the micro-shoots are treated as micro-cuttings. This latter approach generally leads to reduced callus formation at the base of shoots and allows for the concurrent hardening of the rooted shoots (Thorpe and Harry, 1991). Generally, auxin is required for rooting and indole butyric acid (IBA) is often the most effective. This treatment can be given as a pulse e.g., 4 h for *P. rubens* (Lu et al., 1991) or overnight with *P. engelmannii* (Patel and Thorpe, 1986). Commercial rooting powder, containing IBA was most effective for *P. glauca* and *P. mariana* (Rumary and Thorpe, 1984). About 50-80% rooting was achieved with these spruces.

Thus, under the conditions outlined above about 65 plantlets of *P. rubens* would be formed in14 m, 50 plantlets each of *P. glauca* and *P. mariana* after 25 weeks, and about 40 plantlets of *P. engelmannii* after 30 weeks. Clearly these rates of plantlet production are too low for use in mass clonal propagation of these species. As a result most activity from the late 1980s has concentrated on developing micropropagation protocols via somatic embryogenesis.

## 3. Micropropagation by Somatic Embryogenesis

Embryogenic tissue of a conifer was first observed in a spruce by Hakman et al. (1985) with excised embryos from immature seeds of *P. abies*. Embryogenic tissue and somatic embryos have subsequently been obtained from all the native Canadian spruces.

### 3.1 Culture Initiation

Embryogenic tissue in the spruces is generally initiated on AE medium with auxin, usually 2,4-dichlorophenoxyacetic acid (2,4-D) and a cytokinin ($N^6$-benzyladenine, BA) (Table 2). Embryogenic tissues have been initiated from both immature and mature embryos. It has also been possible to induce embryogenic tissue from excised cotyledons of young seedlings of both *P. glauca* and *P. mariana* (Attree et al., 1990; Lelu and Bornman, 1991).

*Table 2* Conditions required for somatic embryogenesis in Canadian spruces.

| Species | Explants | Salt Formulation + Sucrose (%) | Growth Regulators Induction (μM) | Growth Regulators Development (μM) + Sucrose (%) | References |
|---|---|---|---|---|---|
| P. engelmannii | Immature embryos | AE[1], 1-3% | 10 2,4-D + 5 BA | 1 ABA + 1 IBA, 3.4% | Webb et al., 1989 |
| P. glauca | Immature embryos | AE, 1% | 10 2,4-D + 5 BA | 5 BA,0.5 2,4-D or no PGR, 6% | Hakman & Fowke, 1987 |
| | Immature embryos | AE, 1-3% | 10 2,4-D + 5 BA | 1 ABA + 1 IBA, 3.4% | Webb et al., 1989 |
| | Immature embryos | AE, 1% | 10 2,4-D or picloram + 5 BA | 0.5 2,4-D, 3% | Lu & Thorpe, 1987 |
| | Immature embryos | AE, 1% | 10 2,4-D + 5 BA | 12 ABA, 3% | Attree et al., 1989 |
| | Mature embryos | 1/2 LM,1% | 10 2,4-D +5 BA | 5 K + 1 2,4-D, 6% | Tremblay, 1990 |
| | Seedling | MS, 3% | 4.5 BA +0.5 NAA - 21.5 NAA + 0.45 BA | 11 ABA + 1 2,4-D | Lelu & Bornman, 1990 |
| P. glauca x engelmannii complex | Immature embryos | AE, 3% | 5 2,4-D + 2 BA | 30-40 ABA + 0.1-10 IBA, 3.4% | Roberts et al., 1990 |
| | Immature embryos | AE, 3% | 10 2,4-D + 5 BA | 1 ABA + 1 IBA | Webb et al., 1989 |
| | Immature embryos | AE, 3% | 5 2,4-D + 2 BA | 40-60 ABA + 1 IBA, 3.4% | Webster et al., 1990 |
| | Immature embryos | AE, 3% | 5 2,4-D + 2 BA | 40 ABA + 1 IBA, 3.4% | Flinn et al., 1991 |
| P. mariana | Immature embryos | AE, 3% | 10 2,4-D + 5 BA | 5 BA + 0.5 2,4-D or no PGR, 1% | Hakman & Fowke, 1987 |
| | Immature embryos | 1/2LM, 1% | 10 2,4-D + 5 BA | 7.5 ABA, 6% | Tremblay & Tremblay, 1991 |
| | Mature & immature embryos | 1/2AE or AE, 3% | 9 2,4-D + 4.5 BA | None reported | Tautorus et al., 1990 |
| | Seedling parts | MS, 3% | 4.5 BA +0.5 NAA - 21.5 NAA + 0.45 BA | 11 ABA + 1 2,4-D + 0.5 BA + K | Lelu & Bornman, 1990 |
| | Seedling parts & mature embryos | AE, 3% | 9 2,4-D,+ 4.5 BA | 12 ABA | Attree et al., 1990 |
| P. rubens | Mature embryos | AE, 3% | 10 NAA + 4.5 BA | 20-50 ABA, 5% | Harry & Thorpe, 1991 |
| | Mature embryos | 1/2 LM, 1% | 10 2,4-D + 5 BA | 7.5 ABA + 6% | Tremblay & Tremblay, 1991 |
| P. sitchensis | Immature embryos | AE, 3% | 10 2,4-D + 5 BA | 40 ABA + 1 IBA, 3.4% | Roberts et al., 1991 |
| | Immature embryos | 1/2MS, 3% | 50 2,4-D,+ 25BA +25 K | 5 ABA, 3% | Krogstrup et al., 1988 |
| | Mature embryos | MS, 3% | 50 2,4-D +25 BA + 25 K | 5 ABA, 3% | Krogstrup et al., 1988 |
| | Mature embryos | 1/2AE, 1% | 10 2,4-D + 5 BA | None reported | von Arnold & Woodward, 1988 |

[1]AE (von Arnold & Eriksson, 1981); MS (Murashige & Skoog, 1962); LM (Litvay et al., 1981).

### 3.2 Somatic Embryo Development and Maturation

Generally spruce embryogenic tissue, consisting of early stage filamentous embryos (embryonal cell masses) transferred to hormone-free medium shows poor embryo development, with only a few cotyledonary embryos being formed. However, the addition of abscisic acid (ABA) at concentrations up to 60 μM greatly improves embryo development (Table 2). However, although such embryos often appear morphologically normal, they may not develop properly subsequently.

Partial drying of the embryos of *P. glauca x engelmannii* (Roberts et al., 1990) and *P. rubens* (Harry and Thorpe, 1991) enhanced their subsequent performance, as this treatment maintained the shoot meristem organization (Kong and Yeung, 1992) and reduced hyperhydricity (Roberts et al., 1993). As well, somatic embryos (SEs) of *P. glauca* survived drying to a moisture content below those of zygotic embryos excised from mature seed (Attree et al., 1991). Desiccation tolerance was achieved by culturing immature somatic embryos in medium containing ABA and the non-permeating non-plasmolysing osmotica polyethylene glycol (PEG) 4000 or dextran 6000. These treatments promoted the accumulation of storage lipids (Attree et al., 1992) and protein (Misra et al., 1993) and resulted in polypeptide and fatty acid compositions comparable to zygotic embryos.

### 3.3 Germination and Conversion to Plantlets

While germination is described as the emergence of the radicle from the basal cells of the embryo, conversion is concerned with the development of normal plantlets (somatic seedlings or emblings) possessing both a root and shoot. Spruces which had been treated with ABA and underwent some desiccation treatment germinated (in the light in vitro), and produced vigorous somatic seedlings in *P. rubens* (Harry and Thorpe, 1991), *P. glauca* (Attree et al., 1991, 1995), *P. glauca x engelmannii* (Webster et al., 1990), and *P. sitchensis* (Roberts et al., 1991). Treatment of the SEs of *P. glauca* with ascorbate also improved the quality of shoot meristems and enhanced the germination and conversion rates (Stasolla and Yeung, 1999).

### 3.4 Mass Propagation and Artificial Seed

Current methods for maturation of spruce SEs involve the use of semi-solid medium or supports over liquid media. Cultures have also been maintained in liquid shake flasks (Dunstan et al., 1995). For scale-up, bioreactors have been tried, but limited success was achieved with *P. mariana* and *P. glauca x engelmannii* (Tautorus et al., 1992; 1994). For large-scale propagation, a pilot production system capable of delivering 50,000 to 250,000 somatic seedlings of the latter hybrid species as a single crop has been developed (Cervelli et al., 1994), but is not yet fully operational.

Artificial seeds of SEs of conifers have been produced using hydrated gel coatings, desiccated coating and desiccated naked embryos (Attree and Fowke, 1993). Partially dried SEs have been stored for several months without loss of viability, while dried embryos have survived storage for over one year at -20°C without cryoprotectants (Attree and Fowke, 1993). Sodium alginate-encoated SEs of *P. mariana* and *P. glauca x engelmannii* have germinated and were established in the soil (Lulsdorf et al., 1993). Cryopreservation methods for embryogenic tissues of *P. glauca* (Kartha et al., 1988), *P. mariana* (Klimazewska et al., 1992), *P. sitchensis* (Find et al., 1993), and *P. glauca x engelmannii* (Cyr et al., 1994) have been developed, and plantlet regeneration has been achieved.

### 3.5 Acclimatization, Nursery and Field Performance

Acclimatization conditions are extremely important to maximize survival and to provide vigorous somatic seedlings (Roberts et al., 1993). Both *P. mariana* and *P. glauca* emblings acclimatized under high humidity and low light conditions survived transfer to ex vitro conditions, where they survived overwintering (Attree et al., 1990). *P. sitchensis* emblings were also successfully transferred to soil and exhibited normal growth and development when compared with seedlings (Krogstrup, 1990). Similarly, nursery trials on emblings of *P. glauca x engelmannii* under operational conditions showed that at the end of the first season heights were equivalent to control seedlings (Webster et al., 1990; Grossnickle et al., 1994). As well, they had similar growth rates and the same root and shoot morphology. Following frozen storage, stock quality evaluation indicated that the field performance potential of seedlings and emblings would be comparable, if planted in sites under limiting cold or drought conditions (Grossnickle and Major, 1994a). On a reforestation site, emblings and seedlings had similar summer seasonal water relations patterns and gas exchange responses (Grossnickle and Major, 1994b). Emblings were comparable, if not superior, to seedlings in resistance to winter injury, and survival after the first winter was 91% and 87%, respectively. Both types had similar incremental height and diameter growth over two growing seasons.

## 4. Concluding Thoughts

Although it is possible to regenerate spruce plantlets by both organogenesis and somatic embryogenesis, it is clear that only the latter approach can possibly be utilized for mass clonal propagation and reforestation of spruces in Canada. As indicated in this article, good progress is being made towards this goal, and it should become operational before the end of this first decade of this new century.

### References

Attree, S.M. and L.C. Fowke. 1993. Embryogeny of gymnosperms. Advances in synthetic seed technology in conifers. Plant Cell Tiss. Organ Cult. 35:1-35.

Attree, S.M., D.I. Dunstan and L.C. Fowke. 1989. Plantlet regeneration from embryogenic protoplasts of white spruce (*Picea glauca*). Bio/Technology 7:1060-1062.

Attree, S.M., S. Budimir and L.C. Fowke. 1990. Somatic embryogenesis and plantlet regeneration from cultured shoots and cotyledons of seedlings germinated from stored seed of black and white spruce (*Picea mariana* and *Picea glauca*). Can. J. Bot. 68:30-34.

Attree, S.M., D. Moore, V.K. Sawhney and L.C. Fowke. 1991. Enhanced maturation and desiccation tolerance of white spruce (*Picea glauca* (Moench.) Voss) somatic embryos. Effects of a non-plasmolysing water stress and abscisic acid. Ann. Bot. 68:519-525.

Attree, S.M., M.K. Pomeroy and L.C. Fowke. 1992. Manipulation of conditions for the culture of somatic embryos of white spruce for improved triacylglycerol biosynthesis and desiccation tolerance. Planta 187:395-404.

Attree, S.M., M.K. Pomeroy and L.C. Fowke. 1995. Development of white spruce (*Picea glauca* (Moench.) Voss) somatic embryos during culture with abscisic acid and osmoticum, and their tolerance to drying and frozen storage. J. Exp. Bot. 46:433-439.

Bornman, C.H. 1983. Possibilities and constraints in the regeneration of trees from cotyledonary needles of *Picea abies in vitro*. Physiol. Plant. 57:5-16.

Cervelli, R.L., F. Webster, T. Edmonds, B. Sutton and M. Scott. 1994. Pilot production and scale-up delivery of spruce somatic seedlings. 1994 Biological Sciences Symposium TAPPI, pp.191-197.

Cyr, D.R., W.R. Lazaroff, S.M.A. Grimes, G.Q. Quan, T.D. Bethune, D.I. Dunstan and D.R. Roberts. 1994. Cryopreservation of interior spruce (*Picea glauca engelmannii* complex) embryogenic cultures. Plant Cell Rep. 13:574-577.

Dunstan, D.I., T.E. Tautorus and T.A. Thorpe. 1995. Somatic embryogenesis in woody plants. In: In Vitro Embryogenesis in Plants (T.A. Thorpe, ed.), Kluwer Acad. Publ., Dordrecht, The Netherlands.

Farrar, J.L. 1995. Trees in Canada, Fitzhenry and Whiteside Ltd. and the Canadian Forest Service. pp 95-107.

Find, J.I., F. Floto, P. Krogstrup, J.D. Moller, J.V. Norgaard and M.M.H. Kristensen. 1993. Cryopreservation of an embryogenic-suspension culture of *Picea sitchensis* and subsequent plant regeneration. Scand. J. For. Res. 8:156-162.

Flinn, B.S., D.R. Roberts and I.E.P. Taylor. 1991. Evaluation of somatic embryos of interior spruce. Characterization and development regulation of storage proteins. Physiol. Plant. 82:624-632.

Grossnickle, S.C., J.E. Major and R.S. Folk. 1994. Interior spruce seedlings compared with emblings produced from somatic embryogenesis. I. Nursery development, fall acclimation and over-winter storage. Can. J. For. Res. 24:1376-1384.

Grossnickle, S.C. and J.E. Major. 1994. Interior spruce seedlings compared with emblings produced from somatic embryogenesis. II. Stroke quality assessment prior to field planting. Can. J. For. Res. 24:1385-1396.

Grossnickle, S.C. and J.E. Major. 1994. Interior spruce seedlings compared with emblings produced from somatic embryogenesis. III. Physiological response and morphological development in a reforestation site. Can. J. For. Res. 24:1397-1407.

Hakman, I., L.C. Fowke, S. von Arnold and T. Eriksson. 1985. The development of somatic embryos in tissue cultures initiated from immature embryos of *Picea abies* (Norway spruce). Plant Sci. 38:53-59.

Hakman, I. and L.C. Fowke. 1987. Somatic embryogenesis in *Picea glauca* (white spruce) and *Picea mariana* (black spruce). Can. J. Bot. 65:656-659.

Harry, I.S. and T.A. Thorpe. 1991. Somatic embryogenesis and plant regeneration from mature zygotic embryos of red spruce. Bot. Gaz. 152:446-452.

Kartha, K.K., L.C. Fowke, N.L. Leung, K.L. Caswell and I. Hakman. 1988. Induction of somatic embryos and plantlets from cryopreserved cell cultures of white spruce (*Picea glauca*). J. Plant Physiol. 132:529-539.

Klimaszewska, K., C. Ward and W.M. Cheliak. 1992. Cryopreservation and plant regeneration from embryogenic cultures of larch (*Larix* x *eurolepis*) and black spruce (*Picea mariana*). J. Exp. Bot. 43:73-79.

Kong, L. and E.C. Yeung. 1992. Development of white spruce somatic embryos. II. Continual shoot meristem development during germination. In Vitro Cell. Dev. Biol. 28P:125-131.

Krogstrup, P. 1990. Effect of culture densities on cell proliferation and regeneration from embryogenic cell suspensions of *Picea sitchensis* Plant Sci. 72:115-123.

Krogstrup, P., E.N. Eriksen, J.D. Møller and H. Roulund. 1988. Somatic embryogenesis in sitka spruce (*Picea sitchensis* (Bong.) Carr.). Plant Cell Rep. 7:594-597.

Lelu, M.-A.P. and C.H. Bornman. 1990. Induction of somatic embryogenesis in excised cotyledons of *Picea glauca* and *Picea mariana*. Plant Physiol. Biochem. 28:785-791.

Litvay, J.D., M.A. Johnson, D. Verma, D. Einspahr and K. Weyrauch. 1981. Conifer suspension culture medium development using analytical data from developing seeds. Institute of Paper Chemistry, IPC Technical Paper 115.

Lu, C.-Y. and T.A. Thorpe. 1987. Somatic embryogenesis and plantlet regeneration in cultured immature embryos of *Picea glauca*. J. Plant Physiol. 128:297-302.

Lu, C.-Y. and T.A. Thorpe. 1988. Shoot bud regeneration in subcultured callus of Engelmann spruce. In Vitro Cell Dev. Biol. 24:239-242.

Lu, C.-Y., I.S. Harry, M.R. Thompson and T.A. Thorpe. 1991. Plantlet regeneration from cultured embryos and seedling parts of red spruce (*Picea rubens* Sarg.). Bot. Gaz. 152:42-50.

Lulsdorf, M.M., T.E. Tautorus, S.I. Kikcio, T.D. Bethune and D.I. Dunstan. 1993. Germination of encapsulated embryos of interior spruce (*Picea glauca engelmanii* complex) and black spruce (*Picea mariana* Mill.). Plant Cell Rep. 12:385-389.

Misra, S., S.M. Attree, I. Leal and L.C. Fowke. 1993. Effect of abscisic acid, osmoticum, and desiccation on synthesis of storage proteins during the development of white spruce somatic embryos. Ann. Bot. 71:11-12.

Mohammed, G.H. and W.E. Vidaver. 1988. Root production and plantlet development in tissue-cultured conifers. Plant Cell Tiss. Organ Cult. 14:137-160.

Mohammed, G.H., D.I. Dunstan and T.A. Thorpe. 1986. Influence of nutrient medium upon shoot initiation on vegetative explants excised from 15- to 18-year-old *Picea glauca*. N.Z. J. For. Science 16:297-305.

Murashige, T. and F. Skoog. 1962. A revised medium for rapid growth and bioassays with tobacco tissue cultures. Physiol. Plant 15:473-497.

Patel, K.R. and T.A. Thorpe. 1986. *In vitro* regeneration of plantlets from embryonic and seedlings explants of Engelmann spruce (*Picea engelmanii* Parry). Tree Physiol. 1:289-301.

Roberts, D.R., B.C.S. Sutton and B.S. Flinn. 1990. Synchronous and high frequency germination of interior spruce somatic embryos following partial drying of high relative humidity. Can. J. Bot. 68:1086-1090.

Roberts, D.R., W.R. Lazaroff and F.B. Webster. 1991. Interaction between maturation and high relative humidity treatments and their effects on germination of sitka spruce somatic embryos. J. Plant Physiol. 138:1-6.

Roberts, D.R., F.B. Webster, B.S. Flinn, W.R. Lazaroff and D.R. Cyr. 1993. Somatic embryogenesis of spruce. In: Synseeds (K. Redenbaugh, ed.), CRC Press, Boca Raton, USA. pp. 428-450.

Rumary, C. and T.A. Thorpe. 1984. Plantlet formation in black and white spruce. I. *In vitro* techniques. Can. J. For. Res. 14:10-16.

Schenk, R.V. and A.C. Hildebrant. 1972. Medium and techniques for induction of monocotyledonous and dicotyledonous plant cell cultures. Can. J. Bot. 50:199-204.

Stasolla, C. and E.C. Yeung. 1999. Ascorbic acid improves conversion of white spruce somatic embryos. In Vitro Cell Dev. Biol. Plant 35:316-319.

Tautorus, T.E., S.M. Attree, L.C. Fowke and D.I. Dunstan. 1990. Somatic embryogenesis from immature and mature zygotic embryos, and embryo regeneration from protoplasts in black spruce (*Picea mariana* Mill.). Plant Sci. 67:115-124.

Tautorus, T.E., M.M. Lulsdorf, S.I. Kikcio and D.I. Dunstan. 1992. Bioreactor culture of *Picea mariana* Mill. (black spruce) and the species complex *Picea glauca-engelmanii* (interior spruce) somatic embryos. Growth Parameters. App. Microbiol. Biotechnol. 38:46-51.

Tautorus, T.E., M.M. Lulsdorf, S.I. Kikcio and D.I. Dunstan. 1994. Nutrient utilization during bioreactor culture, and evaluation of maintenance regime on maturation of somatic embryos of *Picea mariana* and the species complex *Picea glauca-engelmanii*. In Vitro Cell Dev. Biol. 30P:58-63.

Thorpe, T.A. and S. Biondi. 1984. Conifers. In: Handbook of Plant Cell Culture, Vol. 2 (WL Sharp, DA Evans, PV Ammirato, Y Yamada, eds.), Macmillan, New York, USA, pp. 435-470.

Thorpe, T.A. and I.S. Harry. 1991. Clonal propagation of conifers. Plant Tissue Culture Manual (K Lindsey, ed.), Kluwer Acad. Publ., Dordrecht, The Netherlands, C3:1-16.

Thorpe, T.A. and K.R. Patel. 1984. Clonal propagation: Adventitious buds. In: Cell Culture and Somatic Cell Genetics in Plants, Vol. 1. (IK Vasil, ed.), Acad. Press, New York, USA, pp 49-60.

Thorpe, T.A., I.S. Harry and P.P. Kumar. 1990. Application of micropropagation to forestry. In: Micropropagation: Technology and Application (P Debergh and RH Zimmerman, eds.), Kluwer Acad. Publ., Dordrecht, The Netherlands, pp. 311-336.

Toivonen, P.M.A. and K.K. Kartha. 1988. Regeneration of plantlets from *in vitro* cultured cotyledons of white spruce (*Picea glauca* (Moench.) Voss). Plant Cell Rep. 7:318-321.

Tremblay, F.M. 1990. Somatic embryogenesis and plantlet regeneration from embryos isolated from stored seeds of *Picea glauca*. Can. J. Bot. 68:236-242.

Tremblay, L. and F.M. Tremblay. 1991. Effects of gelling agents, ammonium nitrate, and light on the development of *Picea mariana* (Mill) B.S.P. (black spruce) and *Picea rubens* Sarg. (red spruce) somatic embryos. Plant Sci. 77:233-242.

von Arnold, S. and T. Ericksson. 1981. *In vitro* studies of adventitious shoot formation in *Pinus contorta*. Can. J. Bot. 59:870-874.

von Arnold, S. and S. Woodward. 1988. Organogenesis and embryogenesis in mature zygotic embryos of *Picea sitchensis*. Tree Physiol. 4:291-300.

Webb, D.T., F. Webster, B.S. Flinn, D.R. Roberts and D.D. Ellis. 1989. Factors influencing the induction of embryogenic and caulogenic callus from embryos of *Picea glauca* and *P. engelmanii*. Can. J. For. Res. 19:1303-1308.

Webster, F.B., D.R. Roberts, S.M. McInnis and B.C.S. Sutton. 1990. Propagation of interior spruce by somatic embryogenesis. Can. J. For. Res. 20:1759-1765.

# IN VITRO CULTURE OF JAPANESE BLACK PINE (*PINUS THUNBERGII*)

Katsuaki Ishii and Emilio Maruyama
Forestry and Forest Products Research Institute, P.O. Box 16, Tsukuba
Norinkenkyudanchi-nai, Ibaraki, Japan 305-8687. E-mail: katsuaki@ffpri.affrc.go.jp

**Abstract.** In vitro culture from mature embryos of Japanese black pine (*Pinus thunbergii*) plus trees was developed. For initial adventitious buds induction, Wolter and Skoog (1966) medium with 3 mg/l BAP was the best among screened media. For elongation of shoots, a half strength Lepoivre (LP) medium containing 5 g/l activated charcoal was the best. Rooting percentage was 0-50 % on the modified Root Induction Medium of Abo El-Nil (RIM). However there were big clonal differences in rooting percentages. Habituated plantlets were potted out to the field and growth performance was checked until 8 years. In vitro plantlet regeneration from 6-7 years old black pine which were selected as resistant clones using pine wood nematode inoculation test was tried. Induced brachyblast buds obtained by topping the trees were the best for explants and the half-strength LP medium containing 2.25 mg/l BAP and half-strength LP medium containing 5 g/l activated charcoal were used in turn repeatedly for subculture in about 1-2 months interval. After subculture of 2 years, more than 5000 shoots were obtained from one clone. Shoots rooted in RIM containing IBA. Somatic embryogenic cells were induced from immature embryos collected from the end of June to mid July in Tsukuba, Japan. The suitable media for induction, maturation and germination of somatic embryos were different in the constituents. For induction, modified 1/2LP or Smith (1996) medium containing BAP and 2,4-D was used. For maturation , EM-3 medium (original) was added with ABA and PEG. Maltose instead of sucrose was effective for somatic embryo maturation. For germination of the matured somatic embryos, hormone free medium was used with or without activated charcoal powder. Regenerated plantlets were habituated and potted out in the vermiculite.

**Key index words.** micropropagation, nematode resistant pine, shoot culture, somatic embryogenesis.

## 1. Introduction

Damage to pine trees caused by pine wood nematodes introduced from abroad is still posing a serious problem in Japan. There are about 100 pine trees selected across the country as resistant to pine wood nematode. However the clonal propagation of pines is usually difficult by traditional cuttings. So the attempt to *in vitro* culture of Japanese black pine (*Pinus thunbergii*) which is one of domestic pines was carried out. Somatic embryogenesis is an attractive in vitro culture technique for both micropropagation and genetic transformation. The production of somatic embryo using bioreactors and encapsulated somatic embryos will offer high efficiency in pine production.

## 2. Materials and Methods

Mature seeds of black pine plus tree Kani No. 1 and nematode resistant clone Tanabe No.54 were used for shoot culture. They were surface sterilized and cultured *in vitro* according to the same procedure of Ishii (1988). Habituated plantlets were potted out then transferred to the field and growth performance was checked until 8 years. Six



*C. Kubota and C. Chun (eds.), Transplant Production in the 21st Century, 205–208.*

to seven years old Japanese black pine which survived nematode inoculation test was topped in May, newly induced brachybrasts were collected in July and surface sterilized using benzalkonium chloride, ethyl alcohol and hydrogen peroxide sequentially and cultured *in vitro* as described by Ishii (1989). Somatic embryogenic cells were induced from immature embryos collected from the adult trees at the end of June to mid July in Tsukuba, Japan. For induction, modified 1/2LP or Smith's medium containing BAP and 2,4-D was used. For maturation , EM-3 medium (Table 1) containing ABA and PEG was used. For germination of the matured somatic embryos, hormone free medium was used with or without activated charcoal powder. Regenerated plantlets were habituated and potted out in the vermiculite.

*Table 1* Composition of EM-3 medium (mg/l).

| | | | | | |
|---|---|---|---|---|---|
| $KNO_3$ | 1,000 | $CuSO_4.5H_2O$ | 2.4 | Activated Charcoal | 2,000 |
| $NaNO_3$ | 60 | $Na_2MoO_4.2H_2O$ | 0.2 | Gelrite | 5,000 |
| $KH_2PO_4$ | 70 | $CoCl_2.6H_2O$ | 0.2 | Glutamine | 1,000 |
| $MgSO_4.7H_2O$ | 500 | $FeSO_4.7H_2O$ | 30 | Asparagine | 500 |
| $CaCl_2.2H_2O$ | 75 | $Na_2EDTA$ | 40 | Arginine | 250 |
| $Ca(NO_3)_2.4H_2O$ | 60 | Maltose | 50,000 | Citrulline | 20 |
| $NaH_2PO_4.2H_2O$ | 160 | myo-inositol | 1,000 | Ornithine | 20 |
| KCl | 750 | Thiamin | 5 | Lysine | 15 |
| $MnSO_4.4H_2O$ | 20 | Pyridoxine | 0.5 | Alanine | 10 |
| $H_3BO_3$ | 40 | Nicotinic acid | 5 | Proline | 10 |
| $ZnSO_4.7H_2O$ | 25 | Glycine | 5 | ABA | 25.5 |
| KI | 1 | PEG | 75,000 | | |

## 3. Results and Discussion

### 3.1 Shoot culture from mature embryos

For initial adventitious buds induction, Wolter and Skoog's medium (1966) with 3 mg/l BAP was the best among screened media (Ishii, 1993). For elongation of shoots, a half strength LP medium (Aitken- Christie and Thorpe, 1984) containing 5 g/l activated charcoal was the best. For propagation of shoots, culture on the medium with BAP for 1 month then without BAP but with activated charcoal medium for 1 to 2 months is preferable. With this BAP containing and then BAP free media cycle culture, more than 4000 shoots were obtained during 2 years with one clone. Rooting percentage was 0-50 % on the modified Root Induction Medium (RIM) of Abo El-Nil (Abo El-Nil and Milton, 1982). However there were big clonal differences in rooting percentages (14-50%). Habituated and field planted trees grew in normal healthy shape at least 8 years.

### 3.2 Shoot culture from 6-7 years clone resistant to nematode

Induced brachyblast buds obtained by topping the trees were the best for explants and the half-strength LP medium containing 2.25 mg/l BAP and half-strength LP medium containing 5 g/l activated charcoal were used in turn repeatedly for subculture in about 1-2 months interval. After subculture of 2 years, more than 5000 shoots were obtained from one clone. About 20% shoots rooted in RIM containing IBA. However, after 7 years subculture, the rooting ability of the shoots were dramatically reduced nearly to nil. RAPD PCR did not indicate any DNA level aberration (Goto *et al.* 1998). We expect the possibility of accumulation of BAP in the tissue may affect the physiological state of the shoot for rooting.

### 3.3 Somatic embryogenesis and embrings regeneration

Induction of somatic embryo from immature embryos of black pine was successful. Original EM-3 medium (Table 1) was good for proliferation and maturation. Addition of maltose instead of sucrose was effective for somatic embryo maturation (Table 2). For germination of the matured somatic embryos, hormone free medium with or without activated charcoal was good. Regenerated plantlets were habituated and potted out in the vermiculite (Figure 1).

*Table 2* Effect of sugar kind on somatic embryo maturation in Japanese black pine.

| Sugar kind | Number of mature somatic embryos / petri dish |
|---|---|
| Sucrose | 64 |
| Maltose | 150 |

Maturation media contain 30 g/l sugar, 100 $\mu$M ABA, and 0.2 % activated charcoal.

*Figure 1* Regenerated plants from somatic embryos of Japanese black pine.

## 4. Conclusion

In vitro culture regeneration of Japanese black pine was possible by using shoot culture from mature embryo and 6-7 years old nematode resistant clone. New medium (EM-3) was developed for somatic embryo maturation of Japanese black pine and plant regeneration was achieved through somatic embryogenesis from immature embryo.

**Acknowledgment.** This research was supported in part by a Grant for Research for the Future Program from the Japan Society for Promotion of Science(JSPS-RFTF 96L00603). E.M. is grateful for the STA Fellowship from Science and Technology Agency of Japan.

**References**


Abo El-Nil, A.A. and W. Milton. 1982. Method for asexual reproduction of coniferous trees. United States Patent No. 4353186.

Aitken-Christie, J. and T.A. Thorpe. 1984. Clonal propagation: Gymnosperms. In I.K. Vasil (ed.) Cell Culture and Somatic Cell Genetics of Plants Vol. 1, Academic Press, Inc., London, New York, San Francisco. pp. 82-95.

Goto, S., R. Thakur and K. Ishii. 1998. Detection of genetic stability in long-term micropropagated shoots of *Pinus thunbergii* Parl. using RAPD markers. Plant Cell Reports. 18:193-197.

Ishii, K 1988. In vitro plantlet formation from mature embryo of Japanese Black Pine (*Pinus thungergii*). J. Jpn. For. Soc. 70:278-282.

Ishii, K. 1989. In vitro plantlet regeneration in some species of Japanese conifers. Proc. of the 6th International Congr. of SABRAO (1989). 869-872.

Ishii, K. 1993. Screening of tissue culture conditions of Hinoki cypress and Japanese black pine. Bulletin of the Forestry and Forest Products Research Institute 365:131-167.

Smith, D.R. 1996. Growth medium. United States Patent No. 5565355.

Wolter, K.E. and F. Skoog, 1966. Nutrient requirements of *Fraxinus* callus cultures. Am. J. Bot. 53:236-269

# CONTROL OF THE DEVELOPMENT OF SOMATIC EMBRYO OF JAPANESE CONIFERS BY THE DENSITY OF EMBRYOGENIC CELLS IN LIQUID CULTURE

Shinjiro Ogita[1], Hamako Sasamoto[1] and Takafumi Kubo[2]
[1]Cell Manipulation Laboratory, Bio-resources Technology Division, Forestry and Forest Products Research Institute, Ibaraki, 305-8687 Japan. E-mail: s_ogita@hotmail.com
[2]Department of Environmental and Natural Resource Science, Faculty of Agriculture, Tokyo University of Agriculture and Technology, Tokyo, 183-8509 Japan.

**Abstract.** The morphogenetic changes of embryogenic tissue at different cell densities were investigated in Japanese conifers, i.e. *Larix leptolepis*, *Picea jezoensis*, and *Cryptomeria japonica*. We found that cell lines generated and maintained in liquid suspensions could take on different morphologies and requirements for somatic embryo development and maturation. Cell density within liquid suspension played an important role in embryogenic tissue proliferation and development. We found that in a high cell density culture of *L. leptolepis*, the embryogenic tissue continued to proliferate as small spherical cell aggregates; however, no further development occurred. On the other hand, in a low-density cell culture, development of embryogenic tissue into somatic embryos were promoted. In addition, we have established a micro-culture method for culturing single aggregates of embryogenic tissue of all three species. Individual aggregates could be identified and picked up using a micro-manipulator and cultured in 50 μl liquid medium using a 96-well culture plate. In all three species studied, a majority of the cell aggregates actively proliferated within the wells. Maturation of somatic embryos was successful when the newly proliferated cell aggregates were transferred to solid ABA-containing medium.

**Key index words.** cell density, *Cryptomeria japonica*, embryogenic tissue, *Larix leptolepis*, micro-manipulation, *Picea jezoensis*, suspension culture.

## 1. Introduction

In recent years, many successes have been reported using the somatic embryogenesis technique in the micropropagation of gymnosperms, especially *Picea* and *Larix* species (Dunstan et al., 1995). In order to improve the yield of somatic embryos for various types of studies and the potential for scale-up production, liquid suspension culture techniques are generally recommended. Recently, with the development of liquid suspension protocols for the maintenance of embryogenic tissue of conifer species (Find et al., 1998; Krogstrup, 1990; Lulsdorf et al., 1992), it becomes possible to study the whole process of somatic embryo development using different methods such as histological, physiological and molecular biological procedures. However, many intricate changes occurring within embryogenic tissues leading to embryo formation remain unclear, especially for the Japanese conifer species. The main objective of our research is to determine how embryogenic tissue proliferate and differentiate within the liquid environment and how these processes are regulated. In this study, we have identified some morphological characteristics



*C. Kubota and C. Chun (eds.), Transplant Production in the 21st Century,* 209–214.

of embryogenic aggregates from three Japanese conifer species, i.e. *Picea jezoensis*, *Larix leptolepis* and *Cryptomeria japonica* and the conditions of cultures that are important to somatic embryo production using the liquid suspension culture method.

## 2. Materials and Methods

### 2.1 Embryogenic tissue induction and maintenance

Embryogenic tissues of all three species, *P. jezoensis*, *L. leptolepis* and *C. japonica* were induced and maintained on the solid LP medium (Aitken-Christie and Thorpe, 1984), the modified Campbell and Durzan (mCD, 1975) medium and the mCD medium respectively. Liquid suspensions of the embryogenic tissues were also generated. The methods were detailed in previous reports (Ogita et al., 1999a, d).

### 2.2 Measurement and adjustment of cell densities

In order to investigate the effects of different cell densities on the proliferation and development of embryogenic tissue, different volumes of cells (0.01 - 1 ml packed cell volume (PCV)) were added to a 10 ml liquid mCD medium, at either light or dark condition. The flasks were then placed on a rotatory shaker with a speed of 100 rpm. The increase in PCV were measured after a 4 week culture period. The PCV was determined by placing embryogenic tissue in conical tubes and centrifuged using a centrifuge (5400, KUBOTA, Tokyo) at 500 rpm for 3 minutes. The experiments were repeated at least three times with triplicate samples.

### 2.3 Selection and micro-culture of a single embryogenic aggregate

An inverted microscope (IMT-2, OLYMPUS, Tokyo) with a micro-manipulator (M-152, NARISHIGE) and a micro-injector (IM-5B, NARISHIGE, Tokyo) were used for cell manipulation. Small embryogenic aggregates were picked up using such a cell manipulation system and placed into a well of a 96-well tissue culture plate (3072, FALCON, Becton Dickinson & Co., Sparks, MD) containing 50 μl of liquid medium per well. The cultures were incubated in an incubator (CPD-1701, Hirasawa Tokyo) at 28℃ in the dark (Ogita et al., 1999d).

### 2.4 Maturation of somatic embryos and plant recovery

Maturation of somatic embryos and plant recovery from embryogenic tissues of *P. jezoensis* and *L. leptolepis* were carried out on the LP and mCD media supplemented with 0, 0.1, 1, 10 and 50 μM of ABA and 30 g $l^{-1}$ sucrose (Ogita et al., 1999b, c).

## 3. Results and Discussion

### 3.1 Effects of abscisic acid on somatic embryo maturation in relation to the morphology of embryogenic tissues

Abscisic acid (ABA) has been found to be one of the most important factors for conifer somatic embryo maturation (Gupta and Grob, 1995). Different concentrations of ABA are needed for different conifer species (Tautorus et al., 1991). In our studies, we found that different cell lines generated from the same species can also have different ABA requirements for maturation. One of the cell lines of *L. leptolepis* had only small embryogenic aggregates and was white in color, and the other line had large embryogenic aggregates and was red in color. As shown in a previous paper (Ogita et al., 1999c), a high concentration (50 μM) of ABA was effective in rearing the embryogenic aggregates of the "white" cell line into mature somatic embryos and a low concentration (0.1 μM) was sufficient for embryo maturation in the "red" cell line.

In normal practice, once an optimal concentration of ABA is determined for a given species, the amount of ABA used will be "fixed" for any future experiments. However, our data from two cell lines of *L. leptolepis* clearly showed a necessity to optimize the concentrations of plant growth regulators used in relation to the morphology of embryogenic tissues. Based on these results, we have generated and maintained embryogenic suspension cell lines with reproducible morphological and developmental characteristics. These stable cell lines enable us to study the proliferation and developmental processes of embryogenic aggregates in liquid suspension cultures.

### 3.2 Liquid suspension culture of Japanese conifer species

#### 3.2.1 Effect of cell densities on proliferation and development of embryogenic aggregates

The cell density used in a given culture system can have dramatic effects on cell proliferation (Bhojwani and Razdan, 1996). In *L. leptolepis* cell density also plays an important role in regulating the developmental pattern of the embryogenic tissue. In the maintenance cultures, the embryogenic tissue was consisted of cell aggregates of different sizes. Small, dense cytoplasm-rich cells (20 - 40 μm in diameter) were found to associate with a number of elongated, vacuolated cells (50 - 200 μm in length). When the maintenance cultures were transferred into medium with different cell densities, the embryogenic tissue continued to proliferate and dramatic morphological changes could be seen. In a high-density culture (0.5 - 1 ml PCV/ 10 ml culture medium), the embryogenic tissue continued to proliferate with an increase in 4.3 - 5.3 ml PCV after a 4 week culture period; however, no further development occurred (Table 1).

*Table 1* Effects of cell densities in the multiplication of embryogenic tissue of *L. leptolepis*.

| Cell density[Z] at the time of culture (PCV) | Cell density after 4 weeks of culture (PCV) | Morphology after 4 weeks of culture |
|---|---|---|
| 0.1% (0.01 ml) | 10.3% (1.03 ml) | developed[Y] |
| 0.5% (0.05 ml) | 15.7% (1.57 ml) | developed |
| 1% (0.1 ml) | 18.7% (1.87 ml) | developed |
| 5% (0.5 ml) | 43.3% (4.33 ml) | small and spherical[X] |
| 10% (1.0 ml) | 52.7% (5.27 ml) | small and spherical |

[Z]Cell density represents PCV in 10 ml of liquid medium. One ml of PVC contains approximately 1.5 – 2.0 x $10^4$ cell clusters.
[Y] See *Fig. 1* (b) for further information.
[X] See *Fig. 1* (a) for further information.

The embryogenic tissue appeared as small aggregates and were very uniform in shape and size. Few elongated cells were present in association with the cytoplasm-rich cells that made up the aggregates. The vacuolated cells when present were not as elongated (50 -100 μm in diameter) as those found in low-density cultures. Thus, the embryogenic aggregates tended to take on a spherical shape (Fig. 1 a). At low cell culture densities (0.01 - 0.1 ml PVC/ 10 ml culture medium), an increase in PCV was also observed as in the high-density cultures. The final PCV values increased approximately 19 - 100 times after 4 weeks of culture. However, it was noted that the embryogenic tissue had multiplied and continued to develop with well organized embryonal head regions and elongated suspensor cells (Fig. 1 b). It is clear that dramatic morphological changes had occurred between high and low density cultures. Similar cell density effects were observed in *P. jezoensis* (Ogita et al., 1999b) and *C. japonica* (unpublished results) suspension cells. Furthermore, the embryogenic tissue with well organized embryonal heads and suspensors would mature when transferred to the medium containing ABA and they were able to germinate when transferred to a solid medium.

Cell density is one of the important factors which needs to be optimized for a given culture. In *Picea sitchensis*, a 20 % cell density (as a value of sedimented cell volume (SCV)) appears to be the optimum concentration (Krogstrup, 1990). In our study, low cell densities favor further development of the embryogenic tissue into embryonal-suspensor cell masses, while high densities result in the formation of small spherical cell aggregates. In high-density cultures, rapid utilization of medium components might have resulted in some components becoming limiting factors for further growth and development. Experiments are currently underway to identify the essential components that regulate embryogenic tissue morphogenesis.

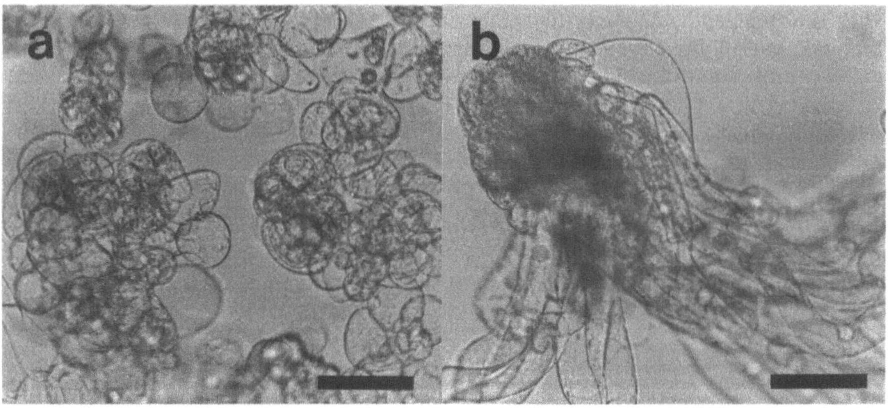


*Fig. 1* Morphology of embryogenic tissue after 4 weeks of culture in *L. leptolepis*. (a) Small spherical aggregates proliferated in a high-density culture. (b) An embryogenic tissue having a well organized embryonal head region and an elongated suspensor region formed in a low-density culture . Scale bars represent 100 μm.

### 3.2.2 A potential application of the cell density effect

One of the potential applications of the observed cell density effect is the generation of large batch cultures of embryogenic aggregates with uniform morphology and developmental stage. As shown in Fig. 2, by culturing small spherical aggregates of *L. leptolepis* at high cell densities (preferably 5 - 10 %)

approximately every 3 - 4 weeks, they continued to proliferate while maintaining their morphology, i.e. small cytoplasm rich cells with the rounded, vacuolated cells. Once the cell density was lowered, further uniform development would take place. This simple manipulation enables us to generate uniform embryogenic aggregates for different experimental studies.

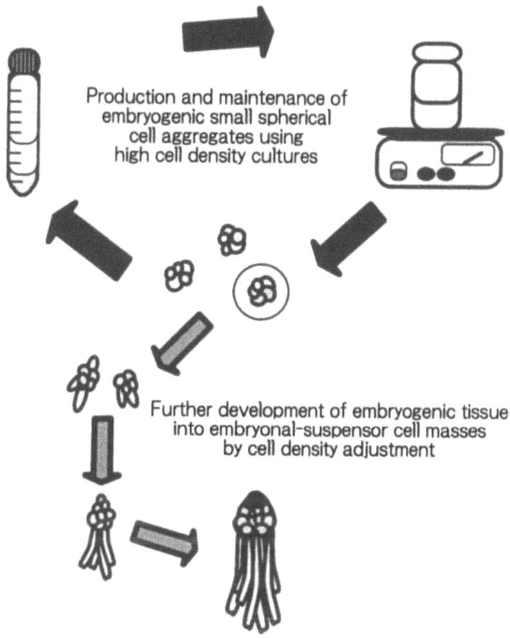


*Fig. 2* An outline of the high - low cell density suspension culture method for the production of uniform somatic embryos in conifer species.

### 3.3 Micro-culture of single embryogenic cell aggregate using micro-manipulation technique

Single embryogenic aggregates of all three species could be easily picked up by using a micro-manipulation technique. After transferring to 50 μl liquid medium in a well of 96-well culture plate, a majority of embryogenic aggregates (more than 80 % in tested) continued to divide and proliferate (Ogita et al., 1999d). Maturation of somatic embryos was successful when the newly proliferated embryogenic tissue was transferred to solid ABA-containing medium.

Furthermore, such a micro-culture method could apply to the high - low cell density suspension cultures. The fact that all the three conifer cell lines developed in our laboratory can be maintained at high cell density and grown at low cell density permits us to develop such a micro-culture method. The advantage of the micro-method is that a single embryogenic aggregate can be propagated to provide a genetically homogenous cell line. If different morphological variants can be identified and cultured separately from one another, it will provide an additional tool for embryogenic cell selection. In conjunction with micro-methods currently developed for amino acid and hormonal analyses in our laboratory, a better insight into the morphogenetic process within the liquid suspension system could be obtained in the near future.

**References**

Aitken-Christie, J. and T.A. Thorpe. 1984. Clonal propagation: Gymnosperms. In: I.K. Vasil (ed.), Cell Culture and Somatic Cell Genetics of Plants, Vol. 1. Academic Press, Orlando. pp. 82-95.

Bhojwani, S.S. and M.K. Razdan. 1996. Cell culture. In: Plant Tissue Culture: Theory and practice, a revised edition. Elsevier, Amsterdam. pp. 63-93.

Campbell, R.A. and D.J. Durzan. 1975. Induction of multiple buds and needles in tissue cultures of *Picea glauca*. Can. J. Bot. 53:1652-1657.

Dunstan, D.I., T.E. Tautorus and T.A. Thorpe. 1995. Somatic embryogenesis in woody plants. In: T.A. Thorpe (ed.), In Vitro Embryogenesis in Plants. Kluwer Academic Publishers, Dordrecht. pp. 471-538.

Find, J.I., J.V. Norgaard and P. Krogstrup. 1998. Growth parameters, nutrient uptake and maturation capacity of two cell-lines of Norway spruce (*Picea abies*) in suspension culture. J. Plant Physiol. 152:510-517.

Gupta, P.K. and J.A. Grob. 1995. Somatic embryogenesis in conifers. In: S.M. Jain, P.K. Gupta and R.J. Newton (eds.), Somatic Embryogenesis in Woody Plants. Vol. 1. Kluwer Academic Publishers, Dordrecht. pp. 81-98.

Krogstrup, P. 1990. Effect of culture densities on proliferation and regeneration from embryogenic cell suspensions of *Picea sitchensis*. Plant Sci. 72:115-123.

Lulsdorf, M.M., T.E. Tautorus, S.I. Kikcio and D.I. Dunstan. 1992. Growth parameters of embryogenic suspension cultures of interior spruce (*Picea glauca-engelmannii* complex) and black spruce (*Picea mariana* Mill). Plant Sci. 82:227-234.

Ogita, S., H. Ishikawa, H. Sasamoto and T. Kubo. 1999a. Somatic embryogenesis from immature and mature zygotic embryos of *Cryptomeria japonica*: embryogenic cell induction and its morphological characteristics. J. Wood Sci. 45:87-91.

Ogita, S., H. Sasamoto and T. Kubo. 1999b. High frequent maturation of somatic embryos from embryogenic suspension cultures of *Larix leptolepis* and *Picea jezoensis*. 110$^{th}$ Mtg. Trans. Jap. For. Soc. 254-255. (in Japanese).

Ogita, S., H. Sasamoto and T. Kubo. 1999c. Maturation and plant recovery from embryogenic cells of Japanese Larch: Effect of abscisic acid in relation to their morphology. J. For. Res. 4:241-244.

Ogita, S. and H. Sasamoto and T. Kubo. 1999d. Selection and micro-culture of single embryogenic cell clusters in Japanese conifers: *Picea jezoensis*, *Larix leptolepis* and *Cryptomeria japonica*. In Vitro Cell. Devel. Biol. Plant 35:428-431.

Tautorus, T.E., L.C. Fowke and D.I. Dunstan. 1991. Somatic embryogenesis in conifers. Can. J. Bot. 69:1873-1899.

# A PRELIMINARY EXPERIMENT ON PHOTOAUTOTROPHIC MICROPROPAGATION OF *RHODODENDRON*

Carmen Valero-Aracama, Sayed M.A. Zobayed and Toyoki Kozai
Faculty of Horticulture, Chiba University, Matsudo, Chiba 271-8510, Japan. E-mail: 99um8301@green.h.chiba-u.ac.jp

**Abstract.** In the present study, nodal leafy cuttings of *Rhododendron* "Hatsuyuki" were cultured for 45 days on sugar-free medium under $CO_2$-enriched and relatively high PPF conditions to enhance growth and multiplication rates of plantlets and to obtain quality transplants ready to be transferred to *ex vitro* conditions. Effects of medium strength (quarter and full) and presence/absence of indole-3-butyric acid were investigated on the growth of plantlets in sugar-free micropropagation. Growth was significantly greater in sugar-free treatments than that in the control, which is the conventional micropropagation with sugar-containing medium under $CO_2$ non-enriched and low PPF conditions, in all measured variables. Among the sugar-free treatments, growth was greatest on plantlets cultured on full strength nutrient solution and absence of indole-3-butyric acid.

**Key index words.** Anderson Rhododendron medium, conventional micropropagation, IBA, medium strength, sugar-free micropropagation.

## 1. Introduction

When woody plants such as *Rhododendron* are conventionally micropropagated, the development sequence is generally divided into four stages (Murashige, 1974; Anderson, 1980) in which sugar, vitamins and growth regulators are generally required in different proportions for each stage. This is a slow process and acclimatization is required after transplanting to *ex vitro* conditions.

In sugar-free micropropagation, using leafy or chlorophyllous explants with photosynthetic ability, high net photosynthetic rates have been reported when $CO_2$ concentration in the culture room and photosynthetic photon flux (PPF) on the culture shelves are relatively high (Kozai, 1991a). Under such conditions, plantlets are forced to rely on their photosynthetic apparatus for their supply of carbohydrates rather than on sugar in the medium (Kozai et al., 1988) and thus, other components like growth regulators and vitamins, necessary for the development of plantlets, are synthesized endogenously.

In the present experiment, sugar-free micropropagation system is proposed as the means to produce high quality *Rhododendron* "Hatsuyuki" plantlets in a short and simplified stage without the necessity of *ex vitro* acclimatization in contrast with the conventional system.

Anderson Rhododendron Medium (Anderson, 1984), characterized by low inorganic nutrient concentration is used for conventional micropropagation of *Rhododendron* and other plants of the family Ericaceae (Ma and Wang 1977; Anderson, 1984; Economou and Read, 1986). Medium strength is full for multiplication stage and one quarter for elongation and root induction stage in conventional micropropagation (Norton and Norton, 1989). For simplification of the micropropagation processes, the effects of medium strength and the necessity of IBA in the sugar-free medium were investigated.



*C. Kubota and C. Chun (eds.), Transplant Production in the 21st Century,* 215–218.

## 2. Materials and Methods

### 2.1 Experimental Set up

Nodal cuttings, each with two nodes and two leaves, excised from photoautotrophically grown *in vitro Rhododendron* 'Hatsuyuki' plantlets were transplanted and cultured in Magenta-type vessels (370 mL; Verde Co., Ltd., Japan) for 45 days. Florialite (vermiculite and cellulose fiber mixture; Nisshinbo Industries Inc., Japan) was used as supporting material imbibed with 50 mL of full or quarter strength Anderson Rhododendron nutrient solution (Anderson, 1984). The pH of the medium was adjusted to 4.5 before autoclaving.

There were four sugar-free treatments; two of them with quarter strength medium and absence or presence of IBA (MA and MP treatments, respectively, hereafter) and other two with full strength medium and absence or presence of IBA (FA and FP, respectively, hereafter). In the sugar-free treatments, the vessels were estimated to have 4.5 air exchanges per hour, measured according to the method described by Kozai et al. (1986), and organic substances like growth regulators and vitamins (*myo*-inositol, thiamine hydrochloride, adenine sulphate $2H_2O$) were excluded from the original nutrient formula. Detailed description of the treatments is shown in Table 1.

The culture room was maintained at 26/24 °C air temperature at photo-/dark periods, respectively, 1500 µmol mol$^{-1}$ $CO_2$, 85% relative humidity, 16 h photoperiod provided with cool-white fluorescent lamps and 100 µmol· m$^{-2}$· s$^{-1}$ PPF on the culture shelves.

Vessels with 0.1 air exchanges per hour containing Florialite with 50 mL quarter strength Anderson's medium and 30 g L$^{-1}$ sucrose were used as control. They were placed under 400 µmol mol$^{-1}$ $CO_2$ and 40 µmol m$^{-2}$ s$^{-1}$ PPF on the culture shelf, maintaining the rest of the culture conditions the same as in the sugar-free treatments.

*Table 1* Description of treatments.

| Treat. Codes | Medium Strength | IBA (mg· L$^{-1}$) | PPF (µmol · m$^{-2}$ s$^{-1}$) | $CO_2$ conc. (µmol· mol$^{-1}$) | No. air exch. (h$^{-1}$) of the vessels | Sucrose in medium (g· L$^{-1}$) | Vita-mins |
|---|---|---|---|---|---|---|---|
| Control | 1/4 | 5 | 40 | 400 | 0.1 | 30 | P* |
| MA | 1/4 | 0 | 100 | 1500 | 4.5 | 0 | A* |
| MP | 1/4 | 5 | 100 | 1500 | 4.5 | 0 | A |
| FA | 1 | 0 | 100 | 1500 | 4.5 | 0 | A |
| FP | 1 | 5 | 100 | 1500 | 4.5 | 0 | A |

*P/A stands for presence/absence of vitamins in the medium, respectively.

After 45 days of culture, net photosynthetic rates per plantlet ($P_p$, µmol h$^{-1}$/ plantlet) were estimated using the following equation (Fujiwara et al., 1987):

$$P_p = k \cdot V \cdot N \cdot (C_{out} - C_{in})/n$$

where $k$ is a conversion factor of $CO_2$ from volume to moles (0.041 mol mL$^{-1}$ at 25°C), $V$ is the volume inside the vessel (mL), $N$ is the number of air exchanges per hour of the vessel (h$^{-1}$), $C_{in}$ and $C_{out}$ are, respectively, $CO_2$ concentration in the vessel and in the culture room (µmol· mol$^{-1}$) and $n$ is the number of plantlets per vessel.

Plantlets were harvested to obtain fresh and dry mass, leaf area and number of unfolded leaves per plantlet on day 45.

## 2.2 Data Analysis

The experimental design was a completely randomized design with four replications and four plantlets per replication. Relation between medium strength and necessity of IBA was tested by the analysis of variance (ANOVA). Each treatment was compared with the control by the *t*-test and significant difference was determined at $P \leq 0.01$.

## 3. Results and Discussion

Fresh and dry mass, leaf area, number of unfolded leaves and net photosynthetic rate were greater in all the sugar-free treatments than in the control (Table 2). Greatest growth in terms of all measured variables was obtained in the FA treatment (sugar-free full strength and absence of IBA) and followed by the FP treatment (sugar-free full strength and presence of IBA).

Medium strength significantly affected fresh and dry mass, leaf area and number of unfolded leaves of the plantlets. In general, plantles cultured on quarter strength medium had less growth and leaves were slightly yellowish compared with those of treatments with full strength medium. On the other hand, absence of IBA in the medium gave greater dry mass and number of unfolded leaves than when IBA was present.

Greater number of unfolded leaves and thus, higher multiplication rates, were obtained in each sugar-free treatment compared with the control. In addition, net photosynthetic rates were higher in sugar-free treatments than that in the control, in which leaves of plantlets became yellowish towards the end of the experiment, probably due to a deficiency of micro-nutrients in the plantlets, caused by low transpiration rates.

Under sugar-free conditions, plantlets use $CO_2$ from the air as unique carbon source to carry on photosynthesis and thus, assimilation of inorganic nutrients is considered important for this process. An insufficient amount of inorganic nutrients becomes a limiting factor for the growth of plantlets. Therefore, full strength Anderson's medium is required for hastening growth of plantlets.

Moreover, in sugar-free treatments, absence of IBA increased the dry mass and number of unfolded leaves per plantlet compared with those of the treatments with IBA in the medium. Hence, the exclusion of IBA from the Anderson's medium formula is proposed. Exogenous growth regulators are not always necessary to obtain a desired plant response. Instead, mobilization, production or breakdown of endogenous growth regulators could be realized through environmental control. For example, levels of endogenous growth regulators in the plantlets are influenced by the duration, quality and intensity of light (Kefeli, 1978) and by chemical environmental factors, such as macro- and micro-nutrients. Control over these variables would contribute to improve this sugar-free system.

Root induction was insignificant in all the treatments and in the control. Understanding the factors that govern root induction and development on sugar-free micropropagation is of theoretical and practical interest. Work is underway in this direction.

*Table 2* Effects of two levels of medium strength and presence/absence of IBA on the plantlet growth on day 45. Means ± SE per plantlet are shown.

| Treat. Codes | Fresh Mass (mg) | Dry Mass (mg) | Leaf Area ($mm^2$) | Number of Unfolded Leaves | NPR ($\mu mol\ h^{-1}$) |
|---|---|---|---|---|---|
| Control | 12.0 ± 4.1 | 4.8 ± 0.2 | 41.4 ± 4.6 | 4.9 ± 1.6 | 0.04 ± 0.00 |
| MA | 15.2 ± 1.3 $^{NS}$ | 6.1 ± 0.3 $^{*}$ | 59.7 ± 5.2 $^{*}$ | 6.8 ± 1.6 $^{*}$ | 1.92 ± 0.05 $^{*}$ |
| MP | 20.3 ± 2.6 $^{*}$ | 5.5 ± 1.9 $^{NS}$ | 65.7 ± 5.9 $^{*}$ | 6.6 ± 1.0 $^{*}$ | 0.58 ± 0.45 $^{NS}$ |
| FA | 10.9 ± 2.7 $^{*}$ | 8.0 ± 0.5 $^{*}$ | 102.5 ± 8.8 $^{*}$ | 12.6 ± 2.3 $^{*}$ | 0.41 ± 0.20 $^{NS}$ |
| FP | 25.8 ± 9.5 $^{*}$ | 6.1 ± 2.0 $^{NS}$ | 85.7 ± 11.4 $^{*}$ | 9.9 ± 3.6 $^{*}$ | 1.23 ± 0.55 $^{NS}$ |
| ANOVA$^{Z}$ | | | | | |
| Strength | ** | ** | ** | ** | NS |
| PGR | NS | ** | NS | * | NS |
| S x PGR | * | NS | NS | ** | NS |

$^{Z}$Significance among treatments; ANOVA = analysis of variance.

$^{NS, *, **}$ Nonsignificantly or significantly different from that of control at $P \leq 0.05$ or $P \leq 0.01$.

**References**


Anderson, W.C. 1975. Propagation of rhododendrons by tissue culture: Part 1. Development of a culture medium for multiplication of shoots. Comb. Proc. Int. Plant Propagators Soc. 25:129-135.

Anderson, W.C. 1980. Mass propagation by tissue culture: principles and techniques. Proceedings of the conference on nursery production of fruit plants through tissue culture-application and feasibility. U.S. Department of Agriculture. Science and Education Administration. ARR-NE-11:1-10.

Anderson, W.C. 1984. A revised tissue cultured medium for shoot multiplication of rhododendron. J. Amer. Soc. Hort. Sci. 109:343-347.

Economou, A. S. and P. E. Read 1986. Microcutting production from sequential reculturing of hardy deciduous azaleas. Proc. Plant Growth Reg. Soc. Am. 8:8-14.

Fujiwara, K., T. Kozai and I. Watanabe 1987. Fundamental studies on environments in plant tissue culture vessels. Measurements of carbon dioxide gas concentration in closed vessels containing tissue cultured plantlets and estimates of net photosynthetic rates of the plantlets. J. Agr. Meteorol. 43:21-30. (in Japanese)

Kefeli, VI. 1978. Effect of light on phytohormones in phytomorphogenesis. Natural plant growth inhibitors and phytohormones. Dr. W. Junk Publishers. The Hague.

Kozai, T., K. Fujiwara and I. Watanabe 1986. Fundamental studies on environments in plant tissue culture vessels. (2) Effects of stoppers and vessels on gas exchange rates between inside and outside of vessels closed with stoppers. J. Agr. Meteorol. 42:119-127. (in Japanese)

Kozai, T., Y. Koyama and I. Watanabe 1988. Multiplication of potato plantlets *in vitro* with sugar free medium under high photosynthetic photon flux. Acta Hort. 230:121-127.

Kozai, T. 1991. Autotrophic micropropagation. In Y. P. S. Bajaj (ed.). Biotechnology in agriculture and forestry 17: High-tech and micropropagation Vol. 1. Springer-Verlag, New York. pp. 313-358.

Ma, S. S. and S. O. Wang 1977. Clonal multiplication of azaleas though tissue culture. Acta Hort. 78:209-213.

Murashige, T. 1974. Plant propagation through tissue culture. Annual review of plant physiology. 25:135-166.

Norton, C. R. and M. E. Norton 1989. Rhododendrons. Biotechnology in agriculture and forestry. Vol. 5. Trees II. In Y. P. S. Bajaj (ed.). pp. 428-451.

# MASS CLONAL PROPAGATION OF *ARTOCARPUS HETEROPHYLLUS* THROUGH IN VITRO CULTURE

Shyamal K. Roy, P.K. Roy, P. Sinha and M.S. Haque
Department of Botany, Jahangirnagar University, Savar, Dhaka, Bangladesh. E-mail: shkroy@juniv.edu

**Abstract.** *Artocarpus heterophyllus* Lam. (Fam. Moraceae) is a large evergreen fruit as well as timber tree. It is grown widely in south and southeast Asia and has been introduced into many other tropical countries. Protocol for mass clonal propagation, through *in vitro* culture of shoot tips and nodal explants of mature trees of *Artocarpus heterophyllus*, has been established. Apical and axillary shoot buds of young sprouts of coppiced branches of 20-year-old elite trees were used as explants. 2-3 cm long shoot tips and nodal segments were cultured aseptically. When the explants were cultured directly on agar gelled MS (Murashige and Skoog) medium with 2.5 mg $l^{-1}$ 6-benzyl adenine, 0.5 mg $l^{-1}$ ∝-naphthaleneacetic acid and 3% sucrose, multiple shoot buds were formed. For sufficient axial growth before transferring into the rooting medium, 100 mg $l^{-1}$ casein hydrolysate and 15% (v/v) coconut milk were required. These shoots continued to proliferate through 24 to more subcultures with an average of 32 shoots per transfer. For rooting the well developed shoots were excised from the culture flask and implanted individually on root induction medium. Within 3 weeks of transfer, 95% rooting was achieved in medium consisting of half-strength MS salts with 1.0 mg $l^{-1}$ each of indole-3-bytyric acid and ∝-naphthalene acetic acid. 85% of the regenerated plantlets survived in the field. The plants are now 4-years old in the field and their growth and vigour are quite satisfactory. The result of the present study show that shoot explants of mature trees of *Artocarpus heterophyllus* are highly potential for mass propagation. Control of the light period and agar concentration in *in vitro* condition is essential to realize full potential for large scale production of quality plantlets in vitro.

**Key index words.** *Artocarpus heterophyllus,* Auxin, Cytokinin, Jackfruit, Plantlet acclimatization, Shoot multiplication.

## 1. Introduction

*Artocarpus heterophyllus* Lam. (Fam. Moraceae) or Jackfruit is a large, evergreen tropical fruit as well as timber tree. It is grown widely in south and south-east Asia and Brazil and has been introduced into many other tropical countries. The tree is of great economic importance for its timber, vegetable (immature fruits), fresh fruit (ripe) and fodder uses. The poor people of the villages extensively utilize the fruit pericarp and seeds as a good source of nutritious food rich in vitamins A & C, carbohydrates and minerals and a considerable amount of protein as well and the leaves and remnants of the fruits as good source of fodder. The timber is excellent for construction of houses and furnitures. Jackfruit is propagated by seeds. Cutting, budding and grafting are not successful.

Trees are being harvested at a faster rate than they are being regenerated (Thorpe et al.,1991). So, there is an urgent requirement for production of large number of improved forest and fruit trees. Hence, *in vitro* techniques are gradually being included in tree improvement programs and considerable progress in the development of tissue culture methodologies for trees has already been made (Bonga and Durzan, 1987). Clonal



*C. Kubota and C. Chun (eds.), Transplant Production in the 21st Century,* 219–225.

propagation of trees offers several advantages for forest renewal and tree improvement programs. It facilitates for the reproduction of elite trees (Thorpe et al., 1991). Clonal multiplication of superior phenotypes and valuable stocks can be used for establishing tree improvement and seed production orchards (Dunstan et al., 1992).

Clonal propagation of *Artocarpus heterophyllus* through in vitro culture has been established using explants from the mature elite trees (Amin, 1992; Roy et al., 1993). Nevertheless, multiplication rate, as achieved so far, is not satisfactory for large scale plantation of selected elite stocks. Thus, the objective of this study was to enhance the clonal propagation potential of *A. heterophyllus in vitro* by optimizing the culture factors.

## 2. Materials and methods

### 2.1 Initiation of aseptic cultures

Young twigs were collected from coppiced shoots on the main trunk of about 20-year-old elite trees of *Artocarpus heterophyllus* growing in the Jahangirnagar University campus were taken in a polyethylene bag, brought to the laboratory and used as plant materials. They were stored in a cool pack at 4°C until processed.

For surface sterilization the shoots were defoliated and apices with 2-3 nodes were taken. They were then first cleaned thoroughly under a continuous stream of running tap water for 1 h, washed with liquid detergent for 15 min and then with a solution of the antiseptic Savlon (A.C. I Pharmaceuticals Ltd., Dhaka, Bangladesh) (5% vol/vol) for 10 min. The shoot apices were then washed repeatedly with distilled water and finally treated with 0.2% $HgCl_2$ for 10 min in a laminar flow cabinet and washed for 3 times with autoclaved double distilled water to remove any trace of $HgCl_2$.

After surface sterilization, shoot apices were excised at the base and divided into pieces as explants of size 25 to 30 mm with one node in each explant. The basal medium used for all the experiments was Murashige and Skoog (1962) mineral formulation (MS) containing standard salts and vitamins, 30 g $l^{-1}$ sucrose and 7 g $l^{-1}$ (unless otherwise stated) bacteriological agar (British Drug House, England). The media were variously supplemented with cytokinin [6-benzyladenine (BA) and Kinetin (Kn)], either individually or in different combinations with auxins, ∝-naphthalene acetic acid (NAA) and indole-3-acetic acid (IAA). The pH was adjusted to 5.8 $\pm$ 0.05 before adding agar, and the media were autoclaved at 1.1 kg. $cm^{-1}$ for 20 min at 120 °C. Cultures were incubated at 25 $\pm$ 1°C with a photoperiods of 16 h (unless otherwise mentioned) at 30 μ mol $s^{-2}$ $m^{-1}$ light intensity of cool white fluorescent light. All cultures were initiated in 150 x 25 mm glass tubes containing 12.5 ml of medium. The cultures were regularly subcultured on fresh medium at four-week intervals in glass tubes or 100-ml flasks. Observations were recorded every 5 days following inoculation and subculturing. All experiments were repeated twice with at least 12 cultures per treatment.

### 2.2 Multiplication of shoots

The primary shoots regenerated from explants cultured in MS basal medium supplemented with 2.5 mg $l^{-1}$ BA and 0.5 mg $l^{-1}$ NAA were isolated and subcultured on the fresh medium of same constituents for 4 weeks and thus the regenerated microshoots were excised several times and cultured in the same medium. For enhancing the number of shoots and axial growth casein hydrolysate (CH) 50-200 mg $l^{-1}$and coconut milk (CM) 5-20% (vol/vol) were also added to the medium mentioned above. To determine optimal photoperiod and agar concentration for maximum number of regeneration the cultures in defined medium were given light period variously (8-20 h per 24 h) and different concentration of agar (0.50-0.80 %) in the medium were tried.

**2.3 Rooting**

In rooting experiments, shoots of 25 to 50 mm long were excised from multiplication cultures and transferred to rooting medium consisting of half strength MS macro and micronutrients ($MS_I$) in the glass tubes. The media were variously supplemented with IBA and NAA.

**2.4 Transplantation**

Rooted shoots from 4-week old cultures on $MS_I$ + 1.0 mg $l^{-1}$ IBA + 1.0 mg $l^{-1}$ NAA were transferred to pots after *in vitro* hardening. For *in vitro* hardening the culture tubes containing plantlets were kept in a room of normal temperature (30 $\pm$ 2°C) and normal day light for 7 days. Plantlets were then taken out from the culture tubes, the agar was washed out from the root and transplanted to small trays (each tray contained 10 plantlets) filled with compost, soil and sand (1:2:1). The trays were kept in shady place, covered with transparent polyethylene sheet and were watered daily. After 3 weeks, when the plantlets were fully acclimatized to outdoor conditions plantlets were again transplanted individually polyethylene bags of 15 cm diameter containing compost and soil (1:1) and they were watered on alternate days. After 3-4 months the plants were planted in the open field.

**3. Results**

**3.1 Initiation of shoot cultures**

The surface sterilization procedure of explants described in materials and methods yielded 90% aseptic cultures. Shoot induction was not found in the MS medium without hormone even after 4 weeks of culture. Addition of cytokinin (either BA or Kn) alone to the basal medium was not found to be fruitful for the induction of multiple shoot. On the MS medium, containing 2.5 mg $l^{-1}$ BA and 0.5 mg $l^{-1}$ NAA, 92% cultures showed induction of shoots within 3 weeks with 2-4 shoots per culture (Table 1). BA - NAA combinations were not efficient to induce shoot. Frequency of shoot induction was strongly influenced by the season in which cultures were initiated. The results obtained and described above refer to the experiments conducted during the rainy season (late June to August). Shoot induction frequency varied considerably during other months. To determine the effect of season on shoot induction, fresh cultures of nodal segments were raised on MS medium supplemented with 2.5 mg $l^{-1}$BA and 0.5 mg $l^{-1}$NAA at regular intervals round the year. The cultures initiated during the rainy season (late June to August) responded both in the frequency of shoot induction and vigour of the regenerated shoots.

**3.2 Shoot multiplication**

Though multiple shoots were induced in MS medium with 2.5 mg $l^{-1}$ BA and 0.5mg $l^{-1}$ NAA, during the first 2-4 cycles of shoot multiplication, almost all the cultures showed dark browning of the medium. Addition of 3000 mg $l^{-1}$ of soluble polyvinyl pyrrolidone (PVP) to the medium overcome the browning problem. At the time of subculture newly formed shoots were separated and transplanted to fresh medium of the same constituent and thus shoots continued to proliferate through several subcultures with an average of 10 shoots per transfer. The length of shoots induced in the medium mentioned above was not sufficient and as such coconut milk (CM) (5-20% v/v) was added to the medium. Addition of 15% (v/v) CM enhanced the number shoot upto 32 per culture. For further development of the medium casein hydrolysate (CH) (50-200 mg $l^{-1}$) was added to the medium. Addition of 100 mg $l^{-1}$ CH to the medium increased the length of shoots. Thus the medium determined for large number (32) of shoot induction with proper length was MS basal medium+2.5 mg $l^{-1}$ BA + 0.5mg $l^{-1}$ NAA + 15% (v/v) CM + 100 mg $l^{-1}$ CH.

*Table 1* Effect of growth regulators in MS basal medium on shoot induction and number of shoots per culture established from apical and nodal bud explants of *Artocarpus heterophyllus*. Data were recorded after 28 days of culture.

| Growth regulators (mg $l^{-1}$) | Percent of culture induced shoots | No. of shoots per culture |
|---|---|---|
| BA 0.5 | 0 | 0 |
| BA 1.0 | 0 | 0 |
| BA 1.5 | 0 | 0 |
| BA 2.0 | 0 | 0 |
| BA 2.5 | 0 | 0 |
| BA 3.0 | 0 | 0 |
| BA 2.0 + NAA 0.5 | 42.3(4.6) | 2.3(0.5) |
| BA 2.0 + NAA 1.0 | 46.5(8.2) | 2.6(0.8) |
| BA 2.5 + NAA 0.5 | 92.1(6.5) | 3.2(1.6) |
| BA 2.5 + NAA 1.0 | 68.3(5.4) | 2.3(1.9) |
| BA 3.0 + NAA 0.5 | 52.6(5.8) | 2.4(0.7) |
| BA 3.0 + NAA 1.0 | 31.8(4.7) | 2.3(0.8) |

Standard error in parenthesis.

*Table 2* Effect of auxins in $MS_1$ medium on root formation in regenerated shoots of *Artocarpus heterophyllus*. Data were recorded after 30 days of culture.

| Growth regulators (mg $l^{-1}$) | Percent of shoots rooted | Length of leader root per culture |
|---|---|---|
| IBA 0.5 | 38.2(5.3) | 2.1(0.1) |
| IBA 1.0 | 36.5(3.5) | 2.8(0.1) |
| IBA 1.5 | 43.6(4.9) | 2.7(0.1) |
| IBA 2.0 | 25.8(4.1) | 2.5(0.1) |
| NAA 0.5 | 0 | 0 |
| NAA 1.0 | 0 | 0 |
| NAA 1.5 | 0 | 0 |
| NAA 2.0 | 0 | 0 |
| IBA 0.5 + NAA 0.5 | 71.6(6.4) | 5.5(2.6) |
| IBA 1.0 + NAA 1.0 | 85.0(9.3) | 6.4(3.2) |
| IBA 1.5 + NAA 1.5 | 66.8(4.2) | 4.3(3.5) |
| IBA 2.0 + NAA 2.0 | 67.4(5.3) | 4.5(2.9) |

Standard error in parenthesis.

To determine the proper light period of culture the medium mentioned above was kept constant and the light period was changed variously (8-20 h). Four consecutive subcultures were maintained in a light period and it was found that 16 h was the optimal period of light, below which the growth of shoots and leaves were gradually deteriorated and above which the number of shoot induction per culture was decreased but the leaves were broadened vigorously. To study the optimum concentration of the agar in the medium, different concentrations (0.5-0.8%) of agar were tried keeping the other constituents in the medium fixed. After four consecutive subcultures in the medium containing 0.65% agar the number of shoots per culture was found to be increased upto 45 $\pm$ 6 (Fig.1). Medium gelled with 0.5% and 0.8% agar caused a decline in the multiplication rate upto 2 and 3 fold respectively. Thus shoot multiplication medium determined was MS + 2.5 mg $l^{-1}$ BA + 0.5 mg $l^{-1}$ NAA + 15% CM + 100 mg $l^{-1}$ CH gelled with 0.65% agar and the appropriate photoperiod was 16 h.

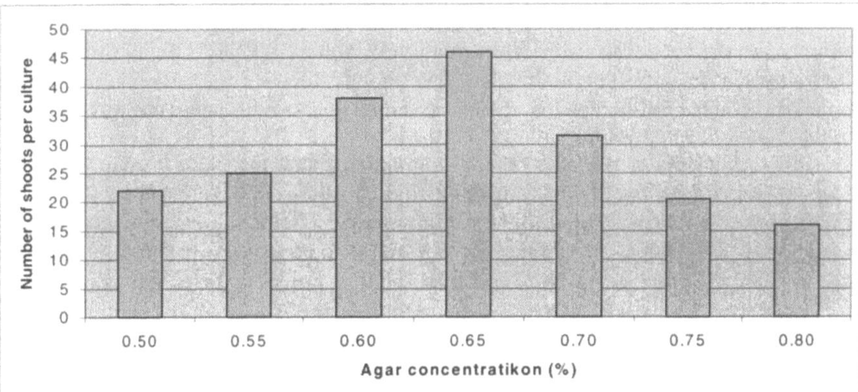


*Fig. 1* Effect of agar concentration on number of shoots developed per culture. Culture medium. MS + 2.5 mg.liter$^{-1}$ BA + 0.5 mg.liter$^{-1}$ NAA + 15% CM + 100 mg.liter$^{-1}$ CH. Photoperiod 16 h.

### 3.3 Rooting

Without auxin treatment, shoots obtained from multiplication cultures failed to root induction. The well developed shoots were excised from the culture flask and implanted individually on root induction medium containing $MS_1$ with different concentrations and combinations of IBA and NAA. 1.0 mg $l^{-1}$ each of IBA and NAA was found to be the best combination of auxins for proper rooting in which 85% shoots rooted. Roots began to emerge from 12th day of culture and within 20th day 85% shoots rooted. Rooting culture was treated with an initial 7-day dark period, but there was no significant difference in root induction between dark- and non- dark treated shoots.

### 3.4 Transplantation

After 4 weeks in rooting medium the rooted shoots were transferred to pots. None of plantlets transferred directly from rooting medium to the pot under natural conditions survived. 80% of the transplanted plants survived if the plants in the rooting in the culture tubes were kept in normal room temperature (as described in materials & methods) for 7 days before transplanting in tray where plants were reared for 3 weeks. At the time of rearing, shoots elongated, leaves expanded and turned deep green and consequently, the plants were much healthy. The plants also grew more vigorously when transplanted to the potting mix. 500 plants have been transferred to the open field where they are still growing with full vigour. Some of the plants are now 4-year- old with 375 cm in height.

## 4. Discussion

Tissue culture from apical and nodal buds is currently being used for the clonal propagation of woody plants. Though it is usually more difficult to establish shoot cultures from mature trees than from juvenile plants (Bonga, 1987) or germinated seedlings (Bonga, 1982), it is the only satisfactory means of preserving the characteristics of a desired clone (Das and Mitra, 1990). The results of the present investigation clearly show that apical and nodal buds of coppiced shoots from selected mature trees of Jackfruit are capable of producing multiple shoots *in vitro* which can be rooted to form complete plantlets. Time of collection of explants is often a key factor in establishing the culture of woody species. According to Bonga (1987), spring and late summer are the best season for bud initiation of woody species. Our findings revealed that rainy season (late June to August) was the best time for the collection of Jackfruit explants, which corroborate the results obtained by Saxena and Bhojwani (1993) in *Dendrocalamus longispathus* culture.

The superiority of BA over other cytokinins for multiple shoot formation in woody fruit plants has been reported (Vieitez and Vieitez, 1980). In the present experiment, BA proved to be a suitable cytokinin for proliferation of shoots in combination with an optimum dose (0.5 mg $l^{-1}$) of NAA. A BA-NAA combination for shoot multiplication was also reported by Lakshmi Sita and Vaidyanathan (1979) and Das and Mitra (1990).

However, we observed the effects of only light period of the culture and agar concentration in culture medium on the number of shoots induction and their growth. Morini et al (1990) studied the effect of different light-dark cycles and observed that the growth of peach shoots was significantly greater with 4 h light and 2 h dark cycle compared to conventional 16 h dark cycle. Whereas, in our experiment we found 16 h light and 8 h dark cycle was the optimum photoperiod for large number of shoots and their proper growth.

Agar concentration in the medium influences several culture parameters and reactions, because it regulates the relative humidity in the culture vessel (Kozai et al, 1992). Tanaka et al. (1992) and Kozai et al. (1993) cultured potato plantlets for 22 days *in vitro* under different relative humidity and found a greater shoot with increasing initial relative humidity. In our experiment we observed that change of agar concentration altered the rate of shoot induction, probably due to change of relative humidity.

For rooting on excised shoots, either single or a combination of tow or three auxins are used routinely. In our experiments a synergistic effect of IBA+NAA was found to be best for root induction. In contrast, only NAA was fruitful for *Eucalyptus citriodora* (Gupta et al., 1981), whereas for rooting *Tectona grandis* (Gupta et al., 1980) microcuttings, three auxins, IBA, NAA and IPA were essential.

The results of the present study show that shoot explants from mature trees of *Artocarpus heterophyllus* could produce multiple shoots *in vitro*. The shoot multiplication rate directly from the shoot explants as observed in the present study was much higher than that reported earlier (Roy et al., 1993). Moreover the potentiality of multiplication continued for a long time. The survivability of the regenerants in the field was also higher. The technique described here would be a promising method of propagation on a commercial scale as well as for the conservation of genetic diversity of an important fruit crop like jackfruit.

### References

Amin, M.N. 1992. *In vitro* enhanced proliferation of shoots and regeneration of plants from explants of jackfruit trees. Plant Tissue Culture 2:27-30.

Bonga, J.M. 1982. Vegetative propagation in relation to juvenility, maturity and rejuvenation In : Bonga J.M.; Durzan, D.J., eds. Tissue culture in forestry Maritnus Nijhoff; The hague. pp 387-412

Bonga, J.M. 1987. Clonal propagation of mature trees: problems and possible and solutions. In: Bonga J.M.; Durzan. D.J. (eds.), Cell and tissue culture in forestry, Vol. 1. Martinus Nijhoff, The hague.

Bonga, J.M.; Durzan, D.J. 1987. Tissue culture in forestry. Vols.I-II. Martinus Nijhoff, Dordracht, The Netherlands.

Das, T.O. and G.C. Mitra. 1990. Micropropagation of *Eucalyptus tereticornis* Smith. plant Cell Tissue Organ Culture 22:95-103.

Dunstan, D.I., D.P. Lashta and S. Kikcio. 1992. Factors affecting recurrent shoot multiplication in *in vitro* cultures of 17 to 20-year-old Douglas fir trees. In Vitro Cell. Dev. Biol. 28P:33-38.

Gupta, P.K., A.L. Nadgir, A.F. Mascarenhas and V. Jagannathan. 1980. Tissue culture of forest trees: Clonal multiplication of *Tectona grandis* L. (teak) by tissue culture. Plant Sci. Lett. 17:259-268.

Gupta, P.K., A.F. Mascarenhas and V. Jagannathan. 1981. Tissue Culture of forest trees: Clonal propagation of mature trees of *Eucalyptus citriodora* Hook. by tissue culture. Plant Sci. Lett. 20:159-201.

Kozai, T., K. Fujiwara, M. Hayashi and J. Aitken-Christie. 1992. The in vitro environment and its control in micropropagation. In: K. Kurata and T. Kozai (eds) Transplant Production System. Kluwer Academic Publishers, Dordrecht, The Netherlands. pp. 247-282.

Kozai, T., K. Tanaka, B.R. Jeong and K. Fujiwara. 1993. Effect of RH in the culture vessel on the growth and shoot elongation of potato (*Solanum tuberosum* L.) plantlets *in vitro*. J. Japn. Soc. Hort. Sci. 62(2):413-417.

Lakshmi Sita, G. and C.S. Vaidyanathan. 1979. Rapid Multiplication of Eucalyptus by multiple shoot production. Curr. Sci. 48:250-352.

Morini, S., P. Fortuna, R. Sciutti and R. Muleo. 1990. Effect of different light-dark cycle on growth of fruit tree shoots cultured *in vitro*. Advances Hort. Sci. 4:163-166.

Murashige, T. and K. Skoog. 1962. A revised medium for rapid growth and bioassay with tobacco tissue culture. Physiol. Plant. 15:473-497.

Roy, S.K., M.S. Islam, J. Sen and S. Hadiuzzaman. 1993. Propagation of flood laterant jackfruit (*Artocarpus heterophyllus*) by *in vitro* culture. Acta horticulturae. 336:273-276.

Saxena, S. and S.S. Bhojwani. 1993. In vitro clonal multiplication of 4-year-old plants of Bamboo, *Dendrocalamus longispathus* Kruz. In vitro Cell Dev. Biol. 29p:135-142.

Tanaka, K., K. Fujiwara and T. Kozai. 1992. Effects of relative humidity in the culture vessel on the transpiration and net photosynthetic rates of potato plantlets *in vitro*. Acta. Hort. 319:59-64.

Thorpe, T.A., I.S. Harry and P.P. Kumar. 1991. Application of micropropagation to forestry. In: Deberg, P.D. and Zimerman, R.H. (eds.). Micropropagation Technology and Application. Kluwer Academic Publishers, Netherlands. pp. 331-336.

Vieitez, A.M. and E. Vieitez. 1980. Plantlet formation from embryonic tissue of chestnut grown in vitro. Phyiol. Plant. 50:127-130.

# PHOTOAUTOTROPHIC GROWTH OF *PLEIOBLASTUS PYGMAEA* PLANTLETS *IN VITRO* AND *EX VITRO* AS AFFECTED BY TYPES OF SUPPORTING MATERIAL *IN VITRO*

Yumiko Watanabe[1,2], Yoshiaki Sawa[1], Nobuharu Nagaoka[1] and Toyoki Kozai[2]
[1]Free Industry Co., Ltd. Taitouku, Tokyo 110-0015, Japan.
[2]Faculty of Horticulture, Chiba University, Matsudo, Chiba 271-8510, Japan.
E-mail: watanabe@free-kogyo.co.jp

**Abstract.** Sasa (*Pleioblastus pygmaea* Miff., cv. Oroshimatiku) has been used as a ground-cover plant because it keeps the ground green throughout the year and prevents soil erosion. Sasa, which rarely flowers, can be micropropagated under artificial light for mass production of transplants. In this study, a micropropagation system for sasa has been developed. The processes are 1) an *in vitro* propagation period, 2) an *in vitro* growth period and 3) an *ex vitro* acclimatization period. The growth of plantlets was examined under six culture conditions: with combinations of sugar-containing or sugar-free medium and one of two types of supporting material (plastic net and vermiculite) or no supporting material. All plantlets were grown under relatively high photosynthetic photon flux (PPF) and $CO_2$-enriched conditions during the growth period. Plantlets grown on sugar-containing medium without supporting material under relatively low PPF and $CO_2$ non-enriched conditions were used as a control. The plantlets grown photoautotrophically with vermiculite during the *in vitro* growth period showed enhanced growth during the *ex vitro* acclimatization period, while the plantlets of the control wilted during that period.

**Key index words.** Acclimatization, mass production, micropropagation, sasa, sugar-free.

## 1. Introduction

There has been an increasing need for planting ground-cover plants on slopes, riverbeds and in parks to prevent soil erosion and to keep the ground green. Sasa is a good choice for a ground-cover plant for three reasons: 1) it extends its rhizomes horizontally from the base of the plantlets below the soil surface to prevent soil erosion, 2) once planted, it has to be mown only a couple of times a year, i.e., it needs little maintenance, and 3) it stays green all through the year. Because sasa rarely flowers, it is difficult to obtain its seeds. Sasa has been conventionally propagated by vegetative methods under natural light. With these methods, it is difficult to produce a large number of transplants. On the other hand, plant tissue culture in a closed system using artificial light could produce a large number of plantlets. Actually there are some kinds of bamboo micropropagated through a shoot propagation (Ravikumar et al., 1998, Sanjay Saxena, 1990). In this study, a micropropagation system for mass production of sasa transplants, which consists of three periods (a propagation period, a growth period and an acclimatization period) has been developed. The production costs should be considered in practical applications of this system. Rooting *in vitro* and acclimatization *ex vitro* can be achieved easily by the photoautotrophic culture method and biological contamination risks *in vitro* are reduced with this method (Kozai, 1991). The photoautotrophic growth of eucalyptus plantlets was enhanced with vermiculite as a supporting material (Kirdmanee et al., 1994). Thus, the photoautotrophic culture method



*C. Kubota and C. Chun (eds.), Transplant Production in the 21st Century,* 226–230.

may not only enhance the growth of sasa plantlets *in vitro* and *ex vitro* but also reduce the production costs. Photoautotrophic growth of sasa plantlets as affected by types of supporting materials during the *in vitro* growth period was examined at the end of the *in vitro* growth period and at the end of the *ex vitro* acclimatization period in the present micropropagation system.

## 2. Micropropagation system, Materials and methods

### 2.1 Micropropagation system

A micropropagation system for mass production of sasa transplants has been developed as follows.

1) *In vitro propagation period*: A propagule consisting of 2 shoots is cultured *in vitro* for propagation.

2) *In vitro growth period*: Multiple shoots are cultured *in vitro* for growing, especially for rooting.

3) *Ex vitro acclimatization period*: Multiple shoots with roots are transplanted *ex vitro* under artificial lights for acclimatization.

### 2.2 Materials

Multiple shoots were induced from axillary buds of sasa (*Pleioblastus pygmaea* Miff.,cv. Orosimatiku) plants. Propagules of Sasa consisting of two shoots were grown to multiple shoots *in vitro* for 30 days on MS (Murashige and Skoog, 1962) medium containing 20 g $L^{-1}$ of sucrose and 2 mg $L^{-1}$ of 6-benzyladenine (Arya et al., 1999). Propagation conditions consisted of 100 $\mu$mol $m^{-2}$ $s^{-1}$ of photosynthetic photon flux (PPF), ambient $CO_2$ concentration (370-400 $\mu$mol $mol^{-1}$) and 1.9 air exchanges per hour of the culture vessel (Kozai et al., 1986). The multiple shoots were transferred to magenta-type vessels (370 mL), containing 40 mL of MS medium, pH 5.8 before autoclaving, and grown for 40 days. Plastic nets (2×2 mm mesh, 55×55 mm/vessel) that had been soaked in 99.8% ethanol for 24 hours and autoclaved vermiculite (10 g/vessel) were used as two separate supporting materials. Each vessel had one plantlet. After the *in vitro* growth period, the plantlets were transplanted and grown for 25 days *ex vitro* in trays (190 mm×145 mm×25 mm) containing commercial soil mixture (Yanmar Agricultural Equipment Co., LTD. Japan). Water was added daily at the rate of 40 mL/tray.

### 2.3 Treatments and culture conditions

There were six treatments consisting of combinations of sugar-containing (20 g $L^{-1}$) or sugar-free medium and with three types of support: plastic net, vermiculite or without supporting material. The culture room conditions during the *in vitro* growth period were PPF of 100 $\mu$mol $m^{-2}$ $s^{-1}$, $CO_2$ concentration of 2000 $\mu$mol $mol^{-1}$ and 1.9 air exchanges per hour of the vessel (on days 0-3), and were PPF of 200 $\mu$mol $m^{-2}$ $s^{-1}$, $CO_2$ concentration of 2000 $\mu$mol $mol^{-1}$ and 4.5 air exchanges per hour of the vessel (on days 3-40). For a control, the plantlets were grown on sugar-containing medium without supporting material under PPF of 100 $\mu$mol $m^{-2}$ $s^{-1}$, ambient $CO_2$ concentration and 1.9 air exchanges per hour of the vessel (conventional method). Therefore, seven treatments were used (Table 1). The PPF and $CO_2$ concentration in each treatment during the *ex vitro* acclimatization period were the same as those in the *in vitro* growth period. Other conditions through the growth and acclimatization periods were 25℃ during the photoperiod and 21℃ during the dark period, a 16-h photoperiod, and 75% relative humidity. White fluorescent lamps were used as the light source.

**2.4 Measurements**

At the end of the *in vitro* growth period, $CO_2$ concentrations inside and outside the vessels, fresh and dry weight of the plantlets were measured. Net photosynthetic rate (NPR) per plantlet was calculated with $CO_2$ concentrations inside and outside the vessels (Fujiwara et al., 1987).

At the end of the *ex vitro* acclimatization period, fresh and dry weight of the plantlets were measured, and relative growth rate (RGR) was calculated using the equation: RGR= [ln (FFW)-ln (IFW)] / 25 days (FFW: final fresh weight on day 25 of the acclimatization period and IFW: initial fresh weight on day 0 of the acclimatization period).

**3. Results and discussion**

NPR and shoot and root dry weights (DW) at the end of the *in vitro* growth period (on day 40) are shown in table 2-1. NPR and root DW were higher in the plantlets grown on sugar-free medium regardless of the type of supporting material as compared to the plantlets grown on sugar-containing medium, and NPR and root DW were lowest in the control. It was also reported that photoautotrophic growth *in vitro* enhanced the growth of *Cymbidium* (Heo et al., 1996). Shoot DW was lower in the plantlets grown with vermiculite as supporting material regardless of sugar concentration, as compared with the plantlets grown without or with plastic net as supporting material. There was no significantly difference between the plantlets of the control and any of the treatments except for the 20V treatment, in which the plantlets had lower shoot DW.

Relative growth rate (RGR) during the *ex vitro* acclimatization period, and shoot and root dry weights at the end of the *ex vitro* acclimatization period (25 days) are shown in Table 2-2. While 50% of the control plantlets wilted, 100% of the plantlets receiving the other treatments survived during the *ex vitro* acclimatization period. RGR and root DW were higher in the plantlets grown with vermiculite as supporting material during the *in vitro* growth period regardless of the sugar concentration than those in the plantlets grown without or with plastic net as supporting material. Shoot and root DW were the highest in the plantlets of 0V, in which the plantlets were grown on sugar-free medium with vermiculite as supporting material during the *in vitro* growth period. In this present experiment, photoautotrophic growth during the *in vitro* growth period enhanced NPR and root DW during the *in vitro* growth period. Moreover, photoautotrophic growth combined with the use of vermiculite as supporting material enhanced the growth of the plantlets during the *ex vitro* acclimatization period, though shoot DW of the plantlets was lower during the *in vitro* growth period. It could be concluded that the high net photosynthetic rate and the extensive roots induced by *in vitro* photoautotrophic growth combined with the use of vermiculite led to the enhancement of the growth of sasa plantlets during the *ex vitro* acclimatization period. Acclimatization *ex vitro* for sasa plantlets can be achieved easily by *in vitro* photoautotrophic culture with vermiculite. This means that a large number of sasa transplants could be produced rapidly by micropropagation when the plantlets are photoautotrophically grown during the *in vitro* growth period. Therefore it is possible that photoautotrophic growth of plantlets with vermiculite reduce the production cost. Additional studies are needed to further reduce costs in order to develop a practical system for the production of sasa transplants.

*Table 1* Description of treatments.

| Treatment | Sucrose ($g\ L^{-1}$) | Supporting material |
|---|---|---|
| 20N | 20 | None |
| 20P | 20 | Plastic net[1)] |
| 20V | 20 | Vermiculite[2)] |
| 0N | 0 | None |
| 0P | 0 | Plastic net |
| 0V | 0 | Vermiculite |
| Control[3)] | 20 | None |

[1)] 55×55 mm/vessel, 2×2 mm mesh.
[2)] 10 g/vessel.
[3)] $CO_2$ concentration, photosynthetic photon flux and number of air exchanges of the vessel were different from those of treatment 20N.

*Table 2-1* Net photosynthetic rate (NPR), shoot and root dry weight (DW) at the end of the *in vitro* growth period (on day 40). Means±standard deviation are shown.

| Treatment | NPR ($\mu molCO_2\ gDW^{-1}\ h^{-1}$) | DW (mg/plantlet) Shoot | Root |
|---|---|---|---|
| Control | 8± 7 | 370± 75 | 4± 3 |
| 20N | 34±16 *[x)] b[y)] | 310±130 NS a | 29±18 * a |
| 20P | 29± 6 ** b | 270±130 NS a | 37±28 ** a |
| 20V | 38±13 ** b | 130± 35 ** b | 21±12 * a |
| 0N | 83±21 ** a | 320± 90 NS a | 71±39 ** a |
| 0P | 70±17 ** a | 380± 32 NS a | 59±39 ** a |
| 0V | 83±23 ** a | 230±110 NS a | 75±60 NS a |
| ANOVA | | | |
| Sugar concentration (C) | **[z)] | NS | * |
| Supporting material (M) | NS | * | NS |
| C×M | NS | NS | NS |

[x)] NS, * and ** indicate non-significant difference, significant difference at p≦0.05 and 0.01, respectively, by t-test compared with conventional treatment.
[y)] The same letters are not significantly different at the 5% level by Duncan's multiple range test.
[z)] NS, * and ** indicate non-significant difference, significant difference at p≦0.05 and 0.01, respectively, by ANOVA.

*Table 2-2* Relative growth rate per day [w] (RGR), shoot and root dry weights (DW) at the end of the *ex vitro* acclimatization period (on day 25). Means ± standard deviation are shown.

| Treatment | RGR ($d^{-1}$) | DW (mg/plantlet) Shoot | Root |
|---|---|---|---|
| Control | — | 220± 72 [x] | 92± 64 |
| 20N | 0.010±0.003bc [y] | 550± 51 b | 180± 55 b |
| 20P | 0.009±0.004bc | 550± 77 b | 240± 43 b |
| 20V | 0.025±0.003a | 570±230 b | 170± 79 b |
| 0N | 0.001±0.002c | 520± 38 b | 130± 50 b |
| 0P | 0.011±0.004b | 530± 53 b | 160± 50 b |
| 0V | 0.031±0.003a | 1020±120 a | 610±110 a |
| ANOVA | | | |
| Sugar concentration (C) | NS [z] | NS | NS |
| Supporting material (M) | ** | NS | * |
| C×M | NS | * | * |

[w] RGR = (ln (final fresh weight)-ln (initial fresh weight))/25 days.

[x] Plantlets that didn't wither were measured.

[y] The same letters are not significantly different at the 5% level by Duncan's multiple range test.

[z] NS, * and ** indicate non-significant difference, significant difference at $p \leqq 0.05$ and 0.01, respectively, by ANOVA.

**References**


Arya, S., S. Sharma, R. Kaur and I.D. Arya. 1999. Micropropagation of *Dendrocalamus asper* by shoot proliferation using seeds, Plant Cell Rep. 18:879-882.

Fujiwara, K., T. Kozai and I. Watanabe. 1987. Fundamental studies on environments in plant tissue culture vessels. (3) Measurements of carbon dioxide gas concentration in closed vessels containing tissue cultured plantlets and estimates of net photosynthetic rates of the plantlets, J. Agric. Meteorol., 43 (1):21-30.

Heo, J.W., C. Kubota and T. Kozai. 1996. Effects of $CO_2$ concentration, PPFD and sucrose concentration on *Cymbidium* plantlets grown *in vitro*, Acta. Hort. 440:560-565.

Kirdmanee, C., Y. Kitaya and T. Kozai. 1995. Effects of $CO_2$ enrichment and supporting material *in vitro* on photoautotrophic growth of *Eucalyptus* plantlets *in vitro* and *ex vitro*, In vitro Cell. Dev. Biol.-Plant 31:144-149.

Kozai, T., K. Fujiwara and I. Watanabe. 1986. Fundamental studies on environments in plant tissue culture vessels. (2) Effects of stoppers and vessels on gas exchnge rates between inside and outside of vessels closed with stoppers, J. Agric. Meteorol. 42(2):119-127.

Kozai, T. 1991. Photoautotrophic micropropagation, *In vitro* Cell Dev. Biol 27:47-51.

Ravikumar, R., G. Ananthakrishnan, K. Kathiravan and A. Ganapathi. 1998. *In vitro* shoot propagation of *Dendrocalamus strictus* nees, Plant cell, Tissue and Organ Culture 52:189-192.

Saxena, S. 1990. In vitro propagation of the bamboo (*Bambusa tulda Roxb.*) through shoot proliferation, Plant Cell Reports 9:431-434.

# EVOLUTION OF CULTURE VESSEL FOR MICROPROPAGATION: FROM TEST TUBE TO CULTURE ROOM

Sayed M.A. Zobayed[1], Fawzia Afreen, Chieri Kubota and Toyoki Kozai
Department of Bioproduction Science, Chiba University, Chiba 271-8510, Japan. Email: S.M.Zobayed@biosci.hull.ac.uk
[1]Present address: Department of Biological Sciences, University of Hull, Hull HU6 7RX, UK.

**Abstract.** To improve the culture conditions for micropropagation, different types of culture vessels and capping systems have been designed. Some of these designs improve the aerial composition in the culture vessel and some for recycling the nutrient medium. This article describes the evolution of different culture vessel and culture systems, with special emphasis on forced ventilation to improve the culture atmosphere and thus to improve the growth and multiplication and also the quality of propagules. By altering the aerial environment of the culture vessel, plantlets can be grown photoautotrophically (sugar free medium) which has many advantages over the photomixotrophic or heterotrophic system. By using forced ventilation and a photoautotrophic culture system, the scaling-up of the culture vessel is possible with high growth rate and survival percentage and with minimum time and space. More recently, this scale-up system has been further extended making the aseptic culture room itself a large culture vessel containing many small sterile trays with plants on the culture shelves and with a common headspace. By using this enlarged system, the production of even more quality transplants was achieved relatively easily.

**Key index words.** forced ventilation, large-scale, natural ventilation, photoautotrophy, temporary immersion.

## 1. Introduction

Plant tissue culture has attained major importance in agriculture and forestry. It plays a unique role to bridge researches with commercial practitioners. The first plants to be propagated by tissue culture technique were the orchids. In 1960, Morel described that by using this technique, within a short time, a multitude of plants can be produced from a single plant; the multiplication rate was distinctly faster than the conventional method (Willmer, 1966). He estimated that as many as 4 million *Cymbidium* plants can be produced per year from a single explant. The success of the orchidologists stimulated growers of other flower and ornamentals to explore a similar applicability to their crops. Thus the method of tissue culture lends itself to scaling up to industrial dimensions.

This article reviews the evolution of different culture vessels and culture systems for the large-scale production, with special emphasis on forced ventilation.

## 2. Evolution of culture vessel

*In vitro* tissue cultures were probably first attempted in a stationary test tube by Loeb in 1897, although it did not prove to be satisfactory. Of the various techniques developed at the early twenty century, the hanging-drop method was the earliest to yield successful results. In this method a small piece of tissue was placed on a cleaned and sterilized coverslip, and then covered with just sufficient nutrient fluid to anchor it, by surface tension, inverted over a sterile hollow-ground slide. With the realization that the



*C. Kubota and C. Chun (eds.), Transplant Production in the 21st Century*, 231–237.

useful life of a hanging-drop culture was not more than a few days, the flask technique was developed by Carrel in 1923, where several pieces of tissue were placed on the bottom of a small flattened flask called Carrel flask. After the development of this technique, different types of vessels such as, roller tubes, tumbler tubes, small flattened rectangular T-flasks, Erlenmeyer flasks, large Roux bottles or petridishes were used extensively (Willmer, 1966) and still now these vessels are being used for the micropropagation purposes. In commercial practice many different types of vessels are now used; vessels are made of various materials such as glass, polypropylene, polyvinylglycine and polycarbonate with different volumes ranging within 15-500 ml. The recent innovation of Magenta vessel (Magenta Corporation, Chicago, IL), Magenta GA-7 vessel (Sigma Co., USA), Phytacon$^{TM}$ (Sigma Co., USA), Phytatray$^{TM}$ (Sigma Co., USA) are specially designed to improve the ventilation in the culture headspace.

**3. Systems for renewing nutrient medium in the vessel**

Steward and Shantz, in 1955 when first used the liquid culture in a specially constructed rotating culture vessel, they noted that the supernatant fluid became turbid, owing to the development and growth of free-floating cells. Then Roberts and Street (1959) developed a system, which allows a continuos flow of sterile culture medium over the growing root culture. Later on, Richez (1965) had worked out a simple apparatus for renewal of the liquid nutrient medium based on the principle of communicating vessels. This renewal of the liquid nutrient had proved to improve the growth of the culture greatly, compared to the shaking of the liquid or rotating motion of the tube. However, Heller in 1965 demonstrates that a mere up-and-down motion of the liquid, without renewal showed the same effect as a true renewal. This is probably the first concept of the temporary immersion system now commonly used in the modern micropropagation.

**4. Photoautotrophic culture system**

Photoautotrophic micropropagation is the propagation of the plantlets using relatively small chlorophyllous (leafy) explants in a sugar-free nutrient medium under pathogen-free condition where they can photosynthesize and produce their own carbohydrates for their growth. However, because of the low $CO_2$ concentrations observed during the photoperiod in the conventional culture system, photosynthesis of *in vitro* plantlets is restricted (Kozai and Iwanami 1988). Mousseau (1986) first concluded that *in vitro* plantlets are apparently photosynthetically active. They also reported that for optimizing photosynthesis $CO_2$ is necessary to be enriched together with light intensity (100-200 $\mu$mol m$^{-2}$ s$^{-1}$). In the same year, Kozai *et al.*, (1988) have successfully developed the photoautotrophic micropropagation system by increasing $CO_2$ concentration and PPF (photosynthetic photon flux) in the culture vessel and omitting sucrose from the nutrient medium. Photoautotrophic micropropagation has many advantages over the conventional system; the benefits are recently listed by Kozai and Zobayed (2000).

**5. Large-scale culture system**

There are two main branches of massive culture: in one the explants are maintained in suspension or liquid medium; in the other they are grown on a gelled or porous solid substrate.

### 5.1 Large-scale systems with liquid culture medium

For the large-scale culture, research was mainly emphasized in the suspension culture by using liquid culture medium. Cultures on a liquid medium have some advantages over gelled or porous solid medium; it allows renewing of the medium easily without manipulating cultures, which is always damaging. Moreover culture conditions are uniform.

Suspension cultures are generally initiated by transferring an established callus tissue to an agitated medium in a culture flask (30-60 ml medium per 250 ml flask). For the large-scale suspension culture different types of apparatus have been designed and used since the early sixties and seventies in the last century. However, it has been realized that the continuos immersion of the cultured explant can seriously hamper the growth and morphology of the cultured plantlets. Recent invention of the temporary immersion system (Etienne *et al.*, 1999) has designed with the intention to avoid this problem. The system has recently been widely used in the filed of micropropagation.

### 5.2 Large-scale systems using porous solid substrate under forced ventilation

#### 5.2.1 Historical background

The benefits of large culture vessel with porous solid substrate for the organ culture have recently been realized. Probably the first system based on this concept was developed by Fujiwara *et al.*, in 1988. In this system, a large culture vessel ( 58 cm long, 28 cm wide and 12 cm high) was used with a forced ventilation system for enhancing the photoautotrophic growth of strawberry (*Fragaria* x *ananassa* Duch.) explants and/or plantlets during the rooting and acclimatization stages. This was a kind of aseptic micro-hydroponic system with a nutrient solution control system. Roche *et al.*, (1996) developed a commercial-scale photoautotrophic micropropagation system for potato microplants in which 100 nodal explants were cultured under natural ventilation in a stainless steel tray containing a block of polyurethane foam (85 x 300 x 25 mm) and enclosed with a polyethylene sleeve. Kubota and Kozai (1992) grew potato plantlets under forced ventilation in a large vessel (2.6 L) containing a multi-cell tray. Recently, Heo and Kozai (1999) developed a similar type of system using a large culture vessel (volume 13 L.), where 20 sweet potato plantlets were cultured photoautotrophically. However, the disadvantage of these large culture vessels commonly faced is the variation in growth of the cultured plantlets mainly due to the non-uniform distribution of $CO_2$ concentration and other environmental factors in the culture headspace. Recently Zobayed *et al.* have developed a large-scale micropropagation technique by using a scaled-up vessel with forced ventilation where uniform growth was achieved (Zobayed *et al.*, 1999a; Zobayed *et al.*, 2000). In this system air distribution pipes are located inside the vessel to distribute $CO_2$ enriched air uniformly in the culture headspace. This scaled-up vessel contained an automatic nutrient supply system which is attached with the vessel and is able to temporarily immersed the root zone of each of the plantlets and thus can improve the root quality.

#### 5.2.2 Beneficial impact of large-culture vessel

Major advantages of the scaled-up vessel

i) By using the forced ventilation system, the relative humidity can be maintained lower (>90%) in a scaled-up vessel than in a usual (small) culture vessel. It is already known that the high percentage of the relative humidity can cause severe physiological disorder (malfunction of stomata, reduction of epicuticular wax formation etc.) in the culture plantlets.

ii) Number of air exchanges between the culture vessel and the outer atmosphere, $CO_2$ concentration inside the culture vessel and the relative humidity can be controlled as required and thus possible to optimize the growth of the cultured plantlets throughout the culture period.

iii) Due to the forced ventilation, the air movement (velocity) inside the vessel is increased which is known to be beneficial for the cultured plantlets.

iv) A considerably large vessel can be employed and hundreds or thousands of plantlets can be cultured without reducing the fresh and dry weight of the plantlets. Thus, the labor cost can be significantly reduced.

v) The automation/robotaization of the micropropagation system is possible by using this type of vessel.

vi) The uniform growth of the micropropagated plantlets can be achieved.

vii) A large culture vessel with nutrient supply system can also make it possible to measure and control the pH, composition and volume of nutrient solution in the culture vessel. The nutrient solution can be circulated in the culture vessel without any manual assistance.

viii)By lowering the relative humidity and enhancing the $CO_2$ concentration and the PPF, the *in vitro* acclimatization of the plantlets can be ensured, thus do not require any specialized *ex vitro* acclimatization.

ix) In a scaled-up system, plug trays with cells containing the artificial substrate [rockwool, Florialite (mixture of vermiculite and cellulose fiber) or vermiculite] can be used to culture the plantlets. Thus each plant can be pulled up from the cell with minimum damage of roots and thus acclimatized *ex vitro* easily.

x) By enhancing the growth (root and shoot growth), shortening the multiplication/rooting cycle, eradicating the *ex vitro* acclimatization, and thus the production cost should reduced significantly.

## 6. Ventilation in culture vessel

It has recently been proved that the ventilation/aeration in the culture vessel has many advantages over the conventional airtight system. Sealing materials in the conventional *in vitro* culture system are screw caps, aluminum foils, transparent films such as parafilm, cling film etc. which are recently known to restrict the air exchanges between the culture vessel and the outer atmosphere. To improve the air exchange and thus the growth and normality of the plantlets, the vessel needs to be ventilated either i) through natural (diffusive) ventilation or ii) by forced ventilation.

### 6.1 Natural ventilation

The natural ventilation system is based on diffusion through the air gap between the vessel and the lid or through a gas permeable film generally attached on the lid or the wall of the vessel. The diffusion rate through the gas permeable film is proportional to the difference in $CO_2$ concentration inside and outside the vessel and the gas conductance of the gas permeable film. Many types of gas permeable film are commercially available nowadays, such as Millipore filter membrane (pore diameter 0.2-0.5 μm; Millipore Corporation), transparent polypropylene sheets (thickness 25 μm; Courtaulds Films, Bridgewater, Somerset, UK), Teflon membranes (Vent Spots; pore diameter 0.5 μm; Millipore Corporation, USA), Suncap closer, (Sigma, Japan) etc.

### 6.2. Forced ventilation

Another method of ventilation is to flush a particular gas mixture directly through the vessel by applying mechanical force, which is called forced ventilation (Kozai *et al.*, 1999). With this system, the gaseous composition ($CO_2$, water vapor, etc.) of incoming air and forced ventilation rate and/or air current speed in the culture vessel can be controlled relatively precisely by use of a needle valve, mass flow controller or an air pump with an inverter.

### 6.2.1 Historical background of forced ventilation system

The concepts of forced ventilation developed for plant micropropagation are only a decade old. Probably Horn *et al.* (1983) first developed a forced ventilation system for the photoautotrophic culture of soybean cell in suspension culture (250 ml flask). In 1988, Fujiwara *et al.*, developed a forced $CO_2$ enrichment technique using a specially equipped growth chamber. The chamber contains a $CO_2$ control unit consisting of a container with pure liquid $CO_2$, an electric solenoid valve with a relay for opening and closing the solenoid, and an infrared type $CO_2$ controller with an air pump for air sampling. Two identical transparent acrylic boxes containing culture vessels were placed in the chamber. In 1989 another apparatus was reported by Walker, *et al.*, to determine the effect of ventilation on Stage II micropropagation of *Rhododendron* 'P.J.M.'. To provide 0, 300 and 1000 $\mu$mol $mol^{-1}$ $CO_2$ gas treatments, gas mixtures were provided from different gas cylinders, and the atmospheric air was supplied by a diaphragm-type air pump. Adkins *et al.* (1989) developed a continuous gas-flow system for the study of callus growth. The system allows for several gases to be mixed and passed through culture tubes containing callus on Miracloth boats placed on a filter paper bridge.

To generate low (30-65%), medium (70-95%) and high (97% and above) relative humidities in the culture vessels Kozai *et al*, (1990) developed a forced ventilation system. In this system a desiccant (silica gel) contained in a flask was used to dehumidify the air which was blown through it by an air pump. Fujiwara *et al.* developed another device in 1993 for experiments on the physical environmental effects on growth and development of cultures. This device was 70 cm wide, 45 cm deep and 70 cm high. The upper part of this device consists of a light source and a culture box containing culture vessels. The lower part is the control box with a control panel. The $CO_2$ is maintained at a certain level by adjusting the flow rates of pure $CO_2$ from the container (volume : 450 ml) and/or incoming air.

Kitaya and Sakami (1993) made a system for $CO_2$ enrichment of chlorophyllous callus by utilizing the respiratory $CO_2$ produced by a crop of mushrooms. A plant tissue culture box was connected to a mushroom culture box using a semi-closed piping (silicone tube) system attached with ethylene absorbent, air pump, solenoid valve, etc. One important feature was that, unlike others, the source of $CO_2$ was free of cost and did not require any gas cylinder.

To control relative humidities in the culture vessels Fujiwara *et al.* (1993) developed another system where RH of the culture boxes were maintained to control the vessels' RH. This was achieved by connecting the box through an inlet pipe to either distilled water or saturated salt solutions in large Erlenmeyer flasks.

In the same year Yue *et al.*, (1993) developed a forced ventilation system in which the RH of the culture vessels could be controlled by adjusting the RH of inlet outdoor air (385-420 $\mu l\ l^{-1}$ $CO_2$) to each of four constant levels, i.e., 100, 91, 78 and 46%. By using ultra-high $CO_2$ with a forced ventilation system, in 1997, Brent *et al.*, enhanced the plantlet growth. By using humidity induced convective through flow (HICT) ventilation (flow rate was 1-2 ml min$^{-1}$ in a 40 ml vessel) Armstrong *et al.* (1997) improved the embryonic callus growth of coconut and controlled the leaf abscission of *Annona squamosa.* Similar types of system were used to improve the growth and physiology of *in vitro* cauliflower (Zobayed *et al.* 1999b). Forced ventilation recently found to improve the growth and physiology of *Eucalyptus* (Zobayed *et al.*, 2000), sweet potato (Zobayed *et al.*, 1999a) and potato plantlets (Zobayed *et al.*, 1999c).

## 7. Micropropagation system by using a pathogen free culture room

More recently, this scaled-up system has been further extended making the aseptic culture room itself a large culture vessel containing many small sterile trays with plants on the culture shelves and with a common headspace (Kozai *et al.*, 1999). By using this enlarged system, the production of even more quality transplants was achieved relatively easily. Basically, the system is a photoautotrophic micropropagation system (sugar free medium and with enriched $CO_2$ and high PPF) in which, instead of using vessels, the plantlets are grown directly in multicelled trays and are placed on a shelf of the aseptic culture room. A robotic transporter transports the plug trays with or without transplants and no person will be allowed to enter the transport production area except for maintenance. This kind of micropropagation system can be considered as a transplant production system using small cuttings under disease-free conditions or a closed vegetative propagation and transplant production system with artificial lighting.

On the basis of such a concept, research and development of a model system are now underway in Chiba University campus in Japan. Detail description of the system and the advantages have been described by Kozai *et al.* (1999). To prevent any accidental introduction of pathogenic microbes entering the system, the plant material, ventilated air, supplies and workers are carefully inspected. The microbes are trapped and killed by using UV-lights or high-powered electromagnetic fields before entering the culture room with ventilated air. The air in the culture room is continuously passed through a micro-porous filter. Introduction of the regulations set by HACCP (Hazard Analysis and Critical Control Point) and ISO-9001 will become increasingly important in this kind of system for quality assurance of transplants.

## 8. Conclusion

It is reasonable to conclude that the design of culture vessel is an important factor to improve the growth, quality and multiplication of plantlets in the micropropagation technology. Only recently, research has been accelerated on the development of the large culture vessels. By using large culture system (large vessel or sterile culture room) described in this article, the quality of transplants can be much improved, resulting in higher yield at a low cost and it is possible to reduce the use of agro-chemicals in the fields or greenhouses for environmental conservation.

### References

Adkins, S.W., T. Shiraishi, and J.A. McComb. 1989. Callus physiology of rice varieties with differing sensitivity to submergence. In: Proceedings of the International Deepwater Rice Workshop, 1987. Bangkok, International rice research institute. pp. 343-350.

Armstrong, J., E.E.P. Lemos, S.M.A. Zobayed, S.H.F.W. Justin and W. Armstrong. 1997. A humidity-induced convective through-flow ventilation system benefits *Annona squamosa* L. explants and coconut calloid. Ann. of Bot. 79:31-40.

Carrel, A. 1923. A method for the physiology study of tissues *in vitro*. J. exp. Med. 38:407.

Etienne, E., C. Teisson, D. Alvard, M. Lartaud, M. Berthouly, F. Georget, M. Escalona and J. C. Lorenzo.

1999. Temporary Immersion for plant tissue culture. In: Altman *et al.* (eds.), Plant Biotechnology and *In vitro* biology in the 21[st] Century. pp. 629-632.

Fujiwara, K., T. Kozai and I. Watanabe. 1988. Development of a photoautotrophic tissue culture system for shoots and/or plantlets at rooting and acclimatization stages. Acta Hort. 230:153-158.

Fujiwara, K., Y. Kitaya, T. Kozai and M. Hayashi. 1993. A simple miniature culture devise with control units of $CO_2$, relative humidity and light intensity. Proc. of the Symposium on Environmental Control and Effect in Plant Tissue Culture, Tokyo. pp. 128-129.

Heller, R. 1965. Some aspects of the inorganic Nutrition of plant tissue cultures. In: P.R. White and A.R. Grove (eds). Proceedings of an International Conference on Plant Tissue Culture. England. pp. 1–18.

Heo, J. and T. Kozai. 1999. Forced ventilation micropropagation system for enhancing photosynthesis, growth and development of sweet potato plantlets. Env. Cont. in Biol. 37:83-92.

Horn, M.E., J.H. Sherrard and J.M. Widholm. 1983. Photoautotrophic growth of Soybean Cells in suspension culture, 1. Establishment of photoautotrophic cultures. Plant Physiol. 72:426-429.

Kitaya, Y. and K. Sakami 1993. Development of $CO_2$ enrichment system for plantlets in vitro using $CO_2$ produced by mushroom. Abstr. of Annual Meeting of Environ. Cont. in Biol. pp. 172-173.

Kozai, T. and Y. Iwanami 1988. Effects of $CO_2$ enrichment and sucrose concentration under high photon fluxes on plantlet growth of carnation (*Dianthus caryophyllus* L) in tissue culture during the propagation stage. J Jap Soc Hortic Sci. 57:279-288.

Kozai, T., Y. Koyama and I. Watanabe. 1988. Multiplication of potato plantlets *in vivo* with sugar free medium under high photosynthetic photon flux. Acta Hort. 230:121-127.

Kozai, T, C. Kubota, S.M.A. Zobayed, Q.T. Nguyen, F. Afreen and J. Heo. 2000. Developing a mass-propagation system of woody plants. In: K. Watanabe and A. Komamine (eds). Challenge of Plant and Agriculture Sciences to the Crisis of Biosphere on the Earth in the 21st Century, USA. pp. 293-306.

Kozai, T., K. Tanaka, I. Watanabe, M. Hayashi and K. Fujiwara. 1990. Effects of humidity and $CO_2$ environment on the growth of potato plantlets in vivo. J. Jap Soc Hortic Sci. (Suppl. 1):289-290.

Kozai, T. and S.M.A. Zobayed. 2000. Acclimatization. In: R. Spier (ed.) Encyclopedia of Cell Technology, (*in press*).

Kubota, C. and T. Kozai. 1992. Growth and Net Photosynthetic rate of *Solanum tuberosum in vitro* under forced ventilation. Hort Sci. 27(2):1312-1314.

Loeb, L. 1897. 'Uber die Enststehung von Bindegewebe, Leucocyten und roten Blutkorperchen aus Epithel und uber eine Methode, isolierte Gewebsteile zu zuchten' M. Stern and Co. Chicago.

Mousseau, M. 1986. CO2 enrichment *in vitro*: Effect on autotrophic and heterotrophic cultures of *Nicotiana tabacum* (var Samsun). Photosynth Res. 8:187-191.

Roche, T.D., R.D. Long, A.J. Sayegh, and M.J. Hennerty. 1996. Commercial-scale photoautotrophic micropropagation: applications in Irish agriculture, horticulture and forestry. In: Kozai, T. (ed.) Acta Hort. 440:515-520.

Walker, P.N., C.W. Heuser and P.H. Heinemann. 1989. Micropropagation: effects of ventilation and carbon dioxide level on Rhododendron 'P.J.M.' Transactions of the Amer. Soc. of Agri. Eng. 32:348-352.

Willmer, E.N. 1966. Cell and tissues in culture, Methods, Biology and Physiology. (Vol 3). Academic press. London. pp. 1-825.

Yue, D., A. Gosselin and Y. Desjardins. 1993. Effects of forced ventilation at different relative humidities on growth, photosynthesis and transpiration of geranium plantlets in vitro. Can. J. of Plant Sci. 73:249-256.

Zobayed, S.M.A., C. Kubota and T. Kozai. 1999a. Development of a forced ventilation micropropagation system for large-scale photoautotrophic culture and its utilization in sweet potato. In Vitro Plant Cell. and Devel. Biol. – Plant. 34:350-355.

Zobayed, S.M.A., J. Armstrong and W. Armstrong. 1999b. Cauliflower shoot-culture effects of different types of ventilation on growth and physiology. Plant Sci. 141/2:221-231.

Zobayed, S.M.A., F. Afreen, C. Kubota and T. Kozai. 1999c. Stomatal characteristics and leaf anatomy of potato plantlets cultured *in vitro* under photoautotrophic and photomixotrophic conditions. In Vitro Plant Cell. and Devel. Biol. – Plant. 35:183-188.

Zobayed, S.M.A., F. Afreen, C. Kubota and T. Kozai. 2000. Mass propagation of *Eucalyptus camaldulensis* in a scaled-up vessel under *in vitro* photoautotrophic condition. Ann. of Bot. 85(5):587-592.

## PHYSIOLOGY OF *IN VITRO* PLANTLETS GROWN PHOTOAUTO-TROPHICALLY

Fawzia Afreen, Sayed M.A. Zobayed, Chieri Kubota and Toyoki Kozai
Department of Bioproduction Science, Chiba University, Matsudo, Chiba 271-8510, Japan. E-mail: afreen@green.h.chiba-u.ac.jp

**Abstract.** Earlier efforts to improve the growth and multiplication of *in vitro* grown plantlets have focused mainly on the composition of the nutrient medium and the use of growth regulators. The physiological characteristics of micropropagated plantlets cultured under conventional photomixotrophic/heterotrophic conditions in airtight vessels are often abnormal. Low rates of photosynthesis and transpiration, poor water and mineral uptake, non-functional stomata, lack of organisation of palisade and mesophyll tissues in the leaves and poor root quality are the major short-comings of plantlets grown under conventional micropropagation conditions. Recent research has revealed that the aerial environment of the culture vessel can significantly affects the growth and quality of the plantlets, the duration of culture and the cost of production. The use of a more suitable supporting material than conventional agar can also improve the root growth and quality as well as shoot growth. Most recently, the use of a photoautotrophic culture system (sugar free medium) with forced ventilation and with fibrous rooting substrates was proved to be the best for improving the growth and quality of the plantlets and for reducing production costs. This article discusses various ways of improving the physiological conditions and quality of micropropagated plantlets with special emphasis on the culture atmosphere and the rooting substrate.

**Key index words.** $CO_2$ concentration, photosynthesis, relative humidity, stomata, transpiration.

### 1. Introduction

One of the major constraints in micropropagation process is the poor aeration of the vessels. The vessels usually have close-fitting lids to avoid the drying out of the rooting medium and to prevent microbes from entering the vessels. Such methods allow at best very restricted gaseous exchange with the outer atmosphere. Thus, the *in vitro* gaseous environment becomes abnormal and is characterized by high relative humidity (>95%), large diurnal fluctuations in $CO_2$, i.e. $CO_2$ concentration is low during the photoperiod and high during the dark period (Tanaka et al., 1992), accumulations of ethylene or other volatile gases and low air movement in the culture vessel (low $CO_2$ and water vapor diffusion coefficients). As a consequence, growth retardation and various physiological abnormalities are observed in the *in vitro* plantlets (Kozai and Smith, 1995) and when these plantlets are transferred *ex vitro* they show poor survival percentage.

To overcome this situation researches have been focused on the increment of the air exchange rate of the culture vessel headspace with the outer environment either by natural ventilation or by forced ventilation. The natural ventilation system is based on diffusion through a gas permeable film generally attached on the lid or at the wall of the vessel. Various kinds of membranes, e.g. Millipore filter membrane (pore diameter 0.2-0.5 μm; Millipore Corporation, Japan), transparent polypropylene sheets (thickness 25 μm; Courtaulds Films, Bridgewater, Somerset, UK), Teflon membranes (Vent Spots; pore diameter 0.5 μm; Millipore Corporation, USA), Suncap closure, (Sigma, USA) etc.



*C. Kubota and C. Chun (eds.), Transplant Production in the 21st Century,* 238–245.

are commonly used for natural ventilation. The natural ventilation system became popular for improving the plant growth to some extent but the constant air exchange rate limits its possibility for its wide application especially in a large culture vessel and with relatively longer culture periods. To overcome this problem another method of ventilation, called forced ventilation, is developed which involves flushing of a particular gas mixture (or air) directly through the vessel by applying mechanical force. The application of forced ventilation has been proved to be successful for improving the growth and physiology of plant species in many cases. Therefore, to improve the growth and morphology of the plantlet a controlled environment is a prerequisite.

This article discusses various ways of improving the physiological conditions and quality of micropropagated plantlets by appropriately controlling the headspace environment.

## 2. Physiology of the *in vitro* plantlets grown in conventional systems

The major shortcomings of the conventional micropropagation are the low transpiration and photosynthetic ability (Kozai and Smith, 1995) associated with low activities of photosynthetic enzymes such as Rubisco (Ribulose-1,5-bisphosphate carboxylase/oxygenase) and abnormal chlorophyll fluorescence responses (Desjardins et al., 1995). Conventionally, micropropagation is carried away using small, relatively airtight culture vessels containing nutrient media with 20–30 g $l^{-1}$ sucrose (as a carbon source for the plantlets) and with photosynthetic photon flux (PPF) of about 30 – 80 $\mu mol\ m^{-2}\ s^{-1}$. Under these conditions plantlets have a significantly higher content of carbohydrates, mostly in the leaves (Kozai and Zobayed, 2000). Thus the leaves have a poorly developed internal structure and become physiologically abnormal and simply act as a storage organ. Anatomical abnormalities observed in the leaves include:

i) Low epicuticular and cuticular wax development (Grout and Aston, 1977; Sutter and Langhans, 1982)
ii) Unorganized palisade and mesophyll layers (Brainerd et al., 1981)
iii) Stomatal malfunction (Pospisilova, 1996)
iv) Low stomatal density (Kirdmanee et al. , 1995) and
v) Presence of large intercellular spaces in mesophyll layer (Ziv, 1986, 1991).

These anatomical abnormalities affect the physiological processes carried out by the leaves. The physiological processes include:

i) Low *in vitro* photosynthetic rate
ii) Low *in vitro* transpiration rate
iii) Low uptake rate of water and minerals and
iv) High water loss during *ex vitro* acclimatization.

In addition to these low chlorophyll contents of the leaves (Grout and Aston, 1977; Lee et al., 1985) and low percent dry matter (Kozai et al., 1992, 1993) are commonly reported.

The anatomical and physiological disorders eventually affect the morphology of the plantlets. Morphological features are:

i) Poor growth of the plantlet
ii) Restricted leaf area
iii) Poor root development
iv) Low rate of multiplication of plantlets
v) Low number of shoots or leaves usable as explants and
vi) The symptoms of hyperhydricity, i.e. the plantlets look `glassy` with thick, translucent, and brittle leaves (Ziv, 1991).

These led to low survival percentage after transplanting *ex vitro*. The water status and the aerial physical environment of the culture vessel during various stages of the culture are considered as the key factor for abnormal plant morphogenesis (Ziv, 1991). The abnormal morphogenesis and malfunctioning of plants in the conventional system emphasize the need for the optimization of the *in vitro* culture condition.

**3. Photoautotrophic micropropagation system**

Kozai et al., (1988) first proposed the photoautotrophic culture of chlorophyllous plants. In photoautotrophic system the environmental condition of the vessel is characterized by high photosynthetic photon flux (PPF), high $CO_2$ concentration and deduction of sucrose, vitamins or amino acids from the nutrient medium. The most important feature of this system, as opposed to conventional system, is that the culture environment can be controlled in the former system. Major features of the photoautotrophic micropropagation system are discussed below.

**3.1 Ventilation**

Major environmental condition in the conventional airtight culture system is the high relative humidity (about 95%) because the culture vessel containing nutrient medium is sealed and the temperature is approximately constant with time. However, the purpose of sealing the culture vessel at the multiplication and rooting stages is not to keep the relative humidity high, but to prevent microbes from entering the culture vessel and to avoid the drying out of the culture medium. Therefore, the major necessary alterations are the reduction of relative humidity and enrichment of $CO_2$ concentration during the photoperiod. These two alterations can be done simply by improving the number of air exchanges in the culture vessel by introducing the natural or forced ventilation. In photoautotrophic micropropagation system the introduction of ventilation (specially the forced ventilation) in the culture vessel reduces the relative humidity at least 10–15% from that of the conventional airtight system. While introducing ventilation in the vessel, care should be taken not to allow the rapid reduction of the relative humidity and thus the possible drying out of the nutrient medium. However, this can be minimized either by humidifying the culture room (using natural or forced ventilation) or by supplying sterile humid air in the culture vessel (using forced ventilation), or by supplying larger volume of culture medium (with reduced nutrient concentrations). Introduction of ventilation in the culture vessel can:

i) Improves the air movement rate in the vessel
ii) Reduces the leaf-boundary layer resistance
iii) Improves the transpiration and net photosynthetic rates of the plantlets to a great extent and

iv) Reduces the accumulation of ethylene and other toxic gases in the culture vessel.

### 3.2 Enrichment of $CO_2$

To ensure the photoautotrophic condition, the $CO_2$ concentration in the culture vessel headspace is needed to be maintained near or above ambient (350–1000 μmol $mol^{-1}$) concentration. There are two different methods to increase the $CO_2$ concentration in the culture vessel; firstly, by increasing the $CO_2$ concentration in the growth chamber and increasing the number of air exchanges between the culture vessel and the outer environment by ventilating either diffusely (natural ventilation) or forcedly. Secondly, by directly supplying sterile and humidified $CO_2$ enriched air in the culture vessel by forced ventilation.
Enrichment of $CO_2$ in the culture headspace is known to,

i) Increase the net photosynthetic rate of the plantlets and thus the dry matter accumulation
ii) Increase stomatal density and
iii) $CO_2$ enrichment together with low relative humidity improves the leaf anatomy.

### 3.3 Photosynthetic photon flux (PPF)

Increasing the PPF for growth improvement of the plantlets is effective only under $CO_2$ enriched or well ventilated conditions. In general the PPF above the vessel should be at least 100-150 μmol $m^{-2}$ $s^{-1}$ or higher compared with a PPF of 30-50 μmol $m^{-2}$ $s^{-1}$ usually used in conventional systems.

### 3.4 Elimination of sugar from the medium

The presence of high concentrations of sugar in the medium is known to reduce the photosynthetic ability of the chlorophyllous plantlets. Alternatively, the lower metabolic activity associated with carbohydrate assimilation is possibly due to the presence of sugar in the medium as well as exposure to low light intensities (Kozai et al., 1987). Low $CO_2$ concentration during the photoperiod can virtually make the environment unsuitable for photosynthesis by the green plants. In the photoautotrophic culture system, chlorophyllous explants are cultured in sugar-free nutrient medium and thus they are able to produce their own carbohydrates by photosynthesis.

## 4. Improvement of growth and morphology of *in vitro* grown plantlets

Growth and morphogenesis of plants *in vitro* are considerably affected by the headspace environment of the culture vessel. Growth of the plants *in vitro* can be improved significantly by proper environmental control using well-designed culture vessel, appropriate supporting medium etc.

### 4.1 Large Culture Vessel

The use of large culture vessels with a $CO_2$ supply system by means of forced ventilation has been proved to be successful for propagating a large number of quality plantlets. In a specially designed large culture vessel (Zobayed et al., 2000) $CO_2$ enriched air is distributed uniformly in the culture vessel headspace, therefore the growth of the plantlets is relatively uniform. This type of vessel not only makes possible to propagate a large number of morphologically and physiologically superior plants in a limited time and space, but also concurrently reduces variation in size and other physiological characteristics. The use of forced ventilation in large culture vessels has successfully improved the growth and physiology of potato (Kubota and Kozai, 1992) sweetpotato (Zobayed et al., 1999a) and *Eucalyptus* (Zobayed et al., 2000) plantlets.

### 4.2 Supporting material

Selection of a supporting material in addition to controlling the aerial physical environment is also important for achieving better growth. The nature of the supporting material exerts considerable influence on the growth and quality of plantlets *in vitro* and thus, can reduce the cost of the micropropagation if an appropriate one is selected. To supply the plants adequately with nutrients the rooting medium must retain nutrients and provide an environment in which the roots can grow and function normally. In the conventional systems the symptoms associated with media quality are mainly as follows: a) Nutrient deficiencies and toxicities, b) root tip death, c) poor root growth, d) poor shoot growth, and e) poor *ex vitro* survival.

The use of a porous, fibrous support, Florialite (mixture of vermiculite and cellulose fiber) enables the plantlet to develop an extensive rooting system, which increase the water and mineral uptake and thus enhance growth, and survival percent (Afreen-Zobayed et al., 1999). In contrast, in conventional system in agar medium the main adventitious root produce sparse, short lateral roots, which might be related to the poor shoot growth of the plantlets (Afreen-Zobayed et al., 1999).

## 5. Physiology of the *in vitro* plantlets grown photoautotrophically

By increasing the PPF and $CO_2$ concentration and decreasing the relative humidity the growth of the plantlets can be enhanced and normal physiological development of the plantlets can be achieved. Anatomical features observed in the leaf include:

- i) Epicuticular and cuticular wax development is prominent (Short et al., 1987; Zobayed et al., 1999b)
- ii) Well organized palisade and mesophyll layers. The palisade cells are more closely packed and has smaller intercellular spaces (Zobayed et al., 1999b)
- iii) Presence of thick palisade layer (Kirdmanee et al., 1995) and
- iv) Normal functional stomata (Zobayed et al., 1999b) and high stomatal density (Kirdmanee et al., 1995).

These anatomical features enable the plant to perform the normal physiological processes. Physiological features include:

- i) Increased net photosynthesis
- ii) Increased *in vitro* transpiration
- iii) High uptake of water and minerals
- iv) Higher diffusive resistance of the leaves
- v) Controlled water loss during *ex vitro* acclimatization and
- v) Elimination of the symptom of hyperhydricity.

In addition to these high chlorophyll contents of the leaves and relatively higher percent dry matter are commonly reported.

The morphological features of the plantlets include:

i) Enhanced growth of the plantlet
ii) Increased leaf area
iii) Enhanced root formation and
iv) Increased multiplication rate of plantlets.

As the plantlets have functional stomata and higher wax deposition, therefore after transplanting *ex vitro* the leaves are able to control the water loss and usually do not wilt rapidly. Thus survival percentage is increased and the growth is faster compared to that of the conventionally grown plantlets. Usually wax deposition on the leaf is a plant's defense mechanism used to protect the plant from desiccation. In conventional system due to the high relative humidity ($\geq$95%), plant does not require to protect itself from desiccation and thus either no or very thin layer of epicuticular or cuticular wax is deposited. Due to reduced amount of wax deposition and nonfunctional stomata the conventionally grown plantlets after transplanting *ex vitro* loss their water very quickly and wilt resulting in low survival percentage.

## 6. Necessity of the normal physiology of the *in vitro* grown plantlets

It is necessary to improve the physiology of micropropagated plantlets. In the conventional system, due to the physiological abnormalities, plants loss significant amount of water immediately after transplanting *ex vitro*. Thus a large number of plants die and even in the survived plants, due to limited number of non-wilted leaves and poorly developed roots, slow growth is reported. On the other hand the photoautotrophic culture system, especially under forced ventilation, ensures the *in vitro* acclimatization during the rooting period. As a result, during the *ex vitro* acclimatization, the leaves do not wilt and survive well. It has been reported that the physiological process of *in vitro* plant is disrupted during the multiplication as well as the rooting stage. In photoautotrophic micropropagation, usually nodal cuttings each with one or two unfolded leaves are used as explants. The larger the leaf area of explant, the greater the initial growth rate of regenerated shoot(s). This is because the leaf is a sole organ, which produces carbohydrates by photosynthesis and their secondary metabolites for growing regenerated shoot(s). To maximize the net photosynthetic rate, it is necessary to use an explant with healthy leaves having normal physiological functions and thus is able to photosynthesize efficiently.

In case of cuttings with multiple shoots, a relatively large shoot with base is used as explant. In the conventional heterotrophic or photomixotrophic conditions, leaves (and stems) are used mainly as a storage organ. The number of explants obtained from one regenerated plant is greater under photoautotrophic than heterotrophic or photomixotrophic (conventional) condition, even when the number of regenerated shoots is the same for both. Many of the multiple shoots regenerated under heterotrophic condition are often too small or hyperhydrated and are not considered as usable explants. It should be noted that plant growth regulators are needed to increase the number of multiple shoots even under the photoautotrophic condition in most cases.

## 7. Conclusions

Considering the above discussion it can be concluded that almost all the physiological processes of the plantlets including two major processes, photosynthesis and transpiration are seriously hampered by the environmental condition in the conventional culture vessel. The recent advancement of the photoautotrophic culture system (sugar free medium) with forced ventilation and with fibrous rooting substrates has brought on a revolution in the field of commercial micropropagation. This technique has proved to be the best for optimizing the environmental conditions and thus improving the growth and physiology of the plantlets and for reducing production costs.

**References**

Afreen-Zobayed, F., S.M.A. Zobayed, C. Kubota, and T. Kozai. 1999. Supporting material affects the growth and development of *in vitro* sweetpotato plantlets cultured photoautotrophically. In Vitro Plant Cell. and Develop. Biol. - Plant 35:470-474.

Brainerd, K.E., L.H. Fuchigami, S. Kwiatkowski and C.S. Clark. 1981. Leaf anatomy and water stress of aseptically cultured 'Pixy' plum grown under different environments. HortSci. 16:173-175.

Desjardins, Y.C., C. Hdider and J. de Riek. 1995. Carbon nutrition in vitro In: J. Aitken-Christie, T. Kozai and M.A.L. Smith (eds.), Automation and Environmental Control in Plant Tissue Culture. Kluwer Academic Publishers, Dordrecht, The Netherlands. pp. 441-471.

Grout, B.W.W. and M.J. Aston. 1977. Transplanting cauliflower plants regenerated from meristem culture. I. Water loss and water transfer related to changes in leaf wax and to xylem regeneration. Hort. Res. 17:1-7.

Kirdmanee, C., Y. Kitaya and T. Kozai. 1995. Effects of $CO_2$ enrichment and supporting material *in vitro* on photoautotrophic growth of *Eucalyptus* plantlets *in vitro* and *ex vitro*: Anatomical comparisons. Acta Hort. 393:111-115.

Kozai, T. and M.A.L. Smith. 1995. Environmental control in plant tissue culture - general overview. In: J. Aitken-Christie, T. Kozai and M.A.L. Smith (eds.), Automation and Environmental Control in Plant Tissue Culture. Kluwer Academic Publishers, Dordrecht, The Netherlands. pp. 910-927.

Kozai, T. and S.M.A. Zobayed. 2000. Acclimatization. In: Spier, R. (eds.), Encyclopedia of Cell Technology (in press).

Kozai, T., H. Oki and K. Fujiwara. 1987. Effects of $CO_2$ enrichment and sucrose concentration under high photosynthetic photon fluxes on growth of tissue-cultured *Cymbidium* plantlets during the preparation stage. In: G. Ducaté, M. Jacob and A. Simeon (eds.), Plant Micropropagation in Horticultural Industries. Presses Universitaires, Liége, Belgium. pp. 47-54.

Kozai, T., Y. Koyama and I. Watanabe. 1988. Multiplication of potato plantlets *in vivo* with sugar free medium under high photosynthetic photon flux. Acta Hort. 230:121-127.

Kozai, T., K. Fujiwara, M. Hayashi and J. Aitken-Christie. 1992. The *in vitro* environment and its control in micropropagation. In: K. Kurata and T. Kozai (eds.), Transplant Production Systems. Kluwer Academic Publishers, Dordrecht, The Netherlands. pp. 247-282.

Kozai, T., K. Tanaka, B.R. Jeong and K. Fujiwara. 1993. Effect of relative humidity in the culture vessel on the growth and shoot elongation of potato (*Solanum tuberosum* L.) plantlets *in vitro*. J. Japan. Soc. Hort. Sci. 62:413-417.

Kubota, C. and T. Kozai. 1992. Growth and Net Photosynthetic rate of *Solanum tuberosum in vitro* under forced ventilation. Hort. Sci. 27(2):1312-1314.

Lee, N., H.Y. Wetzstein and H.E. Sommer. 1985. Effect of quantum flux density on photosynthesis and chloroplast ultrastructure in tissue-cultured plantlets and seedlings of *Liquidamber styraciflua* L. towards improved acclimatization and field survival. Plant Physiol. 78:637-641.

Pospisilova, J. 1996. Hardening by abscisic acid of tobacco plantlets grown *in vitro*. Biol. Plant. 38:605-609.

Short, K.C., J. Warburton and A.V. Roberts. 1987. *In vitro* hardening of cultured cauliflower and chrysanthemum plantlets to humidity. Acta Hort. 212:329-334.

Sutter, E.G. and R.W. Langhans. 1982. Formation of epicuticular wax and its effect on water in cabbage plants regenerated from shoot-tip culture Can. J. Bot. 60:2896-2902.

Tanaka, K., K. Fujiwara and T. Kozai. 1992. Effects of relative humidity in the culture vessel on the transpiration and net photosynthetic rates of potato plantlets *in vitro*. Acta Hort. 319:59-64.

Ziv, M. 1986. *In vitro* hardening and acclimatization of tissue culture plants. In: L.A. Withers and P.G. Alderson (eds.), Plant Tissue Culture and its Agricultural Applications. Butterworths, London. pp. 187-196.

Ziv, M. 1991. Vitrification : morphological and physiological disorders of *in vitro* plants. In: P.C. Debergh and R.H. Zimmerman (eds.), Micropropagation. Kluwer Academic Publishers, The Netherlands, pp 45-69.

Zobayed, S.M.A., C. Kubota and T. Kozai. 1999a. Development of a forced ventilation micropropagation system for large-scale photoautotrophic culture and its utilization in sweet potato. In Vitro Plant Cell. and Develop. Biol. 34:350-355.

Zobayed, S.M.A., F. Afreen-Zobayed, C. Kubota and T. Kozai. 1999b. Stomatal characteristics and leaf anatomy of potato plantlets cultured *in vitro* under photoautotrophic and photomixotrophic conditions. In Vitro Plant Cell. and Develop. Biol. - Plant 35:183-188.

Zobayed, S.M.A., F. Afreen-Zobayed, C. Kubota and T. Kozai. 2000. Mass propagation of *Eucalyptus camaldulensis* in a scaled-up vessel under *in vitro* photoautotrophic condition. Ann. Bot. 85:587-592.

# ENHANCED GROWTH OF *IN VITRO* PLANTS IN PHOTOAUTOTROPHIC MICROPROPAGATION WITH NATURAL AND FORCED VENTILATION SYSTEMS

Quynh Thi Nguyen[1*], Toyoki Kozai[2] and Jeongwook Heo[3]
[1] Institute of Tropical Biology, NCST-VN, 1 Mac Dinh Chi Street, Ho Chi Minh City, Vietnam. E-mail: qtnguyen@netnam2.org.vn
[2] Faculty of Horticulture, Chiba University, Matsudo, Chiba 271-8510, Japan.
[3] Chungbuk National University, Chungbuk 361-763, Korea.

**Abstract.** Recently, photoautotrophic micropropagation has been proved to be one of effective methods of lowering production costs in micropropagation by the reduction of biological contamination, the enhanced growth of *in vitro* plantlets, the increased percent survivals and elimination of acclimatization in the *ex vitro* stage. In naturally ventilated micropropagation, the air outside and inside the culture vessel exchanges through microporous membrane filters covering holes made on cap or upright sides of the culture vessel, or through a gas permeable material replacing the closure completely. In this system, several plant species, such as coffee, banana, Paulownia and sweetpotato, showed no significant difference in their growth when cultured under photoautotrophic conditions as compared with that under photomixotrophic conditions. Furthermore, the photoautotrophic growth of *in vitro* plantlets was significantly enhanced when air-porous supports, such as cellulose plug, vermiculite, Florialite (a mixture of vermiculite and cellulose fibers) or perlite were used instead of agar or gelrite. However, in large-scale production, the control and maintenance of gaseous components, especially carbon dioxide, in the headspace of culture vessels can not be achieved easily in naturally ventilated, small vessels. A forced ventilation system can help to control the ventilation rate and $CO_2$ concentration inside the culture vessel by means of airflow rate meter and $CO_2$ controller. This system may bring more benefits for large-scale micropropagation when large boxes/chambers are used. In case of coffee and sweetpotato, the growth of photoautotrophic plantlets was greater in forced ventilation system as compared with those in a natural ventilation system.

**Key index words.** Acclimatization, forced ventilation, large-scale micropropagation, natural ventilation, photoperiod, quality transplants.

## 1. Introduction

Production of quality transplants at low costs is becoming a critical issue of plant propagation in the 21st century, including production of trees needed for re-forestation, afforestation and desert reclamation, food crops for lessening starvation, plants for manufacturing bio-degradable plastics and hydrogen gas usable as a clean-burning fuel for automobiles, industrial crops, medicinal plants, horticultural plants, and plants used for cleanning up the environmental pollution (Kozai, 1998). Transplants production using micropropagation techniques has more benefits than those using seed or vegetative propagation in terms of genetic uniformity, virus-free or pathogen-free propagules and scheduled production. However, the heterotrophic or photomixotrophic micropropagation, using a sugar-containing culture medium, has become costly due to biological contamination, morphological disorders and low photosynthetic capacity of *in vitro* plantlets. This results in low percent survivals ex vitro and requires acclimatization



*C. Kubota and C. Chun (eds.), Transplant Production in the 21st Century*, 246–251.

prior to the *ex vitro* stage. Photoautotrophic micropropagation, using a sugar-free culture medium, has been studied intensively since the last decade and demonstrated as an effective method for lowering transplant production cost. The cost reduction was achieved due to promoted photosynthetic capability of *in vitro* plantlets, decreased biological contamination and abnormal cultures, and thus, increasing percent survivals in the *ex vitro* stage (Kozai & Jeong, 1993). In large-scale photoautotrophic micropropagation, plantlets can be grown in naturally ventilated, small vessels or in forcedly ventilated large boxes. In this article, the application of these systems are discussed with regards to different plant species including coffee, banana, Paulownia and sweetpotato.

## 2. Photoautotrophic growth of *in vitro* plantlets under natural ventilation systems

### 2.1 Coffee

Nguyen et al. (1999a) showed *in vitro* coffee plantlets could grow photoautotrophically (on sugar-free medium) as fast as photomixotrophically (on sugar-containing medium) in naturally ventilated vessels, i.e., air exchanges through membrane filter discs (Milli-Seal, Millipore, Tokyo, Japan) covering holes on the closure of culture vessel. The ability to develop photoautotrophically of *in vitro* plantlets was mostly affected by $CO_2$ concentration and PPF (photosynthetic photon flux) (Kozai & Jeong, 1993). In case of coffee, these two factors also have critical effects on the growth of *in vitro* plantlets. Single nodal cuttings of *in vitro* coffee (*Coffea arabusta*) plantlets were cultured on half-strength Murashige and Skoog (1962) medium with no sucrose, vitamins and plant growth regulators. These cuttings grew in 370 $cm^3$ Magenta polycarbonate box-type vessels containing 10 g Florialite (a mixture of vermiculite and cellulose fibers with high air porosity, Nissinbo Industries, Inc., Japan) per vessel as a supporting material. The cultures were put under the photoperiod of 16 h $d^{-1}$ and average air temperature of 25 °C. Data of increased fresh weight (IFW), shoot length (SL), leaf area (LA) and net photosynthetic rate ($P_n$) were collected and analyzed using ANOVA and LSD test (Table 1).

*Table 1* Effects of high / low $CO_2$ concentration and PPF on the photoautotrophic growth of *in vitro* coffee plantlets on day 45.

| Treatment Code | IFW (mg) | SL (mm) | LA ($cm^2$) | $P_n$ ($\mu$molh$^{-1}$/plantlet) |
|---|---|---|---|---|
| HH | 390b | 17b | 51b | 14 |
| HL | 760a | 31a | 81b | 15 |
| LH | 220c | 8c | 27c | 5 |
| LL | 200c | 10c | 24c | 4 |

a, b, c within a column show a significant difference at $p = 0.05$. As treatment codes, H and L on the left represent $CO_2$ concentration of 1450 and 450 μmol $mol^{-1}$, respectively. L and H on the right represent PPF of 150 and 350 μmol $m^{-2}$ $s^{-1}$, respectively.

The plant growth was significantly greater in the $CO_2$ enriched condition (1450 μmol $mol^{-1}$) than in the $CO_2$ non-enriched condition (450 μmol $mol^{-1}$). However, under the same $CO_2$ concentration, increasing PPF from 150 to 350 μmol $m^{-2}$ $s^{-1}$ did not increase $P_n$ of *in vitro* coffee plantlets significantly. Although no significant difference in plantlet growth was found at low $CO_2$ concentration either under high or low PPF, coffee plantlets grew significantly faster at $CO_2$ of 1450 μmol $mol^{-1}$ and PPF of 150 μmol $m^{-2}$ $s^{-1}$ (Table 1).

## 2.2 Banana

Nguyen et al. (1999b) showed that there was no significant difference in growth of *in vitro* banana (*Musa* spp.) plantlets under photoautotrophic as compared with those under photomixotrophic condition, and that plantlet growth increased significantly with the increase in photoperiod up to 16 h $d^{-1}$. PPF and ventilation rate also play an important role in the development of *in vitro* banana plantlets. Banana shoots of approximately 1.5 cm long with one unfolded leaf were cultured on half-strength Murashige and Skoog (1962) agar (8g $l^{-1}$) medium without sucrose, vitamins and plant growth regulators. Three levels (50, 100 and 200 μmol $m^{-2}$ $s^{-1}$) of PPF and two levels (1 and 2.5 $h^{-1}$) of number of air exchanges were investigated. The photoperiod was 16 h $d^{-1}$ and air temperature at 25 ± 2 °C. The results showed that under PPF of 50 μmol $m^{-2}$ $s^{-1}$, there was no significant difference in plantlet growth between large and small number of air exchanges. When PPF was raised higher than 50 μmol $m^{-2}$ $s^{-1}$, the effect of ventilation rate on plantlet growth was expressed (Table 2). The high ventilation rate of the vessel resulted in the greater increase in fresh weight, dry weight and shoot length as compared with those under the low ventilation rate, both with PPF of 100 and 200 μmol $m^{-2}$ $s^{-1}$ (Table 2).

*Table 2* Mean increased fresh weight (FW), dry weight (DW) and shoot length (SL) of the banana plantlets on day 28 as affected by number of air exchanges and PPF.

| Treatment | | FW | DW | SL |
|---|---|---|---|---|
| PPF (μmol $m^{-2}$ $s^{-1}$) | No. air exchanges ($h^{-1}$) | (mg) | (mg) | (mm) |
| 50 | 2.5 | 388 c[z] | 28 c | 2 b |
| 50 | 1 | 372 c | 28 c | 3 a |
| 100 | 2.5 | 728 a | 60 a | 3 a |
| 100 | 1 | 514 b | 45 b | 2 b |
| 200 | 2.5 | 703 a | 62 a | 2 b |
| 200 | 1 | 514 b | 43 b | 1 c |
| ANOVA | | | | |
| PPF (A) | | **[y] | ** | ** |
| No. air exchanges (B) | | ** | ** | ** |
| A x B | | ** | ** | ** |

[z] Means in a column followed by same letters are not significantly different at $p \leq 0.05$ by Duncan's multiple range test.

[y] NS, * and ** : Not significant and significant at $p \leq 0.05$ and 0.01, respectively.

## 2.3 Paulownia

*In vitro* Paulownia (*Paulownia fortunei*) plantlets grew more vigorously under photoautotrophic than photomixotrophic condition, especially when supporting materials with high air porosity, such as vermiculite and Florialite, were used (Nguyen et al., *unpublished*). The role of other micro-environmental factors, such as photoperiod and culture vessel ventilation, involved in the enhancement of photosynthetic capacity of in vitro Paulownia plantlets was also demonstrated. Single nodal cuttings of Paulownia plantlets with two opposing leaves on each cutting were used as explants. On half strength MS agar medium without plant growth regulators, the cuttings grew phototautotrophically in 800 mL Phytacon column-type vessels (Sigma Co., USA). There was a significant increase in dry weight, shoot length under high photoperiod and large number of air exchanges (Table 3). Although percent dry matter was not significantly different among treatments, plantlets growing under 2.5 $h^{-1}$ number of air exchanges had percent dry matter lower than those under 1 $h^{-1}$ number of air exchanges in both two photoperiod levels. (Table 3).

*Table 3* Effects of photoperiod and ventilation levels on the increased dry weight (IDW), shoot length (SL) and percent dry matter (DM) of *in vitro* Paulownia plantlets cultured photoautrophically.

| Treatment | | IDW | SL | DM |
|---|---|---|---|---|
| Photoperiod (h $d^{-1}$) | No. of air exchanges ($h^{-1}$) | (mg) | (mm) | (%) |
| 10 | 1 | 22 c[z] | 11 c | 10 |
| 10 | 2.5 | 43 b | 20 b | 9 |
| 16 | 1 | 35 c | 18 b | 10 |
| 16 | 2.5 | 77 a | 36 a | 9 |
| ANOVA | | | | |
| Photoperiod (A) | | ** | ** | NS[y] |
| Ventilation (B) | | ** | ** | ** |
| A x B | | ** | ** | NS |

[z] Means in a column followed by same letters are not significantly different at $p \leq 0.05$ by LSD test.
[y] NS, * and ** : Not significant and significant at $p \leq 0.05$ and 0.01, respectively.

## 3. Forced ventilation systems for photoautotrophic micropropagation

In the natural ventilation method, the $CO_2$ concentration inside the sterile vessels increases with the increase of $CO_2$ concentration in the surrounding atmosphere. However, in large-scale production using several small vessels, concentrations of $CO_2$ and other gaseous components in culture vessels tend to be different from one to another; and thus, the control and maintenance of carbon dioxide at optimal concentration for enhancing plant growth at each growth phase turns to be difficult. This drawback can be overcome by using the forced ventilation method, in which a particular gas mixture is flushed directly through the culture vessel and help to increase $CO_2$ concentration and ventilation rate inside the vessel to a desired level. With forced ventilation method, large culture boxes/chambers can be used in photoautotrophic micropropagtion. Kubota and Kozai (1992) showed a significant increase in growth of potato plantlets under forced ventilation as compared with that under natural ventilation. Heo and Kozai (1997) developed a forced ventilation system for enhancing the uniform growth of *in vitro* sweetpotato plantlets inside a large box (V = 12 L). Sweetpotato plantlets significantly increased shoot dry weight and leaf area when cultured photoautotrophically in this forced ventilation system (Fig. 1).

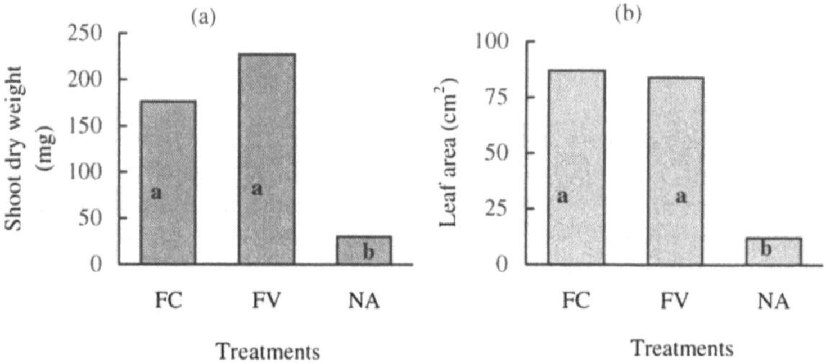


*Figure 1* Shoot dry weight (a) and leaf area (b) of sweetpotato plantlets on day 22 as affected by ventilation method. For treatment codes, F and N on the left represent ventilation types: forced and natural, respectively. C, V and A on the right represent cellulose plug, vermiculite and agar, respectively.

In the same system, *in vitro* coffee (*Coffea arabusta*) plantlets also increased fresh weight, shoot length and leaf area significantly as compared with those in natural ventilation (Table 4).

*Table 4* Increased fresh weight (IFW), shoot length (SL) and leaf area (LA) of coffee (*Coffea arabusta*) plantlets cultured *in vitro* for 40 days.

| Treatment | | | | | |
|---|---|---|---|---|---|
| Ventilation method | Number of air exchanges ($h^{-1}$) | PPF[z] ($\mu$mol $m^{-2}$ $s^{-1}$) | IFW (mg) | SL (mm) | LA ($cm^2$) |
| Natural (control) | 3.9 | 150 | 338 | 11 | 14 |
| Forced | 2.7 | 250 | 260** | 16* y | 15NS |
| Forced | 2.7 | 150 | 116** | 13* | 12* |
| Forced | 5.9 | 250 | 604** | 27** | 23** |
| Forced | 5.9 | 150 | 569** | 36** | 24** |
| *ANOVA* [x] | | | | | |
| Ventilation (A) | | | ** | ** | ** |
| PPF (B) | | | * | ** | NS |
| A x B | | | ** | ** | * |

[z] PPF from day 16 to day 40 of the cultured period.

[y] NS, *, **: nonsignificantly or significantly different from the control treatment at $p \leq 0.05$, respectively, according to t-test.

[x] ANOVA was applied for 4 treatments with forced ventilation method. NS, *, **: nonsignificant or significant at $p \leq 0.05$ or 0.01, respectively.

## 4. Conclusion

Both natural and forced ventilation systems in photoautotrophic micropropagation have obviously been attributed to the full expression of photosynthetic ability of *in vitro* plantlets. However, using forced ventilation method will help to overcome the difficulty in controlling and maintaining the optimal $CO_2$ concentration and ventilation rate inside the culture vessel during the culture period. Thus, productivity of *in vitro* plantlets will be promoted by the faster and uniform plant growth. Automation and computerization, then, can be applied to the forced ventilation system as an attribution to the cost effectiveness of the system.

**References**

Heo, J. and T. Kozai. 1997. Photoautotrophic growth enhancement of in vitro plug plant using a forced ventilation micropropagation system. Abstr. Autumn Meeting of Japan Society of Horticultural Science. (written in Japanese with English abstract) pp. 298-299.

Kozai, T. and B.R. Jeong. 1993. Environmental control in plant tissue culture and its application for micropropagation. *In* Hashimoto Y., Bot G.P.A., Day W., Tantau H.J. and Nonami H. (eds.) The Computerized Greenhouse: Automatic Control Application in Plant Production. Academic Press, Inc., San Diego, U.S.A. pp. 95-116.

Kozai, T. 1998. Transplant production under artificial light in closed systems – quality improvement of seedlings and plantlets by environmental control. Proc. of the 3rd Asian Crop Science Conference, April 27 – May 2, 1998, Taichung, Taiwan. pp. 296-308.

Kubota, C. and T. Kozai. 1992. Growth and net photosynthetic rate of *Solanum tuberosum* in vitro under forced and natural ventilation. HortScience 27:1312-1314.

Murashige, T & F. Skoog. 1962. A revised medium for rapid growth and bioassays with tobacco tissue cultures. Physiol. Plant. 15:473-497.

Nguyen Q.T., T. Kozai and U.V. Nguyen. 1999a. Effects of sucrose concentration, supporting material and number of air exchanges of the vessel on the growth of *in vitro* coffee plantlets. Plant Cell. Tissue & Organ Cult. 58:51-57.

Nguyen Q.T., T. Kozai, K.L. Nguyen and U.V. Nguyen. 1999b. Photoautotrophic micropropagation of tropical plants. In Altman A., Ziv M. and Izhar S. (eds.) Plant Biotechnology and In Vitro Biology in the 21st Century (Proc. The 9th Intl. Congr. of IAPTC, June 14-19, 1998, Jerusalem, Israel). Kluwer Academic Publishers, Dordrecht, The Netherlands. pp. 659-662.

# MICROPROPAGATION OF ORNAMENTAL PLANTS USING BIOREACTOR SYSTEM

Kee Yoeup Paek, Eun-Joo Hahn, Jeongwook Heo and Seong Ho Son[1]
Research Center for the Development of Advanced Horticultural Technology, Chungbuk National University, Chungju 361-763, Korea.
[1]Division of Biotechnology, Forestry Research Institute, Forestry Administration, Suwon 441-350, Korea. E-mail: paekky@cbucc.chungbuk.ac.kr

**Abstract.** Several types of bioreactors were incorporated to verify the effectiveness for mass culture of ornamental plants. Protocorm-like bodies (PLBs) were proliferated from flower stalk of *Phalaenopsis* by using bioreactors. Hyponex medium containing 1% potato and 0.1% activated charcoal supplied by ebb-and-flow method was investigated as the best for PLBs production (17 or more per PLB). The result partially suggested the possibility for mass propagation of PLBs through bioreactor system. Several cultivars of *Lilium* were successfully mass propagated using different types of bioreactors. In the case of 'Marcopolo', the growth of bulblets in balloon type bubble bioreactor (BTBB) was nearly 10 folds faster than that of the solid medium. When 120 single-nodal explants (about 1 cm in length) were cultured for 8 weeks under 16hr light condition, more than 20,000 shoot cuttings of virus-free potatoes could be obtained in 20 L sized BTBB. The percentage of *ex vitro* rooting of the cuttings during weaning stage revealed as more than 95% under 50% shade. Similar results were obtained from the culture of single-nodal explants using bioreactor in chrysanthemum and sweet potato. Based on our above results, we suggest that large-scale cultures of plant cell, tissue and organ using bioreactor system should be quite a feasible approach in terms of increase of multiplication rate and reduction of production cost when compared with conventional plant tissue culture method.

**Key index words.** ballon type bubble bioreactor, ebb-and-flow bioreactor, *Lilium*.

## 1. Introduction

*In vitro* micropropagation methods are still characterized by high labor cost and low degree of automation. The main disadvantages of these techniques are the lack of automatic regulation systems for control of physical and chemical properties of internal vessel and the need to cut regenerated shoot clumps individually by hand (Jones and Sluis, 1991). Therefore, it is important to create new strategies for the development of propagation methods to overcome present limitation of conventionally used techniques (Aitken–Christie, 1991). The application of bioreactors may help to develop more advanced micropropagation methods and production of useful secondary metabolites. Automation of somatic embryogenesis and/or organogenesis through bioreactors has been advanced by several researchers as a possible way of reducing labor costs of micropropagation (Preil et al., 1988; Levin et al., 1988). Bioreactor has traditionally been used for bacterial fermentation and/or for large-scale production of secondary metabolites via plant cell culture (Son et al., 1999a,b; Seon et al., 1998). Since microbial fermentation techniques were firstly reported in studies on growth kinetics of higher plant cell suspensions (Tulecke and Nickell, 1959), major progress has occurred in the area of large-scale liquid culture and in the development of bioreactor process control system. Industrial-scale plant cell cultivation has been accomplished up to a bioreactor



*C. Kubota and C. Chun (eds.), Transplant Production in the 21$^{st}$ Century,* 252–257.

volume of 75,000 L (Ritlershaus et al., 1989). For plant propagation purposes bioreactors with a volume of a few liters are sufficiently large because high numbers of propagules can be obtained in small vessels with special reference to somatic embryogenesis. Although the establishment of somatic embryogenesis has been reported on more than 200 species, just a few cases of the species could be successfully propagated through bioreactor system at the aim for commercialization (Nishimura et al., 1993). This is mainly due to the lack of systematic experiments in bioreactors, where many of interactions between biological and physical factors can be revealed. Bioreactor with computer control systems offers theoretically several advantages over conventional micropropagation procedures due to possibilities of automation, saving labor, and reduction of production costs. Up to now, this assessment could be verified in practice due to increasingly high activities in bioreactor application for plant propagation. Nevertheless, to fine out stable genotypes from vegetatively multiplied horticultural crops seemed to be prerequisite for the application of bioreactor system.

## 2. Bioreactor design for plant cell and tissue culture

In order to achieve effective automated production system via plant cell and tissue culture, original methods for plant production technology had to evolve from a small-scale to large volume. In this point of view, the incorporation of liquid culture system seemed to be more reliable for mass culture of plant biomass and/or plant propagules (Leathers et al., 1995). The basic concept of bioreactor is to provide optimum growth conditions by regulating various chemical and environmental factors. Bioreactor design has been primarily developed for specific requirements of embryogenic or organogenic suspensions. Equipment for universal application does not exist. The optimum bioreactor design for maximum growth and proliferation of differently responding species must be evaluated systematically. Bioreactors for plant propagation can be classified as affected by the type of the agitation system Reactors agitated by air flow such as bubble column and air lift bioreactor. In bubble column bioreactors the dispersion of compressed air at the bottom of the vessel is performed by perforated plates or other devices such as fine or coarse sparger for air distribution. In air lift bioreactors the presence of draught tubes inside the vessel stabilize the liquid circulation patterns, resulting in a well defined flow. The draught tubes divide the vessel into two parts. One of them is gassed, the other serving as loop. The circulation of the liquid is due to the difference in air content between the aerated part and the nonaerated down stream part. Frequently the draught tube is modified by additional baffles. Large air lift tower loop reactors are very common in industrial applications (Preil, 1991). Various modifications have been developed. Aerated mechanically stirred vessels to prevent aggregation of cells or organs such as adventitious root for the production of secondary metabolites now employ the first position. Typical stirrers are propeller and pitched-blade turbine stirrer that cause axial fluid motion. The effects of the number of blades, their internal position, the size of blade in stirrers, and the shape of vessel have to be determined to ensure optimum conditions, especially for culture of shear sensitive cells. In any bioreactor, shearing stress caused by high agitator speed and oxygen limitation favored by low agitator speed must be avoid because both factors are responsible for damage of cells and leaching of intercellular acidic substances decreasing the pit of medium. Prior to describing the most important bioreactor configurations used in the cultivation of plants, it will be useful to examine some of functions of the bioreactors and their consequence to plant cells or organs.

**3. Control of physical and chemical factors**

Methods of bioreactor process control for plant propagation have to be applied from quite well known industrial fermentation technology. Several factors such as temperature, agitator speed, medium pH, dissolved oxygen, redox-potential, $CO_2$ concentration, and medium component can influence the biomass increase and the content of secondary metabolite.

*Temperature*: Temperature inside bioreactor is kept constant at the range of 22 to 25°C during culture by means of internal vessel heat exchanger, by circulation water in double jacket culture vessel, or by keeping the bioreactors in temperature controlled culture room.

*Agitator speed*: To keep the cells permanently in suspension status, usually 50 rpm, or upwards, is required. Setting of cells at agitation speed below 50 rpm can be prevented by the we of large-blade stirrers.

*Medium pH*: Changes of medium pH indicate alterations of internal physical and chemical conditions. These changes appeared to be related to the balance between ammonium and nitrate uptake. Generally ammonium uptake increased with increasing pH while utilization of nitrate decrease with increasing pH. However, the pH can be altered by various factors other than the ammonium-nitrate balance. For example, when regulating the oxygen content periodically between 10 and 80%, the pH changed about 0.2 pH units depending on the volume of medium and the size of culture vessel. The recording precise fluctuations medium pH in computer controlled bioreactor cultures will improve the repeatability of complex biological process like somatic embryogenesis.

*Redox potential*: Changes of redox potential in the medium may give information on technical disturbances during culture. In many cases redox and oxygen registration run in parallel. Therefore, simultaneous interpretation with curves of pH and dissolved oxygen are recommended.

*Effects of accretion*: Determination of dissolved oxygen in liquid medium indicates the amount of oxygen available for cell metabolism. Since oxygen is poorly soluble in medium, its content is rapidly exhausted when air supply is interrupted, especially at high cell or organ density. Oxygen, which is rapidly consumed by logarithmic cell or organ growth, must be continuously supplied to the culture medium to prevent its decrease below the critical level that may differ from culture methods and species to species. In bioreactor experiments with poinsettia cell suspensions, the specific growth rate was inhibited drastically below 10% of air saturation (Preil et al., 1988). In *Lilium* 'Marcopolo' in 20 L BTBB, the concentration of dissolved oxygen in liquid medium rapidly decreased from 8 $mg{\cdot}L^{-1}$ at initial stage to 4 $mg{\cdot}L^{-1}$ after 2 weeks and remained at 4 $mg{\cdot}L^{-1}$ constantly until harvest. It was known that concentration of dissolved oxygen in liquid medium is not only 8 $mg{\cdot}L^{-1}$ under saturated state but also bulblets cultured in liquid medium face up to insufficient oxygen compared to agar medium (Kim, 1999). Takahashi et al. (1992) reported also that bulblet growth could be enhanced by the enforced oxygen concentration. For many suspension cultures it was shown that oxygen influences the metabolic activity of cells. High aeration rates in air lift bioreactors, however, inhibited the growth of *Catharanthus roses* suspensions (Smart and Fowler. 1984). One hypothesis explaining this result based on the fact that plant cell cultures produce a number of volatile compounds that are known to affect the physiology of cells, e.g. $CO_2$, ethylene, and ethanol (Thomas and Murashige, 1979). High aeration rates would result in an increased removal of these gaseous and volatile substances, which are probably necessary to achieve maximal growth rates. Also it was determined that high aeration rate indeed reduced the $CO_2$ concentration in cultures.

*Medium components*: In general the same basal medium is used for growth either on agar or in liquid culture. Analysis of changes in composition of the liquid media during bioreactor culture of *Lilium* may give valuable information on limiting factors. In the initial stage, nitrogen uptake by bulblets was slow but with the lapse of time consumption became very rapidly. In case of salts, almost all of the ions were consumed after 5weeks in culture, although abundant nitrate ion remained until 8weeks later. This tendency was in accord with the sugar consumption. Hence, the nitrogen and carbohydrate roles a keep factors for bulblet formation. Especially, ammonium ion during initial stage of culture is more preferentially used as nitrogen source than nitrate the consumption of chloride, magnesium, potassium and calcium excepting the sulfate ion were relatively low through out the whole culture period.

The limiting growth factor was sugar, rather than remained nutrients. Sugar content in the liquid culture medium of *Lilium* 'Mona' using a 20L non-stirred bioreactor after 2 weeks in culture converted completely into fructose and glucose. Moreover, the remaining monosaccharides in the medium were exhausted after 6 weeks in culture (Lim et al., 1998). Significant preference on sugar uptake between fructose and glucose was not detected. In view of semi-continuous culture, it was desirable to add fresh medium every 4 to 6 weeks interval. In spite of these results and those published papers related to nutrient uptake, there is still a need for detailed investigations on dynamics of uptake of nutritional compounds in order to better manipulation of embryogenesis and organogenesis during bioreactor culture.

## 4. Bioreactor application for mass propagation

A few papers on plant propagation using bioreactors published in the past include the cultivation of cells, somatic embryos, or organogenic propagules like bulblets, corms, nodules, microtubers and shoot clumps. Organogenic plant propagules can be intensively cultivated in bioreactors for end results of producing transplants for mass propagation. For example, bulblets of *Lilium longiflorum* were cultivated on an industrial-scale type of bioreactors having 2,000 L volume for the production of virus-free bulbs (Takahashi et al., 1992). Also several caltivars of *Lilium* were mass propagated in bioreactors such as air lift and/or bubble type bioreactor with two stage culture methods (Kim, 1999). Bulblet formation and subsequent growth was significantly increased in ebb-and-flow type bioreactor with activated charcoal filter. Proliferation of PLBs in *Phalaenopsis* using ebb-and-flow type bioreactor was significantly increased when compared with conventional culture methods. This features strongly suggested the possibility of mass proliferation of PLBs in bioreactors (Lee, 1999). Multiplication of potato plantlets *in vitro* was conducted by inoculating stem segments (1 cm in length) in BTBB. Because the number of nodal segments cultured into the BTBB significantly affected shoot growth, yield of microtuber and/or the number of shoot cuttings at the time of harvest with initial culture of 128 stem segments, average 680 microtubers were obtained by placing a BTBB for 4 weeks at 26°C under 16 hr light condition and another 4 weeks at 18°C under dark. Subjecting potato shoot segments to transient immersion in a BTBB, 12 to 24 cycle/day with immersion period of 30 min was found to be adequate. Shoot cuttings for *ex vitro* rooting were conducted by harvesting shoot cultures reached to 20 cm in height and each shoot divided as node-bud cuttings. Total cuttings obtained from a 20 L BTBB were approximately 20,000. Each cutting planted in 128 cell plug tray filled with peatmoss. More than 95% of cutting was well rooted within 2 weeks and transplanted rooted cutting into 15 cm pot for the production of minituber.

## 5. Large-scale production of somatic embryos through bioreactor system

*In vitro* grown shoots derived from bud culture of *Acanthopanax koreanum* Nakai were used as the source of explants for the induction of callus. The type of tissues from shoot parts was markedly affected for callus induction. Among the tested tissues, node segment revealed as the best type of explant for callogenesis. The pre-embryogenic determined cells (PEDCs) obtained using 2,4-D completely lost the capacity for embryo germination, while callus induced in the presence of NAA and/or IBA showed normal growth compare with that of root sprout. By adding fresh medium at the junction phase between liner growth and late exponential, continuous liner growth could be achieved in bioreactor culture. For a long term culture of *Acanthopanax koreanum* cells, especially fed batch culture, feeding fresh medium when cell density reached to 60% was the best for cell growth compared with that of 70% and 80%. When PEDCs were inoculated into BTBB, 14 L fresh weight of PEDCs could harvest after 6 weeks of culture with 500 mL of inoculum cell density. Old medium may be reused by mixing it with fresh medium instead of replacing to absolute fresh medium in plant cell culture system because of useful compounds related to growth that cells secrete into the old medium. Among the rate test between fresh medium and old medium, the ratio of 1:4 revealed as the best for cell growth increment. By innoculating 0.1 mL of suspended PEDCs onto plastic Petri-dish containing 20 mL of semi-solid medium, more than 5,000 somatic embryos could be obtained. More than 500,000 somatic embryos having different stages were harvested from 10 L sized BTBB after 6 weeks of culture. Further development of these embryos in solid medium and/or eventually in field was successful. The frequency of plant regeneration from the embryos was approximately 93% in the field. The morphology of shoots regenerated by PEDCs mode was similar, and somaclonal variants were not detected by eye.

**Acknowledgements.** The authors would like to thank Lee, J.S. and Kim, Y.S. for providing up-to-date information for this article. This research supported in part by KOSEF through HortTech, Ministry of Science and Technology, and SGRP/HTDP.

### References

Aitken-Christie, J. 1991. Automation, In: P.C. Debergh and R.H. Zimmerman (eds.). Micropropagation. Kluwer Academic Publishers, Dordrecht. pp. 342-354.

Jones, J.B. and C.J. Sluis. 1991. Marketing of micropropagated plants. In: P.C. Debergh and R.H. Zimmerman (eds.). Micropropagation. Kluwer Academic Publishers, Dordrecht. pp. 141-154.

Kim, Y.S. 1999. Development of large scale process for mass production of lily bulblet by bioreactor system. MS thesis in Chungbuk National Univ. Korea.

Leathers, R.R., M.A.L. Smith and J. Aiken-Christie. 1995. Automation of the bioreactor process for mass propagation and secondary metabolism In: J. Aitken-Christie, T. Kozai and M.A.L. Smith (eds.). Automation and Environmental Control in Plant Tissue Culture. Kluwer Academic Publishers, Netherlands. pp. 187-214.

Lee, J.S. 1999. Micropropagation of *Phalaenopsis* by flower stalk-derived axillary bud culture. MS Thesis, Chungbuk National Univ. Korea.

Levin, R., V. Gaba, B. Tal, S. Hirsch, D. De Nola and I.K. Vasil. 1988. Automated tissue culture for mass propagation. Bio/Technology 6:1035-1040.

Lim, S., J.H. Seon, K.Y. Paek, S.H. Son and B.H. Han. 1998. Development of pilot scale process for mass production of *Lilium* bulblets *in vitro*. Acta. Hort. 461:237-241.

Nishimura, S., T. Terashima, K. Higashi and H. Kamada. 1993. Bioreactor culture of somatic embryos for mass propagation of plantsIn ; Redenbaugh, K (ed.). Synseeds-Applications of Synthetic Seeds to

Crop Improvement. CRC Press. London/Tokyo. pp. 175-181.

Preil, W. 1991. Application of bioreactors in plant propagation. In: P.C. Debergh and R.H. Zimmerman (eds.). Micropropagation. Kluwer Academic Publishers, Netherland. pp. 425-445.

Preil, W., P. Florek, U. Wix and A. Beck. 1988. Towards mass propagation by use of bioreactors. Acta Hort. 226:99-105.

Rittershaus, E., J. Ulrich, A. Weiss and K. Westphal. 1989. Large scale industrial fermentation of plant cells: Experiences in cultivation of plant cells in a fermentation cascade up to a volume of 75,000 liters. Bio Engineering. 5:28-34.

Seon, J.H., K.W. Yoo, Y.Y. Cui, M.H. Kim, S.J. Lee, S.H. Son and K.Y. Paek. 1998. Application of bioreactor for the production of saponin by adventitious root cultures in *Panax ginseng*. In: A. Altman, M. Ziv and S. Izhar (eds.). Plant Biotechnology and In Vitro Biology in the 21st Century. Kluwer Academic Plublishers. Netherlands. pp. 329-332.

Smart, N.J. and M.W. Fowler. 1984. An airlift column bioreactor suitable for large-scale cultivation of plant cell suspensions. EXP. Bot. 35:531-537.

Son, S.H., S.M. Choi, D.S. Lee, S.R. Yun and K.Y. Paek. 1999a. Commercial application of mountain ginseng through bioreactor culture system. Proc. Korea-Japan Joint Sym. Transplant Production in Horticultural Plants. Res. Center for the Development of Advanced Horticultural Technology, Changbuk National Univ.

Son, S.H., S.M. Choi, S.R. Kwon, Y.H. Lee and K.Y. Paek. 1999b. Large-scale culture of plant cell and tissue by bioreactor system. J. Plant Bioteh. 1(1):1-8.

Takahashi, S., H. Matsubara, H. Yamagata and T. Morimoto. 1992. Micropropagation of virus free bulblets of *Lilium longiflorum* by tank culture. 1. Development of liquid culture method and large scale propagation. Acta Hort. 319:83-88.

Thomas, D and T. Murashige. 1979. Volatile emissions of plant tissue cultures. I Identification of the major components. In Vitro 15:654-658.

Tulecke, W. and L.G. Nickell. 1959. Production of large amounts of plant tissue by submerged culture. Science. 130:863-864.

# EFFECTS OF MEDIUM SUGAR ON GROWTH AND CARBOHYDRATE STATUS OF SWEETPOTATO AND TOMATO PLANTLETS IN VITRO

Sandra B. Wilson[1], Chieri Kubota[2] and Toyoki Kozai[2]
[1]Department of Environmental Horticulture, Indian River Research and Education Center, University of Florida, 2199 South Rock Road, Fort Pierce, FL 34945, USA.
E-mail: sbwilson@gnv.ifas.ufl.edu
[2]Department of Bioproduction Science, Chiba University, Matsudo, Chiba 271, Japan.

**Abstract.** Carbohydrate status of tomato (*Lycopersicon esculentum* Mill., 'HanaQueen') and sweetpotato (*Ipomoea batatas* (L.) Lam., 'Beniazuma') plantlets was investigated in plantlets cultured for 17 d in vitro photoautotrophically (without sucrose in the medium) or photomixotrophically (with 30 g/L sucrose in the medium) under enriched $CO_2$ conditions (1500 $\mu mol \cdot mol^{-1}$) and 150 $\mu mol \cdot m^{-2} \cdot s^{-1}$ photosynthetic photon flux (PPF). After 17 d of enriched $CO_2$ conditions, there was no significant difference in leaf, stem or root fresh weight among photoautotrophic and photomixotrophic tomato and sweetpotato plantlets, with the exception that the stem fresh weight of photomixotrophic tomato plantlets was higher than that of photoautotrophic plantlets. The addition of sucrose to the medium increased the stem, root and total dry weight of tomato plantlets and increased the root dry weight of sweetpotato plantlets. Tomato plantlets had higher soluble sugar levels (sucrose, glucose, and fructose) than did sweetpotato plantlets, regardless of medium composition. Photomixotrophic plantlets had higher soluble sugar levels that did photoautotrophic plantlets, regardless of plant species. Starch concentrations of photomixotrophic tomato plantlets were higher than that of photomixotrophic sweetpotato plantlets. However, there was no difference in starch concentration of photoautotrophic sweetpotato and tomato plantlets. Regardless of plant species or media composition, carbohydrate status was maintained under enriched $CO_2$ conditions.

**Key index words.** *Ipomoea batatas*, *Lycopersicon esculentum*, micropropagation, photoautotrophic, photomixotrophic.

## 1. Introduction

Plantlets are conventionally cultured photomixotrophically (with sugar and atmospheric $CO_2$ as their carbon sources) in vitro. Sugar in culture medium has been considered the main carbon source for the growth of cells, buds, shoots, and even plantlets. Providing sucrose in the medium has proven beneficial to maintain plant quality and photosynthetic ability during low temperature storage of in vitro broccoli (Wilson et al., 1998a; Wilson et al., 1998b) and hosta (Wilson et al., 2000) plantlets. Capellades et al. (1991) cultured *Rosa multiflora* L. plantlets with four different initial levels of sucrose, and found that plantlets receiving 10 g/L and 30 g/L sucrose in the culture medium had higher net photosynthetic rates and lower starch concentration in leaves than that with 50 g/L sucrose. It was demonstrated that exogenous supply of sugars increases starch and sucrose reserves in micropropagated plants and that appropriate higher levels of carbohydrates can favor plantlet survival upon transfer to ex vitro conditions, improve acclimatization and speed up physiological adaptations (Capellades et al., 1991; Van Huylenbroeck and Debergh, 1996; Wilson et al., 1998b).



*C. Kubota and C. Chun (eds.), Transplant Production in the 21st Century,* 258–265.

However, the absolute requirement of exogenous sugars in tissue culture has been contested. Sugar in the medium has been shown to inhibit photosynthesis and research has revealed that the low net photosynthetic rates of chlorophyllous shoots or plantlets in vitro are largely due to the low $CO_2$ concentration in the vessel during the photoperiod (Kozai et al., 1997). Kozai et al. (1997) report $CO_2$ enrichment and supplemental lighting to significantly reduce the period of acclimatization in the greenhouse, and improve the size and quality of in vitro plantlets.

Although sugar is considered as a carbon source for in vitro cultured plantlets, only a few reports document sugar uptake and metabolism of in vitro plantlets. As mentioned by Desjardins et al. (1995), there are extensive studies on sugar uptake using cell suspension cultures as model systems, but it can not be assumed that responses observed in cultured cells will be the same with tissue cultured plantlets. Quantitative understanding of carbohydrate status is useful for predicting plantlet growth and acclimatization to greenhouse conditions. The objectives of this study were to investigate the influence of initial sucrose concentration of the medium on growth and carbohydrate status of in vitro plantlets grown in enriched $CO_2$ conditions. Tomato and sweetpotato plantlets were selected as model systems for this study, since both plants can be micropropagated either photoautotrophically or photomixotrophically.

## 2. Materials and Methods

### 2.1 Plant material and culture conditions

Single-node stem cuttings each with a leaf were excised from tomato and sweetpotato plantlets and cultured photoautotrophically and photomixotrophically under conditions shown in Table 1.

*Table 1* Experimental culture conditions for tomato and sweetpotato plantlets. Plantlets were cultured photoautotrophically (no sucrose in the medium) or photomixotrophically (30 g/L sucrose in the medium) under otherwise identical culture conditions.

Plant species:
- Tomato (*Lycopersicon esculentum* Mill., 'HanaQueen')
- Sweetpotato (*Impomoea batatas* (L.) Lam., 'Beniazuma')

Culture period: 17 d
Explant: Single node cuttings each with a leaf
- Mean fresh weight per explant: tomato (225 ± 20 mg), sweetpotato (120 ± 20 mg)

Vessel: polycarbonate (volume: 480 mL)
Medium volume: 100 mL/vessel
Basal composition: Murashige and Skoog (1962)
Sugar: 0 or 30 g sucrose/ L
Substrate: Sorbarod cellulose plugs (Baumgartner Papiers SA, Switzerland)
pH: 5.8
No. of air exchanges: 1.6 (days 1-5) and 3.05 (days 6-17) $h^{-1}$
Culture room conditions:
- Air temperature: day (26.8 ± 0.2 °C), night (23.5 ± 0.2 °C)
- Relative humidity: 80%
- $CO_2$ concentration: 1500 ± 10 $\mu mol \cdot mol^{-1}$
- Photosynthetic photon flux: 100 ± 10 (days 1-5); 150 ± 15 (days 6-17) $\mu mol \cdot m^{-2} \cdot s^{-1}$.
- Photoperiod: 16 $h \cdot d^{-1}$
- Light source: cool white fluorescent bulbs

### 2.2 $CO_2$ concentration and gas exchange rate

Three holes (10 mm in diameter) in opposite sides of the vessel lids were covered with 0.5 μm membrane filter disks (Milli-Seal, Millipore K.K., Tokyo) to provide air exchange. At the start of the experiment, two of the air diffusive filters of the vessel were covered with plastic tape to minimize the number of air exchanges. On day 5, the remaining two filters were uncovered to increase the number of air exchanges. $CO_2$ concentrations inside the vessels were monitored after days 0, 5, and 17. The number of air exchanges per hour was determined by the method described by Kozai et al. (1986). The $CO_2$ concentrations inside and outside the vessel were determined during the photoperiod by analyzing gas samples using a gas chromatograph (GC-12A, Shimadzu Co., Kyoto) equipped with a flame ionization detector.

### 2.3 Plant carbohydrate analysis

At day 0 and 17, leaves, stems, and roots were frozen in liquid $N_2$ and freeze-dried for dry weight measurements. To obtain sufficient tissue for carbohydrate analysis, two plantlets were pooled to generate a sample. The procedures for soluble sugar extraction were modified as previously described (Boersig and Negm, 1985; Miller and Langhans, 1989). Leaves, stems, and roots were separately ground, and approximately 50 mg was loaded into glass Pasteur pipettes with glass wool plugs (1 cm) and extracted 3 times with 1.5 mL of 12 methanol : 5 chloroform: 3 water (by volume, MCW). 100 μL sorbitol (10 mg/mL) was added as an internal standard. Distilled water (3.5 mL) was added to samples and aqueous phase was removed and applied to polyethylene columns containing 3 mL 1 methanol : 1 water (v:v, MW) and cation and anion resin (1 mL Amberlite IRA-45 layered with 1 mL Dowex 50-W, Sigma-Aldrich Co.,St. Louis, MO). Soluble carbohydrates were eluted with 1 mL MW twice and evaporated to dryness using a rotary flash evaporator. The dry residue was resuspended in 1 mL HPLC-grade water and filtered through a 0.45 um membrane prior to HPLC injection. Sucrose, glucose, and fructose were separated and detected using an HPLC with a refractive index detector (Hitachi Ltd., Tokyo, Japan) and a Gelpack GL-C611 column (Hitachi Chemical Co., Tokyo, Japan) maintained at 70 °C. Quantification was determined using a D-2000 Chromato-Integrator (Hitachi Ltd., Tokyo, Japan) and a regression equation describing the sucrose, glucose, and fructose calibration lines.

The procedures for starch determination were modified as described (Haissig and Dickson, 1979; Miller and Langhans, 1989). The tissue residue left in the pasteur pipets following soluble sugar extraction was dried overnight at 60 °C, suspended in 4 mL Na-acetate buffer (100mM, pH 4.5) and placed in a boiling water bath for 20 min. After cooling to room temperature, 1.0 mL amyloglucosidase solution (from *Aspergillus niger*, Sigma-Aldrich Co., St. Louis, MO) (50 units/assay in 0.1 M pH 4.2 Na-acetate buffer) was added to each test tube. Samples were incubated for 48 hr at 55 °C with occasional agitation. Glucose determination via a glucose oxidase and peroxidase enzymatic method was completed on a 100 μL sample. Absorbance was determined at 450 nm on a spectrophotometer and starch content calculated based on the regression equation describing the glucose calibration line (0.0 to 0.5 μmol).

### 2.4 Experimental design and statistical analysis

All vessels were arranged in a completely randomized design. Each vessel containing four plantlets was considered a replication and there were three replications per treatment. Data were analyzed by ANOVA and mean separation evaluated using LSD at $P=0.05$.

## 3. Results and Discussion

### 3.1 Plant growth

After 17 d, there was no significant difference in fresh weight among photoautotrophic and photomixotrophic tomato and sweetpotato plantlets, with the exception that the stem fresh weight of photomixotrophic tomato plantlets was higher than that of photoautotrophic plantlets (Table 2). Kozai and Iwanami (1988) also observed enhanced growth of carnation plantlets under a $CO_2$ concentration level of 1000-1500 $\mu mol \cdot mol^{-1}$ and a PPFD level of 150 $\mu mol \cdot m^{-2} \cdot s^{-1}$, regardless of the addition of sugar in the medium. Photomixotrophic tomato and sweetpotato plantlets generally had higher stem, root, and total dry weight than did photoautotrophic plantlets (Table 2). Similarly, Kubota et al. (1998) showed that final dry weight of in vitro tomato and sweetpotato plantlets increased as initial sucrose concentration increased from 0 to 7.5 $g \cdot L^{-1}$. In their experiments, final dry weight continued to increase as sucrose increased to 30 $g \cdot L^{-1}$ for tomato plantlets but remained unchanged for sweetpotato plantlets (Kubota et al., 1998). Deng and Donnelly (1993) showed that both $CO_2$ and sucrose affect in vitro plantlet growth of *Rubus idaeus* L. (red raspberry) independently. In their experiments, in vitro $CO_2$ enrichment significantly increased root count, root length, and total plantlet fresh weight compared with those of plantlets grown under ambient $CO_2$, plantlet height, and percent dry weight were similar among plantlets grown under ambient or enriched $CO_2$ concentrations (Deng and Donnelly, 1993).

*Table 2* Effects of photoautotrophic (PA) and photomixotrophic (PM) conditions on fresh and dry weight of sweetpotato and tomato plantlets after 17 days at 1500 $\mu mol \cdot mol^{-1}$ $CO_2$.

| Treatment | Fresh weight (g) | | | | Dry weight (g) | | | |
|---|---|---|---|---|---|---|---|---|
| | Leaf[z] | Stem | Root | Total | Leaf | Stem | Root | Total |
| Sweetpotato PA | .429 | .077 | .188 | .693 | .038 | .006 | .012 | .056 |
| Sweetpotato PM | .365 | .086 | .191 | .641 | .040 | .009 | .016 | .065 |
| Tomato PA | .516 | .279 | .164 | .959 | .053 | .019 | .011 | .082 |
| Tomato PM | .509 | .386 | .184 | 1.08 | .063 | .042 | .015 | .120 |
| ANOVA[z] | | | | | | | | |
| Species | ** | ** | NS | ** | ** | ** | NS | ** |
| Sucrose | NS | ** | NS | NS | NS | ** | ** | ** |
| Sp * suc | NS | ** | NS | * | NS | ** | NS | * |

[z]ANOVA = analysis of variance.
[NS, *, **] Nonsignificant, significant at $P < 0.05$ or 0.01, respectively.

### 3.2 Carbohydrate analysis

Tomato plantlets had higher sucrose, hexose, and total soluble sugar concentrations than did sweetpotato plantlets, regardless of medium composition. However, total sugar uptake from the media based on dry weight was not significantly different between species (data not presented). Photomixotrophic plantlets had higher sucrose, hexose, and total sugar concentrations than did photoautotrophic plantlets, regardless of plant species. Sucrose concentrations were generally higher than hexose concentrations in stem tissue but lower than hexose concentrations in leaf or root tissue indicating sucrose cleavage and the mobilization of its hexoses. Cournac et al. (1991) showed that the soluble sugar concentrations of *Solanum tuberosum* L. (potato) plantlets cultured photomixotrophically in aerated vessels were higher than those cultured photoautotrophically either with or without $CO_2$ enrichment. Starch concentrations of photomixotrophic tomato plantlets were higher than that of photomixotrophic sweetpotato plantlets. However, there was no difference in starch concentrations of photoautotrophic sweetpotato plantlets verses photoautotrophic tomato plantlets. Photomixotrophic plantlets had higher starch concentrations than photoautotrophic plantlets. This is similar to work of Capellades et al. (1991) who showed that the leaf starch concentration increased when the plantlet was cultured on medium with an elevated sucrose concentration. Sugar in the medium has been reported to reduce the rubisco activity in plants (Desjardins et al, 1995) thereby lowering the net photosynthetic rate (Capellades et al., 1991). Piqueras et al. (1998) reported that micropropagated *Calathea louisae* Gagnep plantlets had higher starch contents in roots and stems compared with leaves, while sucrose concentration was highest in stems, followed by leaves and roots. Providing $CO_2$ enrichment for micropropagated plantlets resulted in highest total soluble sugars in the leaf and root tissue of tomato and in the stem and root tissue of sweetpotato. In vitro tomato and sweetpotato plantlets grown in enriched $CO_2$ conditions without sucrose had sufficient carbohydrate reserves after 17 d in culture.

**Acknowledgments.** Florida Agricultural Experiment Station Journal Series No. R-07421. The authors would like to thank the Japanese Society for the promotion of Science for sponsoring the senior author and Drs. Milton Tignor and Nihal Rajapakse for critically reviewing the manuscript.

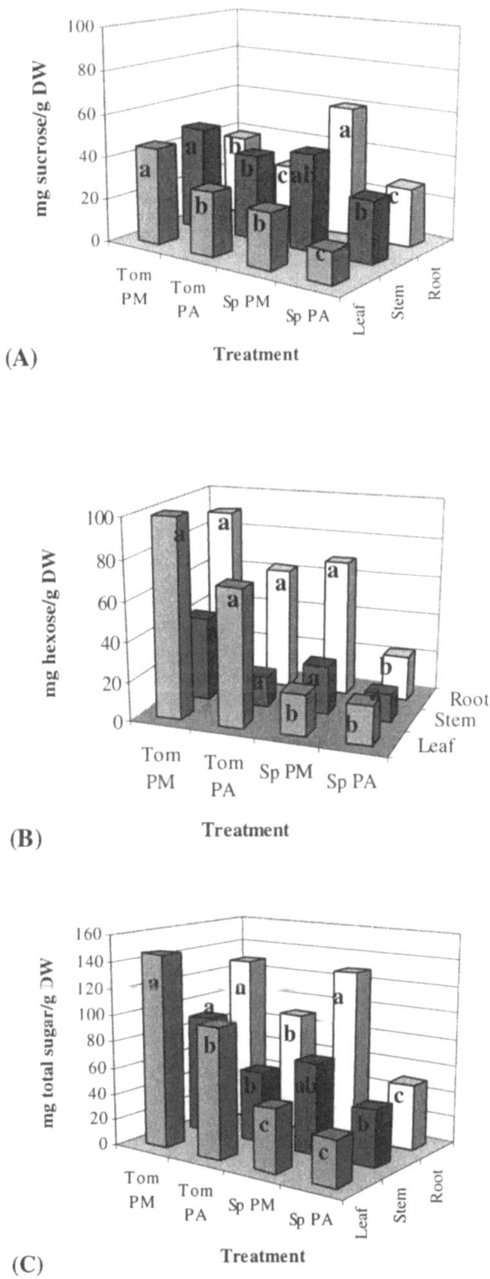


*Fig. 1* Sucrose (A), hexose (B), and total sugar (C) concentrations in leaf, stem and root tissue of tomato (Tom) and sweetpotato (Sp) after 17 d of photoautotrophic (PA) or photomixotrophic (PM) culture.

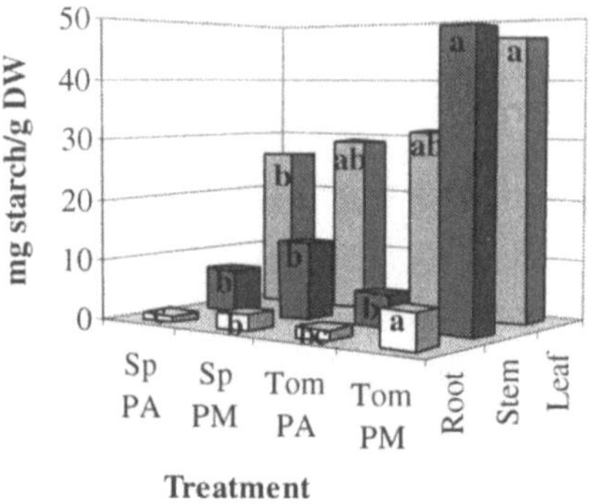


*Fig. 2* Starch concentrations in leaf, stem, and root tissue of sweetpotato (Sp) and tomoato (Tom) plantlets grown photoautotrophically (PA) or photomixotrophically (PM).

**References**


Boersig, M.R. and F.B. Negm. 1985. Prevention of sucrose inversion during preparation of HPLC samples. HortScience 20:1054-1056.

Capellades, M., R. Lemeur and P. Debergh. 1991. Effects of sucrose on starch accumulation and rate of photosynthesis in *Rosa* cultured in vitro. Plant Cell Tissue Organ Cult. 25:21-26.

Cournac et al., 1991. Growth and photosynthetic characteristics of *Solanum tuberosum* plantlets cultivated in vitro in different conditions of aeration, sucrose supply and $CO_2$ enrichment. Plant Physiol. 97:112-117.

Deng, R. and D.J. Donnelly. 1993. In vitro hardening of red raspberry by $CO_2$ enrichment and reduced medium sucrose concentration. HortScience 28:1048-1051.

Desjardins, Y., C. Hdider and J. de Riek. 1995. Carbon nutrition in vitro–regulation and manipulation of carbon assimilation in micropropagated systems. In: J. Aitken-Christie, T. Kozai, and M.A.L. Smith (eds.) Automation and Environmental Control in Plant Tissue Culture. Kluwer Academic Publishers, The Netherlands. pp. 441-471.

Haissig, B.E. and R.E. Dickson. 1979. Starch measurement in plant tissue using enzymatic hydrolysis. Physiol. Plant. 47:151-157.

Kozai, T., K. Fujiwara and I. Watanabe. 1986. Fundamental studies on environments in plant tissue culture vessels. (2) Effects of stoppers and vessels on gas exchange rates between inside and outside of vessels closed with stoppers. J. Agr. Meteorol. 42:119-127.

Kozai, T. and Iwanami, Y. 1988. Effects of $CO_2$ enrichment and sucrose concentration under high photon fluxes on plantlets growth of carnation (*Dianthus caryophyllus* L.) in tissue culture during the preparation stage. J. Japan Soc. Hort. Sci. 57:279-288.

Kozai, T., C. Kubota and B.R. Jeong. 1997. Environmental control for the large-scale production of plants through in vitro techniques. Plant Cell, Tissue and Organ Cult. 51:49-56.

Kubota, C., M. Ezawa, S.B. Wilson and T. Kozai. 1998. Carbon balance of tomato and sweetpotato plantlets cultured with different initial sucrose concentrations in the medium. Jap. Soc. Environ. Cont. Biol.

Miller, W.B. and R.W. Langhans. 1989. Carbohydrate changes of Easter Lilies during growth in normal and reduced irradiance environments. J. Amer. Soc. Hort. Sci. 114:310-315.

Piqueras, A., J.M. van Huylenbroeck, B.H. Han and P.C. Debergh. 1998. Carbohydrate partitioning and metabolism during acclimatization of micropropagated *Calathea*. Plant Growth Regulation. 26:25-31.

Van Huylenbroeck, J.M. and P.C. Debergh. 1996. Impact of sugar concentration in vitro on photosynthesis and carbon metabolism during ex vitro acclimatization of Spathiphyllum plantlets. Physiol Plant. 96: 298-305.

Wilson, S.B., K. Iwabuchi, N.C. Rajapakse and R.E. Young. 1998a. Responses of broccoli seedlings to light quality during low-temperature storage in vitro: I. Morphology and survival. HortScience

33:1253-1257.

Wilson, S.B., N.C. Rajapakse and R.E. Young. 1998b. Responses of broccoli seedlings to light quality during low-temperature storage in vitro. II. Sugar content and photosynthetic efficiency. HortScience 33:1258-1261

Wilson, S.B., N.C. Rajapakse and R.E. Young. 2000. Media composition and light affect storability and post storage recovery of micropropagated hosta plantlets. HortScience (in press).

# PRACTICAL SUGAR-FREE MICROPROPAGATION SYSTEM USING LARGE VESSELS WITH FORCED VENTILATION

Yulan Xiao[1], Jiacong Zhao[1] and Toyoki Kozai[2]
[1]Institute of Environmental Science, Kunming, Yunnan 650032, China. E-mail: ylxiao@ynmail.com
[2]Faculty of Horticulture, Chiba University, Matsudo, Chiba 271-8510, Japan.

**Abstract.** Conventional micropropagation systems using sugar-containing medium under low light intensity and ambient $CO_2$ concentration are widely used for mass-production of plantlets. However, problems such as high % contamination, poor growth and development of plantlets, low % survival at ex vitro acclimatization and high production costs are limiting further commercial use of micropropagation. Our objective in the present study was to develop a practical photoautotrophic (sugar-free in the medium) micropropagation system using a large culture vessel with forced ventilation under high light intensity and $CO_2$ concentration. Explants excised from potato (*Solanum tuberosum* L. cv. Benimaru) and statice (*Limonium latifolium*) plantlets were cultured in sterilized large (125 L) and small (380 mL) vessels. Large vessels with forced ventilation, containing sugar-free perlite as supporting material, and small vessels with natural ventilation, containing sugar-free agar as supporting material, were placed at photosynthetic photon flux (PPF) of ca.120 $\mu$mol $m^{-2}$ $s^{-1}$ and $CO_2$ concentration of ca.1,500 $\mu$mol $mol^{-1}$. Small vessels with sugar-containing agar medium were also placed at PPF of ca. 40 $\mu$mol $m^{-2}$ $s^{-1}$ and $CO_2$ concentration of ca.370 $\mu$mol $mol^{-1}$ as the conventional method. The results showed that the photoautotrophic micropropagation system using the large vessel at $CO_2$ concentration ca.1,500 $\mu$mol $mol^{-1}$ gave the greatest fresh weight, dry weight and leaf area of potato and statice plantlets, and that it gave the highest % survival of 95% after transplanting ex vitro and the highest plantlet quality among the three treatments. This photoautotrophic micropropagation system with forced ventilation has been used for commercial micropropagation of economically important potato and statice plantlets in Kunming, China since 1997, based on the results shown above. The production costs was decreased by about 50% compared with the costs in the conventional method. The average price could be increased by 40%, due to the high quality, compared with the price of plantlets produced in the conventional method, resulting in a great economical profit. This photoautotrophic micropropagation system has an extensive application and will improve the protocol for micropropagation with use of chlorophyllous explants for many plant species in the forthcoming decades.

**Key index words.** $CO_2$ enrichment, culture vessel, potato, statice sugar-free micropropagation.

## 1. Introduction

Technology of plant tissue culture is often applied in micropropagation of flower, forest and fruit crops as well as vegetable and cash crops, foodstuff crops and medicinal plants. Micropropagation is suitable for rapid production of virus-free transplants of the above plants with superior genetic properties. In the world, China is the country with the largest micropropagation laboratory in area, and with the largest number of employees in micropropagation industry (Chen, 1990).



*C. Kubota and C. Chun (eds.), Transplant Production in the 21st Century*, 266–273.

The weather in Kunming, Yunnan, China is cool in summer and not cold in winter. Then, Kunming is often called the city of spring in all seasons with yearly average air temperature of 15 $^0$C. Kunming is an appropriate place for micropropagation industry. In fact, several micropropagation laboratories and the market are located there. According to the statistics, there are 35 tissue culture laboratories in different research institutes and production organizations in Kunming. The laboratories cover 8,000 $m^2$ in floor area with 250 laminar flow benches and 650 workers. The annual production capacity of the laboratories is 35 millions with current production of 8 million plantlets. Major plantlets produced by the means of tissue culture are ornamental flowers, potatoes, fruits, tobacco, medicinal plants, vegetable and perfume plants, etc.

Theoretically, micropropagation should provide high propagation rates compared to the conventional vegetative propagation using cuttings. However, its widespread practical use is still limited mainly by low growth rate and significant losses of plantlets due to microbial contamination, low rooting and survival percents *in vitro* and *ex vitro*.

Thus, it is our objectives to promote the growth *in vitro*, to reduce the microbial contamination, and to increase the percent rooting and survival *in vitro* and *ex vitro*, and thus to decrease the production costs, to improve product quality and quantity, and to make a higher economic profit.

It can be considered that the sugar-free micropropagation makes it possible to solve the above problems. Advantages of sugar-free over conventional (heterotrophic or photomixotrophic) micropropagation systems are said to be: 1) growth and development of plantlets *in vitro* are faster and more uniform, 2) plantlets *in vitro* have less physiological and morphological disorders, 3) biological contamination *in vitro* is less, 4) plantlets show a higher percentage of survival during acclimatization *ex vitro*, and 5) larger culture vessels could be used because of less biological contamination (Kozai, 1997a).

Since 1996, National Foreign Expertise Bureau and Kunming Scientific Commission invited a specialist of sugar-free or photoautotrophic micropropagation once every year and invested to develop a sugar-free micropropagation system suitable for Kunming area.

Our objective in the present study was to develop a practical photoautotrophic micropropagation system using a large culture vessel with forced ventilation and $CO_2$ enrichment at low costs to give a high economic profit.

## 2. Materials and Methods

### 2.1 Forced ventilation micropropagation system

The culture box for forced ventilation micropropagation system was made of plexiglass and glass (volume: 125 L, length: 1,000 mm, width: 500 mm, and height: 250 mm). The length, width and height of the culture box were determined to fit with the length of fluorescent lamps, width of culture shelf and the height of shelf, respectively, with the purpose of making full use of light energy from the lamps and culture space.

The culture box was disinfected or sterilized in the following order: (1) wash the culture box with clean water, (2) clean the culture box with Bromo geraminum, a disinfectant, (3) stifle the culture box with $KMnO_4$ (5 g $m^{-3}$), formaldehyde (10 ml $m^{-3}$) for 12 hours, and (4) clean the culture box with 70% alcohol just before transplanting.

A schematic diagram of the forced ventilation micropropagation system is shown in Fig.1. The pure $CO_2$ from the $CO_2$ container was passed through an airflow rate meter into the gas mixing chamber. Also, the room air was sent by the air pump with a microporous filter into the gas mixing chamber to dilute the $CO_2$ gas and to prevent dusts and microorganisms from entering the gas mixing chamber. The $CO_2$ concentration of the mixed air was controlled at 1,500 μmol $mol^{-1}$. Then, the mixed air was sent into the disinfection box containing 0.1%HgCl solution for making sure that the mixed air was completely sterile. The sterilized $CO_2$ enriched air was sent into the culture vessel.

**2.2 Description of treatments**

In Experiment 1, 2,300 single nodal cuttings each with a leaf of photomixotrophically grown potato (*Solanum tuberosum,* L.) plantlets were used as explants and transplanted in the culture box with a planting density of 4,600 explants per square meter. On the other hand, 15 explants from the same group of cuttings were transplanted in each of the conventional, bottle-type vessel (volume: 380 ml) sealed with a gas-permeable microporus filter (the pore area: 157 $mm^2$, pore diameter: 0.5 μm), with a planting density of 3,000 explants per square meter of culture shelf area.

In Experiment 2, 1,500 plantlets each with 5-6 leaves of photomixotrophically grown statice (*Limonium latifolium*) were used as explants and transplanted in the culture box with a planting density of 3,000 explants per square meter. On the other hand, 10 explants from the same group of plantlets were transplanted in each of the conventional bottle-type vessels, with a planting density of 2,000 explants per square meter of culture shelf area.

In Experiments 1 and 2, MS (Murashige and Skoog, 1962) solution was used as basal medium; 60 mL per the bottle-type vessel, 9,200 and 9,000 mL per the culture box with potato and statice, respectively. Supporting material was agar (6 g $L^{-1}$) for the bottle-type vessels, and perlite for the culture boxes. In the conventional photomixotrophic treatment using the bottle-type vessels, sugar (30 g $L^{-1}$) was added as carbon and energy source for plants. In the photoautotrophic treatment, sugar, vitamins, and other organic compounds were excluded from the medium components.

The experimental conditions, which are common in Experiments 1 and 2 are described briefly in Table 1. The PPF and $CO_2$ concentrations in Table 1 are those on and after day 7. In CSN and FSE, the number of vessels was 100 per treatment. In FLE, only one culture box was used for potato and statice, respectively.

In all the treatments, during the days 0-6, the photosynthetic photon flux (PPF) was 40 μmol $m^{-2}$ $s^{-1}$ with 12 h photoperiod and 12 h dark period and the $CO_2$ concentration of the room air and inlet of the culture box was ca. 370 μmol $mol^{-1}$. The photoperiod and dark period were 16 and 8 h , respectively, on and after day 7. The air flow rate during the photoperiod in FLE was zero during days 0-6 and 15 L $min^{-1}$ on and after day 7.

Throughout the culture period, the temperature and relative humidity of culture room air were 23 °C and 65-70%, respectively. The potato and statice plantlets were harvested on days 20 and 28, respectively. The fresh and dry weights, leaf area, number of leaves per plantlet were measured at the end of experiments. Most of harvested plantlets were transplanted to the soil for acclimatization *ex vitro* and percent survival *ex vitro* were investigated 20 days after transplanting.

**3. Results and Discussions**

**3.1 Effects of ventilation method and vessel size**

The growth of potato and statice plantlets were significantly greater when grown in the culture box with forced ventilation than when grown in the conventional bottle type box with natural ventilation (Tables 2 and 3). The shoot or leaf length, fresh weight, dry weight and leaf area of potato and statice per plantlet were significantly greater in FLE than in FSE and in CSN, respectively.

Percent survival of plantlets after transplanting for acclimatization *ex vitro* in FLE was 95-100 % and was about 10% higher than in FSE and CSN (Table 4), probably due to in the absence of sugar in the medium, higher air exchange rate, lower relative humidity, higher $CO_2$ concentration in FLE than in CSN. In FSE, $CO_2$ concentration was increased to 1,500 $\mu$mol $mol^{-1}$ around the bottles. However, the ventilation of the vessel through the gas permeable filter on the vessel mouth was not high enough to keep the $CO_2$ concentration as high as in FLE. An elevated PPF will not increase the net photosynthetic rate (NPR) under low $CO_2$ concentrations (Kozai, 1997a). The low $CO_2$ concentration in the photoperiod results in the low net photosynthetic rate of plantlets, so that plantlets grow slowly.

The loss of plantlets due to microbial contamination in the photomixotrophic micropropagation generally increases with the increase volume of the vessel. Thus, the large culture box as in FLE can be used only in photoautotrophic micropropagation. One of the key advantages of photoautotrophic micropropagation is that it makes it possible to use large culture vessels with minimum risk of microbial contamination (Kozai, 1999).

### 3.2 Direct production costs of the photoautotrophic and photomixotrophic micropropagations

In the photoautotrophic (sugar-free) micropropagation system using a large vessel with forced ventilation (FLE), the direct production cost per plantlet was 54% in potato or 52% in statice lower compared with the one in the photomixotrophic (conventional) micropropagation system using small vessels with natural ventilation (Table 5).

1) The planting densities in the large box and in the small vessel were basically the same: $4.6\times10^3$ plantlets $m^{-2}$ in potato and $3.0\times10^3$ plantlets $m^{-2}$ in statice. but the number of plantlets produced was different on the same culture layer (0.6 $m^2$), the potatoes were 1500 plantlets (bottle) or 2300 plant (large box), statice were 1000 plantlets (bottle) or 1500 plantlets (large box), the number of plantlets increased 50% more in a large box than in 100 small bottles on the same culture area. Thus production cost per plantlet can be reduced.

2) In the photoautotrophic micropropagation system with large culture vessels, the operation procedures become simpler than conventional photomixotrophic, it saves labor. It does not use the gas-permeable filter membrane, and porous material instead of expensive agar, reduces material cost again. Due to vigorous plantlet the root was white and thick in vitro. The acclimatization process could be simplified or even eliminated (Heo and Kozai 1999). Production costs reduced further.

3) In the photomixotrophic micropropagation system, the loss of plantlets was significantly high due to microbial contamination, high relative humidity, low $CO_2$ concentration, high ethylene concentration, and low PPF in the photoperiod. These in vitro environmental conditions cause physiological and morphological disorders, resulting in low survival percentage in vitro and ex vitro. Low percent survival ex vitro is also caused by the nutritional shift from photomixotrophic in vitro to photoautotrophic ex vitro. In CSN, percent of marketable plantlets was 72% in potato and 62% in statice. On the other hand, in FLE, it was 96% in potato and 91 % in statice.

4) In the photoautotrophic micropropagation system (FLE), the cost of electricity for lamps and air conditioners and the cost for $CO_2$ enrichment were higher than those in

the photomixotrophic micropropagation system (CSN). This is due to higher light intensity and $CO_2$ concentration in FLE than in CSN. However, the increase in cost in FLE is not significant if the equipment is properly designed, installed and controlled (Kozai, 1997b). High $CO_2$ concentration and high intensity promote net photosynthetic rate and then growth of plantlets. Plantlets grown photoautotrophically grow ex vitro vigorously with a high survival percentage ex vitro. Thus, actual production cost per marketable plantlet can be lower in FLE than in CSN.

Kozai (1999) indicated this view explicitly: High production cost per micropropagated plantlet is mostly attributed to the low growth rate, a significant loss of plant in vitro by microbial contamination, poor rooting, low percent survival ex vitro and high labor cost due to intensive manual operations of small plants and small culture vessels.

## 4. Application in production

Based on the experimental results shown above, since 1997, this photoautotrophic micropropagation system with forced ventilation, using the large vessel, the porous supporting material such as vermiculite, perlite, and sand under high light intensity and $CO_2$ concentration has been used in commercial production of micropropagated plantlets in Kunming, China, since 1997. Using this system, we have produced quantities of carnation (*Dianthus caryophyllus*), *Gerbera jamesonii*, *Gypsophila panicudta*, strawberry (*Fragaria chlaensis* var. ananasa), potato, statice, *Platycodon grandiflorum*, *Spathiphyllum,* etc. Their production is successful commercially, the growth and development of plantlets was greater in photoautotrophic system than in conventional photomixotrophic system. Culture cycle was shortened by 25% with increased multiplication rate, and quality and percentage survival of plantlets in vitro and ex vitro was higher in the photoautotrophic system, compared with those in the conventional photomixotrophic system.

Percent rooting in vitro was higher than 93% and percent survival after transplanting ex vitro was higher than 95% in the photoautotrophic system. It was impossible to achieve such high percentages in the conventional, photomixotrophic system.

The production cost in the photoautotrophic system is only about 50% of the cost in the conventional method. The plantlet did not wilt; grew fast and showed high percent survival of 99% after transplanting in the field, so the growers welcomed the products. The price of photoautotrophically micropropagated plantlets could be increased by 40% compared with the price of plantlets produced in the conventional method, resulting in a significant economical profit. A foundation was established for intensive application of this technology for commercial production.

In 1999, we started a demonstration production program of 500,000 plantlets output yearly by this photoautotrophic micropropagation system; the production scale will be enlarged year by year. It is estimated that this system will be used popularly in Kunming or even in many other places in China in the forthcoming decades.

Along with the development agriculture modernization in China, the market needs more the plantlets of improved flowers, fruits, vegetables and other crops. The extensive application of the photoautotrophic micropropagation will improve significantly these plants fast propagation, satisfy the needs of the market, and accelerate the development of agricultural modernization.

Photoautotrophic micropropagation is a vital innovation of conventional photomixotrophic micropropagation, and great progress for plant tissue culture. Its new thought way, new method, providing more wide imagination for plant tissue culture, its extensive application maybe is a revolution of plant tissue culture.

**References**


Chen, W. 1990. The Question and Present Situation of Plant Fast Propagation and Virus-free Young Plants Production in China. Plant Biology Technology and Crop Improvement. Science and Technology Publishing House of China. pp. 213.

Kozai T., C. Kubota and B. R. Jeong. 1997a. Environmental control for the large-scale production of plant through in vitro techniques. Plant Cell, Tissue and Organ Culture 51:49-56.

Kozai, T., T.Q. Nguyen and C. Kubota. 1997b. Environmental control and its effects in transplant production under artificial light J.Kor. Hort. Set. 38(2):194-199.

Kozai, T., C. Kubota, S.M.A. Zobayed, Q.T. Nguyen, F. Afreen-Zobayed and J. Heo. 1999. Developing a mass-propagation system of woody plants. In: Challenge of plant and Agriculture Sciences to the Crisis of Biosphere on the Earth in the 21st Century. Proceedings of the 12th TOYOTA CONFERENCE, Shizuoka, Japan.

Heo, J. and T. Kozai. 1999. Forced ventilation micropropagation system for enhancing photosynthesis, growth and development of sweetpotato plantlets. Environment Control in Biology. 37(1):83-92.


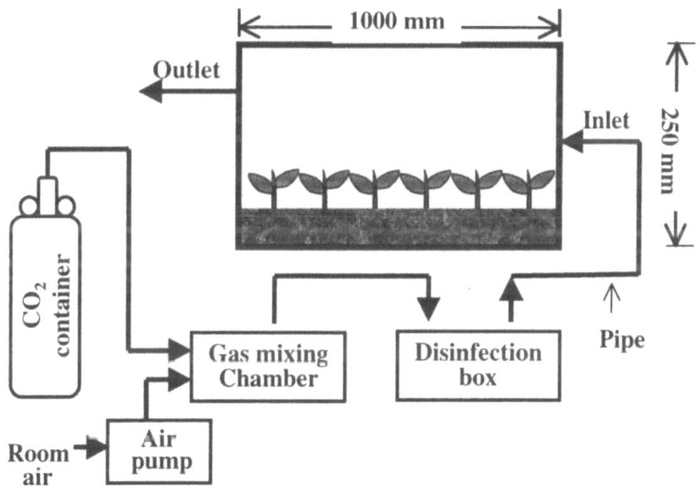


*Fig. 1* Schematic diagram of the forced ventilation micropropagation system.

*Table 1* Description of the treatments.

| Treatment Code[y] | Vessel size | Sugar ($g\ l^{-1}$) | Ventilation method | Supporting material | $CO_2$ con. ($\mu mol\ mol^{-1}$) | PPF[z] ($\mu mol\ m^{-2}\ s^{-1}$) |
|---|---|---|---|---|---|---|
| CSN | Small | 30 | Natural | Agar | 370 | 40±5 |
| FSE | Small | 0 | Natural | Agar | 1,500 | 120±10 |
| FLE | Large | 0 | Forced | Perlite | 1,500 | 120±10 |

[z] Photosynthetic photon flux was measured on the surface of the empty culture shelf.
[y] The letters C, F, S, L, N and E denote sugar-containing, sugar-free, small bottle-type vessel containing agar as supporting material with natural ventilation, large box-type vessel containing perlite as supporting material with forced ventilation, $CO_2$ non-enriched and $CO_2$ enriched, respectively.

*Table 2* Growth of potato plantlets cultured for 20 days.

| Treatment code | Shoot length (mm) | Fresh weight (mg) | Dry weight (mg) | Number of leaves | Leaf area ($cm^2$) |
|---|---|---|---|---|---|
| CSN | 68±8b[z] | 136±30b | 18±4b | 6.2±1.4b | 0.7±0.2b |
| FSE | 61±6b | 113±18b | 15±3b | 6.0±1.1b | 0.7±0.2b |
| FLE | 151±2a | 512±111a | 77±20a | 10.9±1.7a | 8.5±1.4a |

[z] Means±standard deviation per plantlet followed by the same letter in the column are not significantly different at the 1% level by Duncan's multiple range test.

*Table 3* Growth of statice plantlets cultured for 28 days.

| Treatment code | Leaf length (mm) | Fresh weight (mg) | Dry weight (mg) | Number of leaves | Leaf area ($cm^2$) |
|---|---|---|---|---|---|
| CSN | 56±6b[z] | 556±116b | 48±9b | 11.2±2.3ab | 7.8±4.0b |
| FSE | 57±7b | 561±92b | 48±10b | 9.8±2.0b | 18.3±4.3b |
| FLE | 84±13a | 750±163a | 68±15a | 12.8±1.7a | 31.5±11.1a |

[z] Means ±standard deviation per plantlet followed by the same letter in the column are not significantly different at the 1% level by Duncan's multiple range test.

*Table 4* Percent survival of potato and statice plantlets 20 days after transplanting for acclimatization ex vitro.

| Plant species | Treatment code examined | Number of explants *in vitro* | Survived plantlets *in vitro* | Percent survival *in vitro* (%)[z] | Survived plantlets *ex vitro* | Percent survival *ex vitro* (%)[z] |
|---|---|---|---|---|---|---|
| Potato | CSN | 1,500 | 1,315 | 87.7 | 1,086 | 82.6 |
| | FSE | 1,500 | 1,352 | 90.1 | 1,156 | 85.5 |
| | FLE | 2,300 | 2,226 | 96.8 | 2,217 | 99.6 |
| Statice | CSN | 1,000 | 852 | 85.2 | 624 | 73.2 |
| | FSE | 1,000 | 871 | 87.1 | 728 | 83.6 |
| | FLE | 1,500 | 1,441 | 96.1 | 1,370 | 95.1 |

[z] Number of plantlets survived ex vitro were investigated 20 days after transplanting for acclimatization *ex vitro*.

*Table 5* Estimated direct costs of photomixotrophic and photoautotrophic micropropagation of potato and statice plants.

Unit: Yuan

| Items of costs | Cost in photomixotrophic | | Cost in photoautotrophic | |
|---|---|---|---|---|
| | Potato 1500 plantlets | Statice 1000 plantlets | Potato 2300 plantlets | Statice 1500 plantlets |
| Medium | 14.36 | 14.36 | 7.91 | 7.74 |
| Sterilization | 6.00 | 6.00 | 2.00 | 2.00 |
| Gas-permeable films | 6.00 | 6.00 | 0.00 | 0.00 |
| Labor | 45.00 | 45.00 | 36.00 | 36.00 |
| Contamination loss | 8.80 | 7.74 | 0.00 | 0.00 |
| Fluorescent lamps | 16.35 | 23.52 | 28.00 | 42.34 |
| Air conditioner | 3.60 | 5.04 | 7.20 | 10.08 |
| $CO_2$ | 0.00 | 0.00 | 1.04 | 1.68 |
| Air pump | 0.00 | 0.00 | 1.09 | 1.76 |
| Acclimatization | 46.03 | 51.12 | 55.65 | 64.85 |
| Total | 146.14 | 158.78 | 138.89 | 166.45 |
| Plantlets for market | 1086 | 624 | 2217 | 1370 |
| Cost/plantlet | 0.13 | 0.25 | 0.06 | 0.12 |
| Sales price | 0.35 | 0.90 | 0.50 | 1.30 |

# GROWTH AND ACCLIMATIZATION OF CHRYSANTHEMUM PLANTLETS USING BIOREACTOR AND HYDROPONIC CULTURE TECHNIQUES

Eun-Joo Hahn[1], Sun Ja Kim[1], Kee Yoeup Paek[1] and Yong Beom Lee[2]
[1]Research Center for the Development of Horticultural Technology, Chungbuk National University, Cheongju, Chungbuk 361-763, Korea. E-mail: ejhahn@trut.chungbuk.ac.kr
[2]Dept. of Environmental Horticulture, The University of Seoul, Seoul 130-743, Korea.

**Abstract.** Chrysanthemum (*Dendranthema grandiflorum* 'Cheonsu') plantlets were produced in large-scale bioreactors. Among the culture methods, greatest shoot growth was achieved on plantlets in net culture followed by Ebb and Flow culture. Net culture induced highest shoot length, fresh weight, and leaf area. Among $NH_4^+$: $NO_3^-$ ratios, shoot growth was greatest at a $NH_4^+$: $NO_3^-$ ratio of 1:4. Fresh and dry weight, number of leaves, leaf area, and chlorophyll content were all significantly higher than those in the other ratios. High $NO_3^-$ favored plantlet growth but increasing the proportion of $NH_4^+$ significantly restricted plantlet growth. Shoots multiplied in bioreactors were cut by 1.5 cm nodal cuttings and grown hydroponically under various environmental conditions. Optimum aerial environment for chrysanthemum transplants was a photosynthetic photon flux of 200 μmol $m^{-2}$ $s^{-1}$, a 25℃ air temperature, a 70 ± 5% relative humidity, and a $CO_2$ concentration of 1000 μmol $mol^{-1}$. Optimum root zone environment was determined to be a 6.0 solution pH, 1.5 to 2.2 mS $cm^{-1}$ solution EC range, and a 25℃ nutrient solution. The results of our experiment indicated that the large-scale production of high-quality transplants could be achieved by employing bioreactor and hydroponic culture techniques.

**Key index words.** aerial environment, chrysanthemum (*Dendrathema grandiflorum* 'Ch eonsu'), culture methods, $NH_4^+$: $NO_3^-$ ratio, root zone environment.

## 1. Introduction

Micropropagation is a fast multiplication of a selected genotype using in vitro culture technique (Deberg and Read, 1991), with which healthy and uniform plantlets can be mass-produced in shorter period as compared to conventional propagation methods. This propagation method has been valuable, particularly for the propagation of a variety of floricultural crops, e.g., tropical ornamentals, carnation, chrysanthemum, gerbera and so forth (Chu, 1992). And yet, the multiplication efficiency still remains low mainly due to contamination and degeneration of the plantlets during subculture. Low rates of acclimatization and growth after transfer to greenhouse or field conditions also cause the low multiplication efficiency.

Research was carried out to increase multiplication efficiency by improving culture environments. Kozai et al. (1992) reported a significant increase in plantlet growth by increasing photosynthetic photon flux (PPF), $CO_2$ concentration, and the number of air exchanges of the culture vessels with absence of growth regulators and sucrose in the culture medium. Until now, studies have been focusing on inducing vigorous plantlet growth under those improved environments.

To reduce contamination and degeneration of the plantlets, on the other hand, approaches are also required to lessen the subculture frequency. In recent years, various kinds of plantlets have been produced in large-scale bioreactors (Takahashi et al., 1992; Levin et al., 1997). Bioreactor culture is highly effective in mass production of valuable



*C. Kubota and C. Chun (eds.), Transplant Production in the 21$^{st}$ Century,* 274–278.

plantlets since a large amount of plantlets can be obtained from one-time culture.

Acclimatization stage is indispensable for the plantlets to be adapted for ex vitro conditions after transfer to greenhouses or field. Kozai (1989) reported that in most cases, plantlet death during this period was caused by different environmental conditions between in vitro and ex vitro. He suggested improving in vitro environments for increasing acclimatization rate. On the other hand, a new method of acclimatizing in vitro plantlets, called microponics (micropropagation + hydroponics), was introduced, which resulted in significant increases in growth and acclimatization of the plantlets (Hahn et al., 1996).

There have been increasing needs for a scale-up production of high-quality transplants for commercial use in floricultural crops. In this paper, we would like to introduce mass production of chrysanthemum transplants by bioreactor and hydroponic culture techniques.

## 2. In vitro plantlet growth in bioreactors

Bioreactor culture provides advantages of cost reduction, programmable production, and easy control of culture environments over routine tissue culture. With bioreactors, culture period is significantly shortened by obtaining a large amount of virus-free shoots at one-time with little risk of contamination. In addition, vigorous shoots are obtained without shoot degeneration by improving culture conditions such as optimum ratios of $NH_4^+$ to $NO_3^-$, $CO_2$ supply, increasing the number of air exchanges and photosynthetic photon flux, etc.

### 2.1 Growth of chrysanthemum plantlets at different culture methods in bioreactors

Virus-free shoots of chrysanthemum (*Dendranthema grandiflorum* 'Cheonsu') were propagated in vitro and cut by 1.5 cm nodal cuttings. Thirty of them were inoculated in a 5-liter bioreactor containing MS liquid medium (Murashige and Skoog, 1962) supplemented with 30 g $L^{-1}$ sucrose. The nodal cuttings in the bioreactor were grown by three different culture methods: Ebb and Flow culture, soaking culture, and net culture. Cultures were carried out for 8 weeks at 25°C and a PPF of 70 μmol $m^{-2}$ $s^{-1}$ with a 16 h photoperiod.

Plantlet growth was greatest in net culture, showing highest shoot fresh weight, shoot length, leaf area, and stem diameter. No difference in plantlet growth was observed between Ebb & Flow culture and soaking culture but plantlets in soaking culture exhibited hyperhydricity.

### 2.2 Growth of chrysanthemum plantlets culture at various $NH_4^+$: $NO_3^-$ ratios in bioreactors

Nodal cuttings of chrysanthemum were obtained by the same method as the first experiment. Fifteen of them were inoculated in a 1-liter bioreactor containing MS liquid medium. $NH_4^+$: $NO_3^-$ ratio in the culture medium was varied at 0:1 (100% $NO_3^{-1}$-N), 1:4, 1:2 (The basal composition of MS mineral salts), 1:1, 2:1, 4:1, and 1:0 (100% $NH_4^+$-N). Total nitrogen content was maintained at 60 mM. Culture conditions were the same as the first experiment but culture period was 4 weeks.

Plantlet growth was not correlated with the pH, indicating that growth was dependent on the $NH_4^+$: $NO_3^-$ ratio. Shoot growth was greatest at a $NH_4^+$: $NO_3^-$ ratio of 1:4 followed by 1:2. Fresh and dry weight, number of leaves, leaf area, and chlorophyll content were all significantly higher than those in the other treatments. High $NO_3^-$

favored plantlet growth and even 100% $NO_3^-$-N induced greater shoot and root growth than high $NH_4^+$. However, increasing the proportion of $NH_4^+$ significantly restricted plantlet growth, indicating that high $NH_4^+$ decreased carbon use, lowering plant growth (Hdider et al., 1994). Especially, plantlet growth was completely inhibited in 100% $NH_4^+$-N.

**3. Propagation and acclimatization of in vitro plantlets using hydroponic culture technique**

Plantlets propagated in vitro require acclimatization before transfer to greenhouse or field conditions. Acclimatization of the plantlets is generally proceeded by lowering relative humidity and by increasing PPF to produce new leaves having photosynthetic ability similar to that of normal leaves (Ripley and Preece, 1986; Ziv, 1986). On the other hand, acclimatization requires time and complicate process, but the percentage of plantlet survival followed by normal growth is highly dependent on plantlet and environmental conditions during acclimatization (Preece and Sutter, 1991).

To promote propagation, growth, and acclimatization at a time, microponic culture was employed in mass production of chrysanthemum transplants and optimum environments were determined.

**3.1 Optimization of aerial environment for chrysanthemum plantlets in microponic culture**

Chrysanthemum shoots produced in bioreactors were cut by 1.5 cm nodal cuttings, planted in rockwool cubes (2.5 cm×2.5 cm×4 cm), and grown with a nutrient solution (Sonneveld and Straver, 1992). The cuttings were grown for 4 weeks under various aerial environments: air temperature was varied at 15, 20, 25, or 25℃ and relative humidity (RH) at 35 ± 5%, 70 ± 5%, or 90 ± 5%. To observe individual relation and correlation of PPF and $CO_2$ concentration, cultures were varied at 50, 100, 150, 200, or 300 $\mu mol \cdot m^{-2} \cdot s^{-1}$ in PPF and 350, 100, 200, or 300 $mg \cdot L^{-1}$ in $CO_2$ concentration.

Optimum air temperature range for chrysanthemum is generally 15 to 20℃ but higher temperature is required in seedling and rooting stages (Schwarz, 1995). In this experiment, highest fresh weight, shoot length, and number of leaves were obtained at 25℃, in agreement with his observation.

RH was varied at 35 ± 5%, 70 ± 5%, and 90 ± 5% and the cuttings were directly placed under the given RH conditions without gradual changes. Growth was observed in all treatments and a 70 ± 5% RH resulted in the highest growth. Growth under a 35 ± 5% RH was higher than that under a 90 ± 5% RH. When plantlets were transfer to greenhouses or field, humidity should be gradually decreased to that of outside the culture vessel to be adapted for greenhouse conditions (Preece and Sutter, 1991). However, the result proved that gradual decrease of humidity was not required in microponic culture.

Shoot and root growth was affected both by $CO_2$ concentration and by PPF level. Under no $CO_2$ supply, growth significantly increased with increasing PPF levels. Under the $CO_2$ concentrations of 1000 and 2000 mg $L^{-1}$, however, growth was increased but not significantly affected by high PPFs (150 $\mu mol\ mol^{-1}$), indicating that growth was more affected by $CO_2$ concentration than PPF. A high $CO_2$ concentration (3000 $\mu mol\ mol^{-1}$) combined with high PPFs restricted shoot and root growth.

Overall, optimum aerial environment in microponic culture was determined to be a

25℃ air temperature, a 70 ± 5% RH, PPFs of 150 μmol $m^{-2}$ $s^{-1}$, and a 1000 μmol $mol^{-1}$ $CO_2$ concentration.

### 3.2 Optimization of root zone environment for chrysanthemum plantlets in micropo nic culture

The nodal cuttings, obtained as described in the first experiment, were grown for 4 weeks under various root zone environments: 3.0, 4.0, 5.0, 6.0, 7.0, or 8.0 in solution pH, and 0.8, 1.5, 2.2, 2.9, or 3.6 mS $cm^{-1}$ in solution EC. Temperature of the nutrient solution was varied at 15, 20, 25, or 30℃.

pH 5.0 and 6.0 were optimum for growth, in agreement with the observation of Schwarz (1995), but growth in other pH levels were not as significantly interfered as in the case of hydroponic culture. In microponic culture, the cuttings in extreme pHs of 3.0 and 8.0 survived and successfully grew to complete plants although the growth rate was significantly lower than those in other pHs.

Optimum solution EC varies with plant species, season, growth stages, and the quality of water. In general, low EC is required in rooting and seedling stages, but high EC in vegetative stage (Fonteno, 1996). However, growth in microponic culture was greatest in EC 1.5 and 2.2 mS $cm^{-1}$, without showing negative effect of high ECs on the growth.

In the case of nutrient solution temperature, 20℃ and 25℃ resulted in highest shoot and root growth, while 15℃ resulted in the lowest. Shoot and root fresh weight, shoot length, and leaf area at 15℃ were halves of those at 20℃ and 25℃. It seemed that increasing nutrient solution temperature accelerated nutrient uptake by roots, while low solution temperatures suppressed it. Roots in 30℃ were longer but thinner and fragile compared to those in the other treatments.

## 4. Conclusion

There has been remarkable advance in mass propagation of in vitro plantlets by improving culture conditions. Nevertheless, new approaches are required for the large-scale production of transplants; that is, production efficiency should be increased by improving transplant quality and by reducing production cost.

In this regard, it could be possible to mass-produce healthy and uniform transplants by bioreactor culture and microponic culture with shortening the production period, improving the transplant quality, and reducing the production cost.

**References**

Chu, I.Y. E. 1992. Perspective of micropropagation industry. In: Kurata K. and T. Kozai (eds.). Transplant Production Systems. Kluwer Academic Publishers, Dordrecht, Hollands. pp. 137-150.

Debergh, P.C. and P.E. Read. 1991. Micropropagation. In: P.C. Debergh and R.H. Zimmerman (eds.). Micropropagation: Technology and Application. Kluwer Academic Publishers, Dordrecht, The Netherlands. pp. 1-13.

Fonteno, W.C. 1996. Growing media: Types and physical/chemical properties. In: David Wm. Reed (ed.). Water, Media and Nutrition for Greenhouse Crops. Ball Publishing. Batavia, Illinois, USA. pp. 93-122.

Hahn, E.J., Y.B. Lee and C.H. Ahn. 1996. A new method on mass-production of micropropagated chrysanthemum plants using microponic system in a plant factory. Acta Hort. 440:527-532.

Hdider, C.H., L.P. Vezina and Y. Desjardins. 1994. Short-term studies of $^{15}NO_3^{-1}$ and $^{15}NH_4^+$ uptake by micropropagated strawberry shoots cultured with or without $CO_2$ enrichment. Plant Cell Tiss. Org. Cult. 37:185-191.

Kozai, T. 1989. Autotrophic (sugar-free) micropropagation for a significant reduction of production costs.

Chronica Hort. 29:19-20.

Kozai, T., K. Fujiwara, M. Hayashi and J. Aitken-Christie. 1992. The in vitro environment and its control in micropropagation. In: K. Kurata and T. Kozai (eds.). Transplant Production Systems. Kluwer Academic Publishers, Dordrecht, The Netherlands. pp. 247-282.

Levin, R., S. Ran, A. Yekutiel, and A.W. Abed. 1997. A technique for repeated non-axenic subculture of plant tissues in a bioreactor on liquid medium containing sucrose. Plant Tissue Culture and Biotechnology 3:41-45.

Murashige, T. and F. Skoog. 1962. A revised medium for rapid growth and bioassays with tobacco tissue cultures. Physiol. Plant. 15:473-497.

Preece, J.E. and E.G. Sutter. 1991. Acclimatization of micropropagated plants to the greenhouse and field. In: P.C. Debergh and R.H. Zimmerman (eds.). Micropropagation: Technology and Application. Kluwer Academic Publishers, Dordrecht, The Netherlands. pp. 71-94.

Schwarz, M. 1995. Soilless culture management. Springer-Verlag, Berlin, Heidelberg, NewYork. pp. 7-31.

Sonneveld, I.C. and N. Straver. 1992. Nutrient solutions for vegetable and flowers grown in water or substrates. Glasshouse crops research station Naaldwijk, The Netherlands. pp. 33.

Takahashi, S., H. Matsubara, H. Yamagata and T. Morimoto. 1992. Micropropagation of virus free bulblets of *Lilium longiflorm* by tank culture. 1. Development of liquid culture method and large-scale propagation. Acta Hort. 319:83-88.

Ziv, M. 1986. In vitro hardening and acclimatization of tissue culture plants. In: L.A. Withers, and P. G. Anderson (eds.). Plant Tissue Culture and It's Agricultural Application. Butterworths, London. pp. 187-196.

# MASS PROPAGATION OF PINEAPPLE THROUGH *IN VITRO* CULTURE

Shyamal K. Roy, M. Rhaman and S. Hauqe
Department of Botany, Jahangirnagar University, Savar, Dhaka, Bangladesh. E-mail: shkroy@juniv.edu

**Abstract.** A large number of shoots were obtained from dormant axillary buds excised from the crown of pineapple and cultured on MS nutrient medium supplemented with different concentrations and combinations of kinetin, benzyl adenine (BA), α-naphthalene acetic acid (NAA) and indole-3-butyric acid (IBA). Condensed nodal regions of the crowns were used as explants. When the explants were cultured directly on MS medium with 2.5 mg $l^{-1}$ BA + 0.5 mg $l^{-1}$ NAA multiple shoot buds formed. These shoots continued to proliferate through several subcultures with an average of 20 shoots per transfer but the shoots were too small to transfer to rooting medium. When the concentrations of BA and NAA were lowered to 1.25 mg $l^{-1}$ and 0.25 mg $l^{-1}$, respectively, and 1.25 mg $l^{-1}$ kinetin was added to the medium the shoots elongated and the number of shoots increased up to 30 per culture. When the explants of meristem tip of crown were cultured on MS medium supplemented with 1.5 mg $l^{-1}$ NAA and 1.0 mg $l^{-1}$ kinetin callus was initiated within 3 weeks. This callus, when subcultured on MS medium with 1.5 mg $l^{-1}$ kinetin + 0.5 mg $l^{-1}$ NAA, produced large number of shoots. After four weeks these shoots along with calli were transferred to fresh medium of same constituents, where the old shoot buds elongated and many new buds also emerged. Shoots of both the categories rooted well, within two weeks, when they were excised individually and implanted in half strength MS medium with 2 mg $l^{-1}$ IBA. Eighty percent plantlets survived when transferred to open field. The technique described would be a promising method of propagation on a commercial scale and the regeneration technique via callus would be utilized for genetic improvement of the crop.

**Key index words.** *Ananas comosus*, micropropagation, growth regulator, Pineapple.

## 1. Introduction

*In vitro* culture techniques constitute an important component of biotechnology and have the potential not only to improve the existing cultivar, but also for the generation of novel plants and early release of high yielding plants resistant to various diseases, pests, and stresses. Biotechnology can bring new ideas, improved tools and novel approaches to the solution of some persistent, seemingly intereactable problems in food crop production. Given the pressing need to enhance and stabilize food production in Bangladesh in response to mounting population pressures and increasing poverty, there is an urgent need to explore novel technologies that will break traditional barriers.

Fruit is the human's oldest food whose history of cultivation lies back to the earliest known human civilization (Singh, 1985). It has already been proved that *in vitro* cloning is efficient and reliable horticultural technique for mass propagation of a number of fruits. Numerous tropical and subtropical fruit plants have been propagated successfully *in vitro,* but actual commercial production is probably still limited to only a few of them.

Pineapple (*Ananas comosus* (L.) Merr.; Bromeliaceae) is one of the most popular and delicious tropical fruit and is esteemed for its pronouced flavor and nutritive constituents. The fresh fruit is a good source of vitamins A and B and is rich in Vitamin



*C. Kubota and C. Chun (eds.), Transplant Production in the 21st Century,* 279–283.

C. It also contains some minerals, such as iron and phosphorus. It has low sugar and fat contents. Pineapple is always propagated vegetatively using suckers, arising from buds below the ground level, slips, which are borne on the peduncle just below or at the base of the fruit or crowns on top of the fruit.

Microparopagation of pineapple has been reported by many authros (Benega et al., 1995; Cote et al., 1991; Daquinta et al., 1994), but extensive work for large scale propagation in commercial basis has not been reported. Thus, the objective of the present study was to enhance the number of shoots per culture for mass clonal propagation.

## 2. Materials and Methods

Explants were taken from the crowns of *Ananas comosus* var. Madhupur. They were collected in April and May from Madhupur pineapple fields of Tangail district in Bangladesh.

For surface sterilization the shoots were defoliated and apices with 6-8 condensed nodes were taken. They were then first cleaned thoroughly under a continuous stream of running tap water for 1 h, washed with a liquid detergent for 15 min and then with a 5% solution of antiseptic savlon (ACI Co., Bangladesh) for 10 min. The shoot apices were then washed repeatedly with distilled water and finally treated with 0.2% $HgCl_2$ for 10 min in a laminar flow cabinet and washed for 3 times with autoclaved double distilled water to remove any trace of $HgCl_2$. After surface sterilization, shoot apices were excised at the base and divided into pieces of explants with 2-3 nodes in each explant.

The basal medium used for all experiments was Murashige and Skoog (1962) mineral formulation (MS) containing standard salts and vitamins, 30 g $l^{-1}$ agar (BDH, England). The media were variously supplemented with cytokinin [6-benzyladenine (BA) or kinetin] alone or in combinations with auxins [α-naphthalene acetic acid (NAA) or indole-3-butyric acid (IBA)]. The pH was adjusted to 5.8 $\pm$ 0.05 before adding agar, and the media were autoclaved at 1.1 kg $cm^{-2}$ for 20 min at 120°C.

Cultures were incubated at 25 $\pm$ 1°C with a photoperiod of 16 h $d^{-1}$ at 30 μ mol $s^{-1}$ $m^{-2}$ light intensity by cool white fluorescent light. All cultures were initiated in 150 x 25 mm glass tubes containing 12.5 ml of medium. The cultures were regularly subcultured at five-week intervals on fresh medium in glass tubes or 100-ml flasks. Observations were recorded every 5 days following planting and subculturing. All experiments were repeated twice with at least 12 cultures per treatment.

Several weeks after planting the new shoots grew into a small, leafy structure with several axillary branches. At this stage they were recovered aseptically from the culture vessels and the small shoots were separated and subcultured in the medium containing same or different combinations of hormonal supplements for multiplication of shoots. Multiplication was repeated at regular intervals. The optimum hormone levels in the culture medium essential to maximize production of uniform plantlets were selected. After several subculturs the shoots about 3-8 cm in length obtained from the multiplication media were separated aseptically from the culture vessels and transferred to rooting media containing different combinations of hormonal supplements. After developing sufficient root systems the regenerated plantlets were considered ready for transplantation. The roots of the plantlets were gently washed under running tap water to remove agar and plantlets were transplanted to small pots containing garden soil, compost and sand (1:1:1 v/v).

## 3. Results and Discussion

The surface sterilization procedure of the explants described in materials and methods yielded 85% aseptic cultures. The greatest shoot induction was found in a MS medium containing 2.5 mg $l^{-1}$ BA + 0.5 mg $l^{-1}$ NAA (Table 1). Repeated subculture in the same medium produced multiple shoots up to 20 per culture. Addition of cytokinin (either BA or kinetin) alone to the basal medium was not found to enhance the induction of multiple shoots. Though multiple shoots were induced in MS medium with 2.5 mg $l^{-1}$ BA and 0.5 mg $l^{-1}$ (data not shown) NAA, the length of the shoots was not long enough for transferring into the rooting medium. When the concentrations of BA and NAA were lowered to 1.25 mg $l^{-1}$ and 0.5 mg $l^{-1}$, respectively, and 1.25 mg $l^{-1}$ kinetin was added to the medium the shoots elongated and the number of shoots increased up to 30 per culture (data not shown).

*Table 1* Effect of growth regulators in MS medium on shoot induction and number of shoots per culture established from apical and nodal explants of *Ananas comosus* crown. Data were recorded after 35 days of culture.

| Growth regulators (mg $l^{-1}$) | Cultures with induced shoot (%) | Number of shoots per culture |
|---|---|---|
| BA 1.0+NAA 0.2 | 00 | 00 |
| BA 1.0+NAA 0.5 | 00 | 00 |
| BA 1.5+NAA 0.5 | 23 (1.9) | 7.3 (1.2) |
| BA 2.0+NAA 0.5 | 39 (2.3) | 11.7 (2.5) |
| BA 2.0+NAA 1.0 | 68 (6.7) | 13.9 (4.2) |
| BA 2.5+NAA 0.5 | 100 | 20.6 (2.5) |
| BA 2.5+NAA 1.0 | 78 (6.3) | 11.7 (3.2) |
| BA 3.0+NAA 1.0 | 61 (5.2) | 9.6 (3.6) |

Standard errors in parenthesis.

When the explants of meristem tips of crown were cultured on MS medium supplemented with 1.5 mg $l^{-1}$ NAA and 1.0 mg $l^{-1}$ kinetin, callus was initiated within three weeks. This callus, when subcultured on MS medium with 1.5 mg $l^{-1}$ kinetin + 0.5 mg $l^{-1}$ NAA, produced large number of shoots. After four weeks these shoots along with calli were transferred to fresh medium of same constituents, where the old shoot buds elongated and many new buds also emerged.

Without auxin treatment, shoots obtained from multiplication cultures failed to root. The well developed shoots of both categories were excised from the culture flask and implanted individually on a root induction medium containing half strength MS medium with different concentrations and combinations of IBA and NAA. The best concentration of auxin for rooting was found to be 2.0 mg $l^{-1}$ IBA in which 100% shoot rooted.

After 4 weeks in the rooting medium the rooted shoots were transferred to pots. None of the plantlets transferred directly from the rooting medium to the pot under natural condition survived. Eighty percent of the transferred plants survived if the plantlets in the rooting culture tubes were kept in room temperature (32-34°C) for 7 days before transplanting in the pots where plants were reared for thewe weeks. At the time of rearing, leaves elongated, expanded and turned deep green and consequently the plants were much healthy.

Necessity of cytokinin for shoot initiation is well established, but the process appears to be so complex that one cytokinin may depend upon some other factors for functioning as shooting stimulants. These factors may already exist in cultured cells or may need exogenous supply in the culture medium.. Ratio of auxin to cytokinin seems to play an important role. The results of the present investigation clearly showed that shoot tip explants of suckers of *Ananas comosus* plants were capable of producing multiple shoots in the medium with combinations of cytokinin and auxin. This observation on shoot tip culture of suckers was similar to the reports by Mapes (1973). Axillary buds of suckers of *Ananas comosus* exhibited morphogenetic responses which showed similarities with the responses found by other workers (Canlas and Javier, 1994).

*Table 2* Effect of auxins in half strength MS medium on root formation in regenerated shoots of *Ananas comosus*. Data were recorded after 35 days of culture.

| Growth regulators (mg $l^{-1}$) | Rooted shoots (%) | Days required for root induction |
|---|---|---|
| IBA 1.0 | 0 | 0 |
| IBA 1.5 | 42.5 (3.2) | 26 (18) |
| IBA 2.0 | 100 | 12 (0.9) |
| IBA 2.5 | 73.5 (5.2) | 20 (1.3) |
| NAA 1.0 | 0 | 0 |
| NAA 1.5 | 0 | 0 |
| NAA 2.0 | 0 | 0 |
| NAA 2.5 | 0 | 0 |
| IBA 2.0+NAA 1.0 | 52.5 (5.2) | 17 (1.3) |
| IBA 2.0+NAA 1.5 | 59.3 (6.1) | 16 (1.0) |
| IBA 2.0+NAA 2.0 | 41.7 (4.2) | 19 (1.2) |
| IBA 2.0+NAA 2.5 | 42.1 (3.9) | 18 (1.5) |

Standard errors in parenthesis.

The superiority of BA over other cytokinins for multiple shoot formation was reported by Das and Mitro (1990). In the present experiment, BA proved to be a suitable cytokinin for proliferation of shoots in combination with NAA. BA and NAA combination for shoot multiplication was also reported by Lakshmi Sita and Vaidyanathan (1979) and Das and Mitra (1990). For shoot elongation both BA and Kinetin were required in this experiment which corroborate the results obtained by Zepeda and Sagawa (1981).

For rooting on excised shoots, either single or a combination of two ro three auxins are used routinely. In our experiment 2 mg $l^{-1}$ IBA was found to be best for root induction (Table 2).

The result of the present study show that explants from crowns of pineapple could produce multiple shoots *in vitro*. Moreover the potentiality of multiplication continued for a long time. The technique described here would be a promising method of propagation of pineapple on a commercial scale. Furthermore, in this experiment regeneration system through callus induction was also developed which is a prerequisite for genetic transformations.

**References**

Benega, R., M.J. Isidron, M. Hidalgo and C.G. Borroto. 1995. *In vitro* germination and callus formation in pineapple (*Ananas comosus* (L.) Merr.) hybrid seed. Acta Hort. 425:215-220.

Canlas, A.M. and F.B. Javier. 1994. Multiple shoot formation, callus induction and plantlet regeneration of pineapple (*Ananas comosus*). J. Crop Sci. (Philippines). 19:39.

Cote, F., R. Domergue, M. Folliot, J. Bouffin and F. Marie. 1991. *In vitro* micropropagation of pineapple. Fruits Paris, Numero Special *Ananas* 46 : 359-366.

Daquinta, M., R. Castillo, J.C. Lorenzo, I. Lunas, M. Escalona, R. Trujiloy and C.G. Borrota. 1994. Formacion de callos en pina (*Ananas comosus* (L.) Merr.) Revista Brasileira de Fruticulura 16(2):83-91.

Das, T.O. and G.C. Mitra. 1990. Micropropagation of *Eucalyptus tereticornis* Smith. Plant Cell, Tissue and Organ Culture 22:95-103.

Lakshimi Sita, G. and C.S. Vaidyanathan. 1979. Rapid multiplication of *Eucalyptus* by multiple shoot production. Curr. Sci. 48:250-352.

Mapes, M.O. 1973. Tissue culture bromeliads. Proc. Int. Plant. prop. Sco. 23:47-55.

Murashige, F. and K. Skoog. 1962. A revised medium for rapid growth and bioassay with tobacco tissue culture. Physiol. Plant. 15:473-497.

Singh, R. 1985. Fruits (4th edition). National Book Trust, India. pp. 213.

Zepeda, C. and Y. Sagawa. 1981. *In vitro* propagation of pineapple. Hort Science. 16:495-496.

# MICROBIAL CONTAMINATION UNDER PHOTOAUTOTROPHIC CULTURE SYSTEM

Nazrul Islam[1] and Sayed M.A. Zobayed[2]
[1]Aristopharma Ltd., Dhaka 1204, Bangladesh.
[2]Department of Biological Sciences, University of Hull, Hull HU6 7RX, UK. E-mail: S.M.Zobayed@biosci.hull.ac.uk

**Abstract.** Microbial contamination is a serious and an unavoidable problem in the field of plant tissue culture. The use of antibiotic, commercial fungicide, thermotherapy, warm-water treatment and ultrasonic treatment are some of the commonly used method to control *in vitro* contamination. Almost all of these methods are expensive, with a low success rate and thus increased labor cost. We have found that the use of sugar free nutrient medium (photoautotrophic culture system) has significantly reduced the contamination in the multiplication and rooting stages of banana. In the conventional system (medium containing sugar and vitamins), atleast 34% of the vessels were contaminated (which is common in tissue culture of banana in our laboratory); 10% with fungal and 24% with bacterial contamination. However, by using the photoautotrophic culture system only 12% vessels were contaminated; 9% with fungal and only 3% with bacterial contamination. In this paper, we have discussed some common types of contamination in the sugar-containing medium and the possible use of photoautotrophic culture system for reducing the contamination rate in plant micropropagation of banana.

**Key index words.** antibiotic, bacterial contamination, banana rmicropropagation, fungal contamination.

## 1. Introduction

Successful tissue culture of all plants depends on the removal of exogenous and endogenous contaminating microorganisms. Microbial contamination is a serious and an unavoidable problem. Fungi and bacteria are the most common microorganisms to be found on or in plant tissues. Viruses, viroids, mycoplasma, spiroplasma, and rickettsia are also found inside the plants.

To eliminate contamination during *in vitro* propagation, different techniques have been developed in the last few years (Pype et al., 1996). The use of antibiotic, commercial fungicide, thermotherapy, warm-water treatment and ultrasonic treatment are some of the commonly used methods to control *in vitro* contamination. In order to reduce/eliminate contaminating microorganisms antibiotics are often used by many researchers and commercial propagators with mixed success (Cole, 1996).

Banana is one of the important fruit crops and is widely cultivated in many parts of the world. Banana-planting material derived from *in vitro* techniques has been commercially used in some countries as an alternative to conventional planting material since 1985 (Robinson, 1996). One of the major problems of banana (*Musa sapientum* L.) micropropagation especially with the `Sagar` variety is some bacterial and fungal contaminations. Major problem is the bacterial (endogenous) contamination in multiplication stage and thus millions of cultures are thrown away every year.

Therefore, in the present study, attempt was made to culture banana plantlets photoautotrophically during the initiation stage with the aim to reduce microbial contamination. Sterile MS medium (Murashige and Skoog, 1962) with and without



*C. Kubota and C. Chun (eds.), Transplant Production in the 21$^{st}$ Century, 284–288.*

sucrose was exposed for few minutes in a growth room and the percentage of contamination was also studied.

## 2. Materials and Methods

### 2.1 Experiment I: Contamination in sugar containing and sugar-free media

To compare the contamination in MS medium with or without sugar, 1.0 g $l^{-1}$ agar (Sigma, USA) was added to the solution as a gelling agent. For sugar-free treatment sucrose was not added and for the sugar containing treatment (control) 20 g $l^{-1}$ sucrose was added; 5.0 ml medium was pored into each of the sterile plastic petridishes (volume 40 ml). Petridishes were then placed in a growth chamber and the lids were left open for 5 minutes and then sealed by using parafilm. The growth room condition was as follows: 75% relative humidity, 23°C temperature, and 50-60 μmol $m^{-2}$ $s^{-1}$ photosynthetic photon flux. Six days after treatment, each of the vessels were studied in relation to the contamination percentage.

### 2.2 Experiment II: Control of contamination in banana cultures

To control contamination of banana culture following treatments were given:

A. Explants were immersed in sodium hypochloride solution (2%) for 10 mins. and washed in sterile DW for 3 times

B. Treatment A + adding Gentamicin (25 mg $L^{-1}$) in MS medium.

C. Treatment A + Kanamycin (50 mg $L^{-1}$)

D. Treatment A + Erytromycine (50 mg $L^{-1}$)

E. Treatment A + photoautotrophic culture system (sugar-free medium).

For treatments A - D, 20 g $L^{-1}$ sucrose was added in the MS medium (photomixotrophic condition). For treatment E, sucrose was not added in the medium to ensure the photoautotrophic condition. Agar was used as a gelling agent in all the treatments. For treatment B, C and D, specific amount of antibiotic was dissolved in distilled water and was filter sterilized. The solutions were then added directly in the medium after autoclave.

Meristem shoot tips of banana (*Musa sapientum* L. var. Sagar) were taken from vigorous sword suckers of selected healthy mother plants as the source for explants. The shoot tips were trimmed to cubes with 5 mm sides and the above mentioned treatments were given. Explants were inoculated in a glass vessel (volume 60 ml) and only one explant was inoculated per vessel; 20 vessels were prepared for each treatment.

Vessels were kept in a growth chamber with 80 μmol $m^{-2}$ $s^{-1}$ PPF, 23°C air temperature and under ambient $CO_2$ concentration. Transparent polypropylene sheets (thickness 25 μm; Courtaulds Films, Bridgewater, Somerset, UK) were used to seal the vessels in all the treatments. After 60 days of culture, percentage of contamination and survival of the explants were studied.

## 3. Results and Discussion

### 3.1 Experiment I: Contamination in sugar containing and sugar-free media

Only three days after treatment, visible fungal and bacterial colonies were observed in both sugar-free treatment and in the control (sugar containing medium). In general, colony appearance was delayed at least for 48 to 72 h in the sugar-free medium.

However, the size, shape and number of the colonies varied significantly between the two treatments. Large and prominent fungal and bacterial colonies were noticed six days after treatment in sugar containing medium (Fig. 1a). This was probably because of the presence of sugar in the medium, which nourished the microorganisms to grow well. In the sugar-free treatment, the growth of the colonies was poor; especially the bacterial colonies were very small (Fig. 1b).

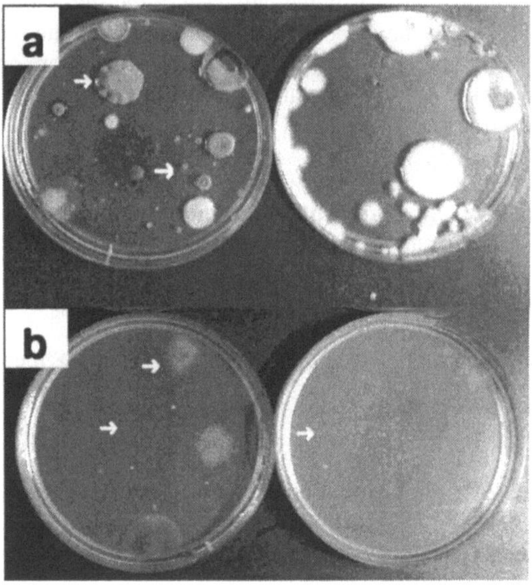


*Fig. 1* Microbial contamination in MS medium a) with 2.0 g $l^{-1}$ sucrose and b) without adding sucrose. Sterile petridishes were uncapped for 5 minutes and placed under growth room conditions. Photographs were taken after 6 days. Arrows indicate bacterial colony.

### 3.2 Experiment II: Control of contamination in banana cultures

In general, cultures subjected to photoautotrophic treatment exhibited low contamination and high percent survival of banana plantlets (Table 1). In all the treatments, percentage of fungal contamination was almost similar. In the photoautotrophic treatment, bacterial contamination was reduced significantly (3%), although fungal contamination was similar as observed in treatments A, B, C and D. However, the growth of microbes in the photoautotrophic treatment was poor and delayed. Among the treatments, highest percentage of contamination (35%) was observed in treatment `A` where explants were immersed in sodium hypochloride solution (2%) for 10 mins. and washed in sterile DW (surface sterilization). A major difference was in the percentage of bacterial contamination, which was at least 8-fold than that of the photoautotrophic treatment. The use of antibiotics (treatments B, C and D) can reduce bacterial contamination (9-12% vessels were contaminated) although the survival percentage of banana explants was reduced.

It should be mentioned that the growth of banana explants was poor in the photoautotrophic treatment compared with that of the sugar containing treatments. This was probably due to the low $CO_2$ concentration in the culture headspace and low air exchange rate in the vessel. However, by increasing the $CO_2$ concentration and air exchange rate, growth can be greater than that of the sugar-containing medium (Kozai et al., 1988, Kozai 1991).

*Table 1* Showing the percentage of bacterial and fungal contaminations in the culture vessel and survival percentage of banana plantlets grown for 60 days in sugar-free and sugar containing media.

| Treatments | Contamination (%) (B+F) | †Survival (%) |
|---|---|---|
| Treatment A | 35 (24+10) | 76 |
| Treatment B | 22 (11+11) | 54 |
| Treatment C | 22 (10+12) | 47 |
| Treatment D | 21 (9+12) | 61 |
| Treatment E | 12 (3+9) | 78 |

†Survival percentage not included contaminated vessels; F = fungal and B = bacterial contamination.

Some of the fungal-species were identified at the end of the experiment. Usually in all the treatments, similar types of fungal contamination were observed; however well-nourished fungal colonies were observed in all the treatments except the photoautotrophic. Following fungal species were identified:

i) *Mucor sp.*
ii) *Candida sp.*
iii) *Penicillium sp.*
iv) *Aspergillus fumigatus*

Therefore, by using photoautotrophic culture system (sugar-free medium) in banana micropropagation it is possible to control bacterial contamination. Reduction of sugar from the medium is known to be beneficial for the growth and physiology of the plant as well as the production costs. On the other hand, use of antibiotics for controlling contamination is a common practice but an expensive business, moreover it can be toxic for plants to *some extend* and thus possibly reduce the plant growth and survival.

## 4. Conclusion

The use of sugar-free medium to control the microbial contamination in the field of micropropagation is a useful technique. By using this culture system, bacterial contamination is possible to control in banana, which is a significant improvement in banana micropropagation. Bacterial contamination usually seriously hampers the propagation cycle and increases the production costs tremendously in the banana micropropagation. In some cases, bacterial contamination is observed even from the explant itself (endogenous). However, further study is necessary to investigate the possibility of controlling fungal contamination in the sugar-free medium and to distingue between pathogenic and non-pathogenic fungal contamination.

### References

Cole, M. 1996. Microbial contaminations and aseptic technique in plant tissue culture, In: A. Taji and R. Williams (eds.) Tissue culture of Australian plants. University of New England, Armidale, Australia. pp. 204-238.

Kozai, T. 1991. Micropropagation under photoautotrophic conditions. In: P.C. Debergh and R.H. Zimmerman (eds.). Micropropagation technology and application. Kluwer Academic Publishers. pp. 447-469.

Kozai, T. and Y. Iwanami. 1988. Effects of $CO_2$ enrichment and sucrose concentration under high photon fluxes on plantlet growth of carnation (*Dianthus caryophyllus* L) in tissue culture during the propagation stage. J Jap Soc Hortic Sci. 57:279-288.

Pype J., K. Everaert and P. Debergh. 1996. Contamination by micro-arthropods in tissue cultures. In: A.C. Cassells and B. Hayes (eds.). Proceedings of second international symposium on bacteria and bacteria-like contamination of plant tissue cultures. University college Cork, Ireland. pp. 21.

Robinson, J.C. 1996. Establishing A Plantation In: J.C. Robinson, (eds.) Banana and Plantains. CAB International, Oxon, UK. pp. 104-127.

## Author Index


Adelberg, J. 131
Afreen, F. 3, 231, 238
Arai, K. 145

Bishop, D. 131
Bostick, M. 131

Chintakovid, W. 98
Choi, K.Y. 166
Chun, C. 3, 20, 145, 149, 154

Enomoto, H. 183

Fang, W. 108
Fleisher, D.H. 53

Giacomelli, G.A. 72

Hahn, E.-J. 252, 274
Haque, M.S. 219
Harry, I.S. 197
Hauqe, S. 279
Hayashi, Y. 38
He, D. 154
Heins, R.D. 59
Heo, J. 246, 252
Hiruta Y. 67
Honma, T. 119
Hoshi, T. 38

Ibaraki, Y. 63
Ishii, K. 205
Ishii, M. 172
Islam, A.F.M.S. 149
Islam, N. 284
Ito, T. 125, 172
Iwabuchi, K. 63, 67

Jao, R.C. 108
Jeong, B.R. 102

Kan, H. 119
Kania, S.T. 72
Kawai, M. 83
Kida, H. 125
Kim, H.-H. 137
Kim, S.J. 274
Kim, Y.H. 91
Kirdmanee, C. 78
Kitaya, Y. 83
Kiyota, M. 83
Kozai, T. 3, 20, 28, 38, 98, 137, 145, 149, 154, 215, 226, 231, 238, 246, 258, 266
KrishnaRaj, S. 160
Kubo, T. 209
Kubota, C. 3, 47, 231, 238, 258
Kurata, K. 63, 67

Lee, Y.B. 166, 274
Lok, Y.H. 154

Maruo, T. 125, 172
Maruyama, E. 205
Miyajima, H. 119
Morohoshi, N. 191
Mosaleeyanon, K. 78
Murase, H. 33
Murata, Y. 183
Murch, S.J. 160

Nagaoka, N. 226
Nguyen, Q.T. 246
Nhut, D.T. 114

Ogita, S. 209
Ohyama, K. 3, 28
Okabe, K. 145
Omura, Y. 145

Paek, K.Y. 166, 252, 274
Pollock, R. 131

Rhaman, M. 279
Roy, P.K. 219
Roy, S.K. 219, 279

Sakami, K. 149
Sakurai, N. 183
Sasamoto, H. 209
Sawa, Y. 226
Saxena, P.K. 160
Seyama, K. 67
Shibuya, T. 83
Shimizu, H. 59
Shinohara, Y. 125, 172
Sinha, P. 219
Son, S.H. 252

Takagaki, M. 149, 183
Takamura, T. 114

| | |
|---|---|
| Tanaka, M. | 114 |
| Thorpe, T.A. | 197 |
| Ting, K.C. | 53 |
| Tsuchiya, H. | 119 |
| Tsuji, M. | 125 |
| Tsuruyama, J. | 83 |
| Udagawa, Y. | 183 |
| Valero-Aracama, C. | 215 |
| Watanabe, H. | 114 |
| Watanabe, Y. | 226 |
| Wilson, S.B. | 258 |
| Xiao, Y. | 266 |
| Yabusaki, M. | 63 |
| Yamazaki, A. | 119 |
| Yoshida, S. | 178 |
| Yoshinaga, K. | 28 |
| Zhao, J. | 266 |
| Zobayed, S.M.A. | 215, 231, 238, 284 |

MIX
Papier aus verantwortungsvollen Quellen
Paper from responsible sources
FSC® C105338

If you have any concerns about our products,
you can contact us on
**ProductSafety@springernature.com**

In case Publisher is established outside the EU,
the EU authorized representative is:
**Springer Nature Customer Service Center GmbH**
**Europaplatz 3, 69115 Heidelberg, Germany**

Printed by Libri Plureos GmbH
in Hamburg, Germany